Optimization and Decision-Making in the Renewable Energy Industry

Figen Balo
Fırat University, Turkey

Ayşe Topal
Nigde Omer Halisdemir University, Turkey

Ezgi Demir
Sumer Robotics Engineering and Consultancy Ltd., UK

Alptekin Ulutaş
Sivas Cumhuriyet University, Turkey

A volume in the Advances in Environmental Engineering and Green Technologies (AEEGT) Book Series

IGI Global
PUBLISHER of TIMELY KNOWLEDGE

Published in the United States of America by
IGI Global
Engineering Science Reference (an imprint of IGI Global)
701 E. Chocolate Avenue
Hershey PA, USA 17033
Tel: 717-533-8845
Fax: 717-533-8661
E-mail: cust@igi-global.com
Web site: http://www.igi-global.com

Library of Congress Cataloging-in-Publication Data

Names: Balo, Figen, 1972- editor. | Topal, Ayse, 1983- editor. | Demir, Ezgi, 1991- editor. | Ulutas, Alptekin, 1983- editor.
Title: Optimization and decision-making in the renewable energy industry / Figen Balo, Ayse Topal, Ezgi Demir, and Alptekin Ulutas, editor.
Description: Hershey PA : Engineering Science Reference, [2022] | Includes bibliographical references and index. | Summary: "The main objective of this book is to analyze renewable energy sources using current mathematical methods and techniques, giving an introduction of trends and current mathematical methods to the masses and providing advanced knowledge on renewable energy sources for both academic and practical professionals"-- Provided by publisher.
Identifiers: LCCN 2021054738 (print) | LCCN 2021054739 (ebook) | ISBN 9781668424728 (hardcover) | ISBN 9781668424735 (paperback) | ISBN 9781668424742 (ebook)
Subjects: LCSH: Renewable energy sources. | Energy industries--Mathematical models. | Energy industries--Decision making. | Mathematical optimization.
Classification: LCC TJ808 .O67 2022 (print) | LCC TJ808 (ebook) | DDC 333.79/4--dc23/eng/20220105
LC record available at https://lccn.loc.gov/2021054738
LC ebook record available at https://lccn.loc.gov/2021054739

This book is published in the IGI Global book series Advances in Environmental Engineering and Green Technologies (AEEGT) (ISSN: 2326-9162; eISSN: 2326-9170)

British Cataloguing in Publication Data
A Cataloguing in Publication record for this book is available from the British Library.

All work contributed to this book is new, previously-unpublished material.
The views expressed in this book are those of the authors, but not necessarily of the publisher.

For electronic access to this publication, please contact: eresources@igi-global.com.

Advances in Environmental Engineering and Green Technologies (AEEGT) Book Series

ISSN:2326-9162
EISSN:2326-9170

Editor-in-Chief: Sang-Bing Tsai, Zhongshan Institute, University of Electronic Science and Technology of China, China & Wuyi University, China; Ming-Lang Tseng, Lunghwa University of Science and Technology, Taiwan; Yuchi Wang, University of Electronic Science and Technology of China Zhongshan Institute, China

MISSION

Growing awareness and an increased focus on environmental issues such as climate change, energy use, and loss of non-renewable resources have brought about a greater need for research that provides potential solutions to these problems. Research in environmental science and engineering continues to play a vital role in uncovering new opportunities for a "green" future.

The **Advances in Environmental Engineering and Green Technologies (AEEGT)** book series is a mouthpiece for research in all aspects of environmental science, earth science, and green initiatives. This series supports the ongoing research in this field through publishing books that discuss topics within environmental engineering or that deal with the interdisciplinary field of green technologies.

COVERAGE

- Green Technology
- Water Supply and Treatment
- Contaminated Site Remediation
- Electric Vehicles
- Policies Involving Green Technologies and Environmental Engineering
- Cleantech
- Alternative Power Sources
- Radioactive Waste Treatment
- Pollution Management
- Industrial Waste Management and Minimization

IGI Global is currently accepting manuscripts for publication within this series. To submit a proposal for a volume in this series, please contact our Acquisition Editors at Acquisitions@igi-global.com or visit: http://www.igi-global.com/publish/.

The Advances in Environmental Engineering and Green Technologies (AEEGT) Book Series (ISSN 2326-9162) is published by IGI Global, 701 E. Chocolate Avenue, Hershey, PA 17033-1240, USA, www.igi-global.com. This series is composed of titles available for purchase individually; each title is edited to be contextually exclusive from any other title within the series. For pricing and ordering information please visit http://www.igi-global.com/book-series/advances-environmental-engineering-green-technologies/73679. Postmaster: Send all address changes to above address.

Titles in this Series

For a list of additional titles in this series, please visit: http://www.igi-global.com/book-series/

Improving Integrated Pest Management for Crop Protection and Food Security
Muhammad Haseeb (Florida A&M University, USA) Lambert H.B. Kanga (Florida A&M University, USA) and Jawwad A. Qureshi (University of Florida, USA)
Engineering Science Reference • © 2022 • 315pp • H/C (ISBN: 9781799841111) • US $195.00

Prevention and Management of Soil Erosion and Torrential Floods
Slobodan Milutinović (University of Niš, Serbia) and Snežana Živković (Faculty of Occupational Safety, University of Niš, Serbia)
Engineering Science Reference • © 2022 • 335pp • H/C (ISBN: 9781799884590) • US $195.00

Applications of Nature-Inspired Computing in Renewable Energy Systems
Mohamed Arezki Mellal (M'Hamed Bougara University, Algeria)
Engineering Science Reference • © 2022 • 326pp • H/C (ISBN: 9781799885610) • US $225.00

Physiology, Genomics, and Biotechnological Applications of Extremophiles
Aparna B. Gunjal (Dr. D.Y. Patil Arts, Commerce, and Science College, Pune, India) Rebecca Thombre (The University of Kent, UK & Blue Marble Space Institute of Science, USA) and Javid A. Parray (Department of Higher Education, Government Degree College, Srinagar, India)
Medical Information Science Reference • © 2022 • 473pp • H/C (ISBN: 9781799891444) • US $345.00

Optimal Planning of Smart Grid With Renewable Energy Resources
Naveen Jain (College of Technology and Engineering, Udaipur, India) Jai Kumar Maherchandani (College of Technology and Engineering, Udaipur, India) Navneet Kumar Agrawal (College of Technology and Engineering, Udaipur, India) and Trilok Gupta (College of Technology and Engineering, Udaipur, India)
Engineering Science Reference • © 2022 • 294pp • H/C (ISBN: 9781668440124) • US $225.00

IGI Global
PUBLISHER of TIMELY KNOWLEDGE

701 East Chocolate Avenue, Hershey, PA 17033, USA
Tel: 717-533-8845 x100 • Fax: 717-533-8661
E-Mail: cust@igi-global.com • www.igi-global.com

Table of Contents

Detailed Table of Contents

Renewable energy sources are clean energy sources and have a much lower environmental impact than other energy sources. Energy production from biomass allows the storage of energy raw materials, unlike other renewable energy sources. Some factors and criteria affect the site selection of biomass power plants. Determining the significant levels of these conflicting criteria is vital for investors in choosing a biomass power generation facility location. Location selection of biomass power plants is a multi-criteria decision problem for decision makers. The relative importance of the criteria includes quantitative evaluations and is an uncertain process. In this study, a comprehensive literature study determined the criteria to be considered in selecting biomass energy production facility locations. The fuzzy analytical hierarchy process was used to determine the importance levels of the criteria to be used to select the bio-mass power plant establishment location.

Consumption of renewable energy sources for countries shows a rising trend. Providing progress in the renewable energy field, countries try on related regulations and accurate investments according to renewable energy consumption and generation. European

Union (EU15) countries play an essential role increasing renewable energy efficiency, which is share of Europe in total energy usage. In this chapter, deterministic and stochastic methods were used to examine whether the renewable energy efficiencies of EU15 countries and Turkey are sensitive to different data envelopment analysis and stochastic frontier analysis models using renewable energy consumption and generation parameters. The chapter presents how the renewable energy efficiency results of related countries change with different optimization models in the context of deterministic and stochastic framework, and it proposes a new method to find a common solution for the different results of different optimization models.

The aim of this study is to determine the criteria used in the evaluation of renewable energy resources with the extremely high strategic importance that Turkey has and to find their degree of importance. For this purpose, an application has been made to find which criteria come to the fore and the weights of these criteria in order to evaluate Turkey's renewable energy resources. Five main criteria (technical, economic, environmental, social, and political) and a total of 22 sub-criteria related to these criteria were included in the scope of the study. Fuzzy SWARA method, one of the multi-criteria decision-making methods, has been used in the study. According to the results of the analysis, the most important criterion among the main criteria was "environmental criteria." As a result, it has shown the importance of the priority criteria for Turkey and environmental evaluation criteria, which are important for the common future of humanity in parallel with the results in the world.

The terms "sustainable development" and "sustainability" have become more popular because of the critical troubles faced through mankind such as growing humankind effect on ecology and the risk of energy source depletion. Solar energy is one of the most rapidly developing sources of sustainable energy available today. The solar panel is the foundation of a photovoltaic system. In this study, the COPRAS and ENTROPY methods have been applied to select the most effective solar panel (100W) for a solar farm design. The five various solar panel brands have been evaluated, and professionals' choices have been dependent on the most important characteristics of the solar panels. From the top panel firms worldwide, the solar panel data utilized in this chapter is obtained. The most effective panel selection has been analyzed to affect the solar panel property potency employing in scales, in conjunction with several competing solar panels from which corporations must choose the top requirements, using COPRAS and ENTROPY techniques.

The use of fossil fuels has decreased compared to the past due to the gradual depletion of fossil fuels and the greenhouse gases emerging in their use. As a result of this decrease, many countries have turned to alternative energy sources instead of fossil fuels. The most popular energy sources among these alternatives are renewable energy sources. Compared to fossil fuels, renewable energy sources cause little or no harm to nature. While renewable energy sources differ according to the countries, it is necessary to determine the most optimal renewable energy source for countries. In the literature, the most optimal renewable energy source selection has been made many times, and while this selection is made, multi-criteria decision-making (MCDM) methods are generally used. In these studies, the most optimal renewable energy source was selected by considering multiple criteria. In this chapter, criteria that are frequently used in the literature will be evaluated with the Plithogenic PIPRECIA method. In addition, the most important criterion will be determined.

The electric vehicle segment is gaining momentum around the globe, and Morocco will not be the exception in this regard. The present study serves to look into the question of the current and future electricity needs of this segment of the means of transport. The main contribution is preparing the necessary adaptations in the frame of electricity production capacity at the national level. This chapter aims to highlight the enablers to be seized and the main barriers to be overcome by the use of an integrated SWOT-PESTEL analysis in combination with the analytical hierarchy process. First, the SWOT-PESTEL framework is dedicated to identifying the main criteria that enable and hinder the viability of implementing electric freight vehicles (EFV) in Morocco from a sustainability perspective. Afterwards, the quantification process of the output is realized through the application of the AHP method.

Biomass energy is an essential and sustainable type of energy for today and the future because it is produced from renewable sources and is environmentally friendly compared to fossil-based energy. Biomass energy is the only renewable energy source that creates social and economic impact together. It creates added value, provides employment, and creates new tax opportunities in many fields from agriculture to industry, from the transportation sector to the banking insurance sector. While other renewable energy sources cannot be stored, the energy obtained from biomass can be stored. In this aspect, energy production from biomass stands out from other renewable energy sources. One of the crucial handicaps in evaluating biomass as an energy source is the collection of biomass. Biomass is produced in different amounts in different village centers, while country roads are suitable for trucks of different sizes. In this chapter, the multi-capacity vehicle routing problem is modeled for biomass collection in village centers at different production capacities.

Global climate change is one of the most challenging problems of today's world and its effects have become more noticeable day after day. The magnitude of climate change is closely related to our carbon footprint, so replacing the resources such as petroleum, coal, nuclear energy by which humankind generates their energy requirements with new ones is essential. The usage of renewable energy resources is one of the effective ways to decrease CO2 emissions and environmental pollution.

Biomass energy is one of the promising future energies as a renewable resource. Therefore, many requirements should be considered and evaluated carefully to produce and sustain a successful biomass energy system. This chapter presents a review of academic research attempting to face the biomass energy sector's problems using multi-criteria decision-making (MCDM) methods. Related articles in the international journals from 2010 to 2021 are collected and reviewed to answer the following questions: (1) Which methods are mainly used? (2) Which problems attract the most attention?

Renewable energy resources have become popular in energy policies as sustainable development in the energy field requires the transition to clean or renewable energy resources such as solar, wind, and hydro to mitigate global warming. Renewable resources play a more significant role in the energy future of Turkey. However, despite renewable energy resources being cleaner and causing fewer environmental problems, the renewable energy selection problem is a complex task due to the involvement of various conflicting factors and uncertainty. Therefore, multi-criteria decision-making methods are commonly used to handle this complexity successfully. In this chapter, the studies focused on renewable energy resource selection problem in Turkey with multi-criteria decision-making methods were reviewed. Findings suggest that the number of studies increased due to the growing importance of renewables. Also, AHP, TOPSIS, and ANP have risen to the top of the literature as the most extensively used approaches.

Because of the effects of nuclear and fossil-based energy on the environment, economics, and security in the world, the need for alternative energy sources has grown steadily and dramatically during the last years. An increasing attraction in renewable power sources, due to rising energy expenses and country-level tax inducement, is driving the research to advance a sequence of improving unified resolutions and novel energy generation equipment. The novel wind turbine installation and the novel wind farm building are critical procedures for long time energy generation. In this chapter, a comprehensive analysis, which combines ARAS and ENTROPY methods,

Preface

We have no chance but to speed the changeover to sustainability of the world due to rising energy costs, climate change, and fossil fuel expenditure. We should expect increased strain on energy supplies from increasing population in the future decades, as well as a growing threat to power provides from climate change. Energy systems patterns are evolving, as highlighted in a latest protocol published through the International Agency of Renewable Energy, and the transition to more maintainable provide mechanisms may be unavoidable. Long-term power mechanisms sustainability is challenged by rising resource constraint and population. Many scholars had previously analyzed and investigated sustainability in the power industries in an isolated way and fragmented. Latest researchers have revealed that power systems are extremely interrelated, and that enhancing system operation while maintaining sustainability can't be achieved just through research on individual energy systems. Additionally, energy planning has become more complicated in the sustainable development's contemporary period due to the inclusion of numerous standards such as economic, social, environmental, and technical considerations. More worldwide investigation is required to better understand energy system behavior and build the technical enablers required to develop system efficiency.

Renewable energy, when comparing to conventional energy, tackles the global warming problem, and the previous ten years was a watershed moment for the renewable energy industry due to notable advancements in the area. Because the application of any sustainable energy design requires a significant investment of diverse elements, like permits and labor from both local and central governments, there may be a critical need for a decision-making module to determine and understand the renewable energy type that best fits and supplies the most efficient findings for a specific area where each noted criterion has different precedencies.

Multi criteria decision making, which replaces the dependence and uncertainty on "gut instinct" through professionals with metric computation architectural to facilitate the decision-making operation, can be demonstrated to be an asset in this case.

This, in turn, places significant innovations on decision-makers' ability to discretely and independently maximize energy sources, particularly in remote settings. Furthermore, because of topographical constraints relating to sustainable energy sources, which are usually scattered in nature, energy design happens more difficult. Even at fragmented levels of electrification, decision making process plays a critical part in the planning of such systems through taking into account numerous objectives and criteria. Multiple-criterion decision-making is operational research's a branch that deals with determining the best outcomes in complicated cases involving multiple indicators, competing criteria, and objectives. This tool is gaining popularity in the field of energy planning because it allows authorized persons to make judgments while concurrently considering overall objectives and criteria.

In this book, energy industry availability in the sector and the outputs of some value critical researches conducted to gain a more understanding of multi criteria decision making implementation in the area. The latest multi criteria decision making technics based on scientific techniques and developments for renewable energy systems decision-making, operation, and design are presented. It also provides an overview of latest several multi-criteria decision making techniques, as well as advances made in this field through examining renewable energy implementations over multi-criteria decision making methodologies and next expectations. Furthermore, the multi-criteria decision making techniques were used as to compare regional and worldwide of latest equipment to conduct a comprehensive evaluation in the field of renewable energy. This book also tried to draw attention to the consequences of different approaches so that renewable energy systems can operate with optimum performance.

In this reason, this book compiles a lot of material that will be useful to engineers, scientists, and researchers working on renewable energy development and research. The editors would like to convey their heartfelt gratitude to the contributors for contributing their experience and knowledge to this book.

THE CHALLENGES

As the world's fossil fuel reserves dwindle, increasing emphasis is being placed on capturing power from renewable energies derived from alternative natural sources. Renewable energy sources such as solar, wind, biomass, and hydropower have become well-established in both industrial and home contexts.

Problems With Power Quality

To provide network efficiency and stability, consistently excellent power quality is required. The power supply's quality allows the system to operate with cheap costs and high dependability. On the other hand, low energy quality, can have important results for the energy mechanism and also technological processes. It has the potential to valuation in equipment failure and very high costs. Frequency disturbance, less power factor, voltage & current harmonics, conduction line transits, and voltage variation are overall instances of quality issues of power.

Power Availability

Power production depend on renewable sources that are uncontrollable through human is one of the most important concerns in the area of nature-sourced power. For instance, wind energy is depending on wind usability: whether the wind speed is too less, the wind turbine won't turn, emerging in no energy flux to the network. Moreover, a lot of wind can damage to the generator part, requiring the careful balance's maintenance in order to maintain consistent power generation, sun-powered electric is formed solely when sun light is existent and cuts off at darkness: Because of the variability in power production in renewable energy industries, integration is getting more difficult.

Location of the Resource

The majority of renewable power facilities that feed into the network require a lot of location. Renewable power resources are typically dictated through site, which can be discouraging to users. To begin with, some renewable power resources are easy unavailable in certain areas. Second, in terms of efficiency and cost, the distance between the grid and the renewable power resource is critical. Furthermore, renewable power resources are affected by climate, geographic location, and weather, so one kind of power production mayn't be ideal for the location.

The Issue of Cost

One of the biggest obstacles to the renewable energy's improvement is the great start-up expense of assemblage. Though a coal facility expenses about $6 per MW to construct, it is well knowledge that solar and wind power facilities also expense a lot of cost. Additionally, storage resolutions for produced power are expensive and supply an important impediment in terms of MW production.

Obstacle to Information

While there has been some improvement in this field, there is still a dearth of understanding and knowledge about the benefits and importance of renewable energy. Capital and investment appropriations have made renewable energy applications conceivable. Government officials should provide clear advice and guidance to applicants and potential recipients on how to qualify for renewable power incentives.

SEARCHING FOR A SOLUTION

This book serves as a resource for those conducting decision-making analysis for academical aims as well as researchers interested in learning more about multi-criteria decision modeling issue solutions.

Optimization and Decision-Making in the Renewable Energy Industry

- Illustrates the utilize of analysis and modeling, as well as case researches to help researchers grasp.
- Provides a comprehensive overview of developing market development and research trends in optimization and modeling utilizing multi-criteria decision modeling for many technological industries.
- Provides a step-by-step, complete technique for applying multi-criteria decision modeling to a wide range of scenarios.
- Provides a view of the multi-criteria decision modeling technique's primary lines of thought.
- Describes how businesses can utilize these approaches to their benefit in order to attain sustainability.
- Provides a solid foundation in give a roadmap for dealing with emerging market issues.
- Enables a range of approaches to strategic decision-making in various applications.

ORGANIZATION OF THE BOOK

The book is organized into 10 chapters. A brief description of each of the chapters follows:

Chapter 1 conducts a comprehensive literature study to determine the criteria to be used in location selection for biomass energy production facility. It identifies the importance levels of the criteria to be used to select the bio-mass power plant by using Fuzzy Analytical Hierarchy Process.

Chapter 2 uses deterministic and stochastic methods to examine whether the renewable energy efficiencies of EU15 countries and Turkey are sensitive to different Data Envelopment Analysis and Stochastic Frontier Analysis models using renewable energy consumption and generation parameters. The chapter presents how the renewable energy efficiency results of related countries change with different optimization models in the context of deterministic and stochastic framework and it proposes a new method to find a common solution for the different results of different optimization models.

Chapter 3 determines the criteria used in the evaluation of renewable energy resources with extremely high strategic importance that Turkey has, and to find their degree of importance with Fuzzy SWARA method. For this purpose, an application has been made to find out which criteria come to the fore and the weights of these criteria in order to evaluate Turkey's renewable energy resources. 5 main criteria as technical, economic, environmental, social and political and a total of 22 sub-criteria related to these criteria were included in the scope of the study.

Chapter 4 presents an analysis to select the most effective solar panel (100W) for a solar farm design by using COPRAS and Entropy methods. The authors obtain the solar panel data used in this chapter from the top panel firms worldwide.

Chapter 5 presents that the use of fossil fuels has decreased compared to the past due to the gradual depletion of fossil fuels and the greenhouse gases emerging in their use. As a result of this decrease, many countries have turned to alternative energy sources instead of fossil fuels. Compared to fossil fuels, renewable energy sources cause little or no harm to nature. The authors select the most optimal renewable energy source with the Plithogenic PIPRECIA method.

Chapter 6 investigates the question of the current and future electricity needs of this segment of the means of transport. The main contribution is preparing the necessary adaptations in the frame of electricity production capacity at the national level. It aims to highlight the enablers to be seized and the main barriers to be overcome using an integrated SWOT-PESTEL analysis in combination with the Analytical Hierarchy Process.

Chapter 7 states that biomass energy is an essential and sustainable type of energy for today and the future because it is produced from renewable sources and is environmentally friendly compared to fossil-based energy. It creates added value, provides employment, and creates new tax opportunities in many fields from agriculture to industry, from the transportation sector to the banking insurance sector. It argues that one of the crucial handicaps in evaluating biomass as an energy source is the collection of biomass. Biomass is produced in different amounts in different village centers, while country roads are suitable for trucks of different sizes. Authors in this chapter model the multi-capacity vehicle routing problem for biomass collection in village centers at different production capacities.

Chapter 8 presents a review of academic research attempting to face the biomass energy sector's problems using multi-criteria decision-making (MCDM) methods. It reviews related articles in the international journals from 2010 to 2021 to answer the following three questions. (i) Which methods are mainly used? (ii) Which problems attract the most attention.

Chapter 9 argues that despite renewable energy resources being cleaner and causing fewer environmental problems, the renewable energy selection problem is a complex task due to the involvement of various conflicting factors and uncertainty. It states that multi criteria decision-making methods are commonly used to handle this complexity successfully. This chapter reviews the studies focused on renewable energy resource selection problem in Turkey with multicriteria decision making methods. Its findings suggest that the number of studies increased by the time due to the growing importance of renewables. Also, AHP, TOPSIS and ANP have risen to the top of the literature as the most extensively used approaches.

Chapter 10 states that an increasing attraction in renewable power sources, due to, rising energy expenses and country-level tax inducement are driving the research to advance a sequence of improving unified resolutions and novel energy generation equipment. The novel wind turbines' installation and the novel wind farms' building are critical procedures for longtime energy generation. In this chapter, a comprehensive analysis, which combines ARAS and ENTROPY methods, is structured to choose appropriate turbines when improving a wind power plant.

Figen Balo
Firat University, Turkey

Ayse Topal
Nigde Omer Halisdemir University, Turkey

Ezgi Demir
Sumer Robotics Engineering & Consultancy Ltd., United Kingdom

Alptekin Ulutas
Sivas Cumhuriyet University, Turkey

Chapter 1
A Fuzzy Multi-Criteria Decision-Making Method for Selection of Biomass Power Plant Location

Uğur Atici
Sivas Cumhuriyet University, Turkey

Ömer Faruk Gürcan
Cumhuriyet University, Turkey

Meral Güldeş
https://orcid.org/0000-0001-6310-2504
Engineering Faculty, Cumhuriyet University, Turkey

Cenk Şahin
Cukurova University, Turkey

ABSTRACT

Renewable energy sources are clean energy sources and have a much lower environmental impact than other energy sources. Energy production from biomass allows the storage of energy raw materials, unlike other renewable energy sources. Some factors and criteria affect the site selection of biomass power plants. Determining the significant levels of these conflicting criteria is vital for investors in choosing a biomass power generation facility location. Location selection of biomass power plants is a multi-criteria decision problem for decision makers. The relative importance of the criteria includes quantitative evaluations and is an uncertain process. In this study, a comprehensive literature study determined the criteria to be considered in selecting biomass energy production facility locations. The fuzzy analytical hierarchy process was used to determine the importance levels of the criteria to be used to select the bio-mass power plant establishment location.

DOI: 10.4018/978-1-6684-2472-8.ch001

INTRODUCTION

Humanity is faced with the depletion of resources as a result of socio-economic development. The development of non-polluting renewable energy should be encouraged for resources to be long-term and sustainable (Xue, Ding, Zhao, Zhu, & Li, 2022). Developing energy efficiency and renewable energy sources is one of the effective solution methods in preventing climate change (Adua, Zhang, & Clark, 2021). The environmental effects of fossil fuels and the depletion of resources have necessitated the development of many policies to use renewable energy sources (Lerkkasemsan & Achenie, 2014). Effective energy management is possible with the efficient use of renewable energy sources (Ilbahar, Kahraman, & Cebi, 2021).

The renewable energy market is increasing around the world. Corporate companies make the majority of the supply and demand in the renewable energy market. As production and supply expertise in renewable energy sources increases, energy costs decrease (O'Shaughnessy, Heeter, Shah, & Koebrich, 2021). Energy policies should be developed to regulate the energy supply processes of renewable energy sources. Energy policies implemented in West Africa have resulted in widely access to electricity (Maman Ali & Yu, 2021). Renewable energy positively impacts economic development (Q. Wang, Dong, Li, & Wang, 2022).

Cities and centres of mass production settlement have an important place in the transition to renewable energy sources. Alternative plans are needed to transition to renewable energy sources (Hoicka, Conroy, & Berka, 2021). The most crucial problem experienced in using renewable energy sources is the technical difficulties in storing energy obtained from the sun and wind. Integration of renewable energy sources with the grid, the inability to store energy is one of the crucial handicaps in disseminating renewable energy sources (Ben Yosef et al., 2021). Contrary to other renewable energy sources, storing the gases obtained from biomass is technically more economical. However, biomass has a low energy density compared to other energy sources, and transportation costs are high (Cheng, Zhang, & Wang, 2020). Contrary to renewable energy sources, the most crucial advantage of obtaining energy from biomass is that the gases obtained from biomass can be stored. Thus, the grid can be fed with energy when needed.

Obtaining energy in biomass power plants depends on feeding with bio-waste at the location where the power plant is installed. Therefore, it is necessary to collect, store and transport waste from waste production centres of different scales to the power plant. The transportation of the feedstock, minimizing the pollutant emissions, and logistics costs should be considered in the placement of biomass power plants (Zhao & Li, 2016). Potential raw material sources for biomass are agricultural, forest, animal, urban, and industrial wastes. In this respect, another vital issue that increases biomass input is biomass diversity. (Cebi, Ilbahar, & Atasoy, 2016). Biomass is in a wide range. Forest wastes (Viana, Cohen, Lopes, & Aranha, 2010), agricultural products (Guo & Zhu, 2021), domestic wastes (food), industrial wastes (beet, olive pomace) can be counted as biomass input.

The location selection problem is one of the essential issues during the establishment of new enterprises. The priorities of the decision-makers should also be taken into account in the facility location selection problem. Facility location selection criteria vary according to the field of activity of the enterprise. The qualitative and quantitative priorities of the decision-makers are essential in terms of facility development, cost-effectiveness, and national policies. In this study, the FAHP technique was used to consider the qualitative and quantitative evaluations of the decision-makers in the biomass plant site selection.

In this study, an expert team was formed to determine the criteria, compare the criteria with each other, and weight the alternatives in terms of criteria. The expert team consists of three personnel. The first member of the team is an entrepreneur with ten years of experience investing in renewable energy. The second member is a manager with a doctoral degree who has worked in various positions in companies in renewable energy sources and is still the general manager. The third member is a consultant with seven years of experience in public policy.

In multi-criteria decision-making, it is essential to consider the qualitative and quantitative evaluations of the decision-maker. Linguistic expressions of decision-makers and uncertainty are among the factors that complicate the decision-making process. Adopted AHP has selected uncertainty in the decision-making process. FAHP method is selected uncertainty in the decision-making process. Tree expert personnel determined the criteria for selecting biomass power plant location. The hierarchical structure between criteria and alternatives was created. The criteria are compared according to their importance. The alternatives are ranked in order of importance by using pairwise comparison matrices. Indecision and uncertainty situations are modeled during the decision-making process. Thus, the aggregated results for each alternative were obtained, and the alternatives' priority was determined.

This study provides an appropriate evaluation tool for selecting alternative biomass power plants. The criteria that must be considered to select the biomass energy production facility are determined. Environmental, safety and economic criteria are the most important criteria. The power plant should be checked whether the facility location is within the flight safety zone, built on non-urban land and routes. Docking centres must have covered floods and landslides Etc. Bio-power plant and storage areas should be established at a distance that will not adversely affect industrial-scale production

In multi-criteria decision making, it is essential to consider the qualitative and quantitative evaluations of the decision-maker. Linguistic expressions of decision-makers and uncertainty are among the factors that complicate the decision-making process. Adopted AHP is selected uncertainty in decision-making process. FAHP method is selected uncertainty in decision-making process. Tree expert personnel determined the criteria for selecting biomass power plant location. The hierarchical structure between criteria and alternatives was created. The criteria are compared according to their importance. The alternatives are ranked in order of importance by using pairwise comparison matrices. Indecision and uncertainty situations are modelled during the decision-making process. Thus, the aggregated results for each alternative were obtained, and the alternatives' priority was determined.

BACKGROUND

Different facility location selection approaches have been used related to the facility selection problem of biomass plants. This study focuses on the studies in the literature using the MCDM method. The criteria to be evaluated in the study were selected based on the literature. Perpiña et al. (2013) defined three main criteria as economic, social, and environmental criteria of biomass plant location selection and 13 different related sub-criteria (Perpiña, Martínez-Llario, & Pérez-Navarro, 2013). However, many different criteria have been evaluated in the studies conducted in the literature in the last ten years. The profit, price, etc. should be considered when investing in renewable energy sources since renewable energy generation have some risk in terms of pricing, (Tsao, Vu, & Lu, 2021). Certification systems in renewable energy propose that biomass be sustainable (Fehrenbach et al., 2008). The physical properties biomass, location area, transportation infrastructure (highway, railway, sea), raw material logistics, raw material type (all logs, chips, etc.), biofuel efficiency, proximity to the source of biowaste at industrial scale, proximity of other biomass power plants have primary priority in the selection of the establishment location (Stephen, Mabee, & Saddler, 2010). Since the wastes to be used in the power plant feeding will be collected from places close to the power plant, other power plants in the target region affect the amount of input. For this reason, the positions and capacities of renewable energy sources relative to each other are important in site selection (Ben Yosef et al., 2021). Physiography, crop types, vegetation biomass waste is other criteria that should be evaluated in determining the location of the biomass power plant to be established (Perpiña et al., 2013).

In biomass plants, the input is fed with a wide range of raw material sources. Forest remains are considered among the essential biomass resources. In this context, criteria such as proximity to forest areas, short-term cultivation of vacant lands, stakeholder participation, attractive electricity price tariffs, appropriate harvesting, and use of transportation technologies should be considered in selecting the plant location (Prasad & Raturi, 2021). The planned total electricity capacity, storage capacity, and power plant type are essential in choosing the biomass plant establishment location (Bojić, Đatkov, Brcanov, Georgijević, & Martinov, 2013). Another criteria taken into consideration when choosing a biomass plant location is the size of the biomass supply area (Vera, Carabias, Jurado, & Ruiz-Reyes, 2010). The life cycle of a biomass power plant is linked to crucial phases that involve the storage and transportation of biomass and its conversion to biowaste or bioproducts. The risks of the production stages should be considered in selecting the establishment location (El-Halwagi et al., 2013).

The location selection of biopower plants is similar to the facility location selection problem, which is widely researched in the literature. While the criteria specific to biomass power plants are used in location selection, the same methodology is used to solve the problem (Karatop, Taşkan, Adar, & Kubat, 2021). A new multi-criteria decision-making model consisting of fuzzy Stepwise Weight Assessment Ratio Analysis (SWARA) and Combined Compromise Solution (CoCoSo) method been used by Ulutaş et al. (2020). It is recommended to use multi-criteria decision-making techniques when the criteria in selecting the facility location are based on subjective judgments, and the importance weights of the alternatives are determined. The Fuzzy Technique for Order Preference by Similarity to an Ideal Solution (TOPSIS) method was used by Chu (2002) to determine the weights of subjective attributes (Baykasoğlu, Kaplanoğlu, Durmuşoğlu, & Şahin, 2013; Chu, 2002). The fuzzy group decision-making method was used by Ertuğtul (2011) in selecting facility (Ertuğrul, 2011). AHP method was used by Badri (1999) to solve the facility location selection problem (Badri, 1999).

Different problems of biomass plants have been discussed in the literature (Ilbahar, Kahraman, & Cebi, 2022). The primary purpose of the circular economy is to reduce energy waste, eliminate and re-use waste. The Gross Domestic Product (GDP) has been determined with fuzzy logic by Petkovic et al. (2020) (Petković et al., 2021). The fuzzy model was used by Böhler et al. (2020) to control the criteria of the biomass combustion model (Böhler, Krail, Görtler, & Kozek, 2020). Geological structure, groundwater, Etc., were analyzed based on GIS by Karakuş et al. (2020) in determining the solid waste storage area. Analytical Hierarchy Process (AHP) and Simple Addition Weighting (SAW) multi-criteria decision-making methods were used in the analysis (Karakuş, Demiroğlu, Çoban, & Ulutaş, 2020).

In the biomass plant location selection, the feed supply points are located in different regions. Geographical information system-based multi-criteria decision making (GIS-MCDM) and Fuzzy Decision-Making Trial and Evaluation (F-MCDM) technique were used by Jeong and Ramirez (2018) to determine suitable and suitable location for biomass plants. Three different criteria have been established (Jeong & Ramírez-Gómez, 2018). It has been reported by Guo and Zhu (2021) that diversification of the input is more important instead of using a single type of raw material for biomass in choosing a biomass plant location (Guo & Zhu, 2021). It was reported by Lerkkasemsan and Achenie (2014) that it is vital to establish and operate the biomass power plant cost-effectively, and fuzzy modelling was used to estimate the production (Lerkkasemsan & Achenie, 2014). Biomass-based hydrogen production multi-actor multi-criteria decision-making method was proposed by Ren et al. (2013) (Ren, Fedele, Mason, Manzardo, & Scipioni, 2013). Wang et al. (2019) used FAHP to determine the most suitable alternative locations and TOPSIS to identify potential locations (C. N. Wang, Tsai, & Huang, 2019). Multi-criteria decision-making based on the Hurwitz algorithm was used in the location selection of regional landfills by Curčic et al. (2011). Fuzzy logic was used by Rezk et al. (2022) to simulate methane production through biomass gasification (Rezk, Inayat, Abdelkareem, Olabi, & Nassef, 2022).

Voivontas et al. (2001) examined estimating the production amount of renewable energy sources (Voivontas, Assimacopoulos, & Koukios, 2001). Diversification of energy markets dependent on imports of renewable energy sources is an important alternative. VIKOR (VlseKriterijumska Optimizacija I Kompromisno Resenje), one of the multi-criteria decision-making techniques, was used by Cristóbal (2011) to select the investment area among renewable energy sources (San Cristóbal, 2011). Multi-criteria decision-making techniques were used by Scot et al. (2012) to select the most suitable alternative among bioenergy plans and to meet the conflicting needs of stakeholders (Scott, Ho, & Dey, 2012). Mourmouris and Potolias (2013) used the multi-criteria decision-making technique to support the renewable energy market development (Mourmouris & Potolias, 2013). AHP was used by Perpiña et al. (2013) to determine the suitable alternative plant location (Perpiña et al., 2013).

In the literature, linear programming and multi-criteria decision-making techniques have been used to select biomass facility locations. However, the number of studies that use qualitative and quantitative criteria together, considering the priorities of decision-makers in decision-making processes, is quite limited. The novelty of this study is that it uses the qualitative and quantitative priorities of the decision-makers in the selection of biomass facility locations and shows how the method will be applied in the Sivas example.

MAIN FOCUS OF THE CHAPTER

Renewable energy sources are clean energy sources. They are compatible with their habitat, and their impact on the environment is neutral. Renewable energy sources consist of biomass, hydroelectric, solar, wind, marine, and geothermal energy. These resources are converted into electricity or motion power to produce power, heat, or mechanical energy by using renewable energy technologies. One of the disadvantages of renewable energy, which prevents the widespread use of renewable energy sources, is its dependence on weather conditions and its shortcoming to store and send energy when necessary. Biomass energy is stored within the organism, unlike other renewable energy sources such as solar or wind or and harvested as needed. Biomass energy has disadvantages as well as advantages.

One of the crucial issues that will allow the advantages to outweigh is the selection of the locations of the facilities where this energy will be obtained. Appropriate site selection for the biomass plant is essential because the amount transported and collected should be balanced. Otherwise, the shipping cost will cause a significant loss of profit. The site selection of biomass power plants is affected by various criteria. Identifying the significant levels of these conflicting criteria is vital for investors in choosing a biomass power plant location. Location selection of biomass power plants is a multi-criteria decision problem for decision-makers. The relative importance of the criteria involves quantitative assessments and is an uncertain process.

In decision-making problems, considering the qualitative and quantitative evaluations of the decision-maker is essential in terms of cost and national regulation. Linguistic expressions of decision-makers and uncertainty are common problems in plant location selection. Fuzzy sets are widely used to tackle this problem.

The number of studies that use qualitative and quantitative criteria together, considering the priorities of decision-makers in decision-making processes, is quite limited. The qualitative and quantitative priorities of the decision-makers in the selection of biomass facility locations should be considered. The main focus of this study is to provide an appropriate evaluation tool for selecting alternative biomass power plants. For this purpose, an expert team was formed to determine the criteria, compare the criteria, and weigh the alternatives in terms of criteria. The criteria for selecting biomass power plant location was determined. The hierarchical structure between criteria and alternatives was created. The importance of criteria was determined. The alternatives are ranked in order of importance by using pair wise comparison matrices. The aggregated results for each alternative were obtained, and the alternatives' priority was determined. The criteria that must be considered to select the biomass energy production facility are presented. Environmental, safety and economic criteria are the most important criteria. The power plant should be checked whether the facility location is within the flight safety zone, built on non-urban land and routes. Docking centers must have covered floods and landslides.

CASE STUDY AND DATA

The current number of animals in Sivas 2021 was obtained from the Provincial Directorate of Agriculture. The number of cattle in 2021 is presented in Table 1.

25% of the bovine stock is raised in enterprises close to the centre of Sivas. The distribution of the number of cattle by urban area is shown in Figure 1. The wastes considered as biomass in Sivas province are vermicompost, slaughterhouse wastes and animal faeces. Vermicompost is the process of composting organic residues by worms. In this process, organic wastes are fermented by microorganisms in the environment and then subjected to an accelerated humification and detoxification process as they pass through the digestive system of earthworms. (Lim, Wu, Lim, & Shak, 2015). Vermicompost is used for the product obtained due to the organic waste and/or waste composting process in which worms are used. (Kale, Mallesh, Kubra, & Bagyaraj, 1992).

Table 1. Number of cattle in 2021

Residential area	Quantity	Residential area	Quantity
Akıncılar	10515	Kangal	22323
Altınyayla	13899	Koyulhisar	18451
Divriği	16574	Sivas Center	78258
Doğanşar	5417	Şarkışla	27405
Generic	23051	Suşehri	21177
Gölova	2494	Ulaş	16436
Gürün	1825	Yıldızeli	4621
Hafik	16622	Zara	28247
İmranlı	7406		

Wastewater from slaughterhouses and meat processing processes often contains suspended solids, oil and grease, and floating matter (Salminen & Rintala, 2002). When blood from cattle slaughter is discharged into the sewer, the sewage pollution load is significantly increased. This is why blood recovery is encouraged by governments (S. Wang, Jena, & Das, 2018). Blood and blood meal obtained from blood are necessary biomass inputs (Hollander & Wright, 1980).

Cow manure is another important source of biomass obtained in agriculture (Fan, Zhang, Guo, Xing, & Fan, 2006). The critical problem to be encountered in implementing the disposal methods of animal waste is the ability to economically deliver a sufficient amount of farm animal manure to the central units. Using cow dung as a biomass input instead of disposal reduces costs in agriculture (Pattanaik, Duraivadivel, Hariprasad, & Naik, 2020).

Figure 1. Distribution of the number of cattle by urban area

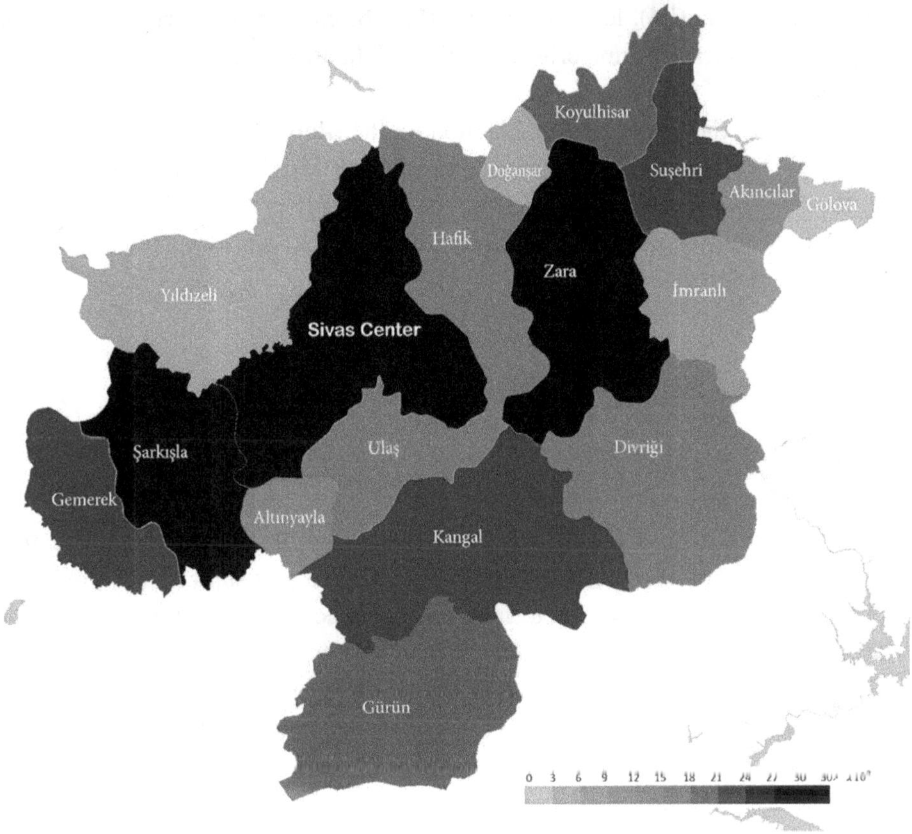

Figure 2. The selection of bio power plant location criteria and sub-criteria

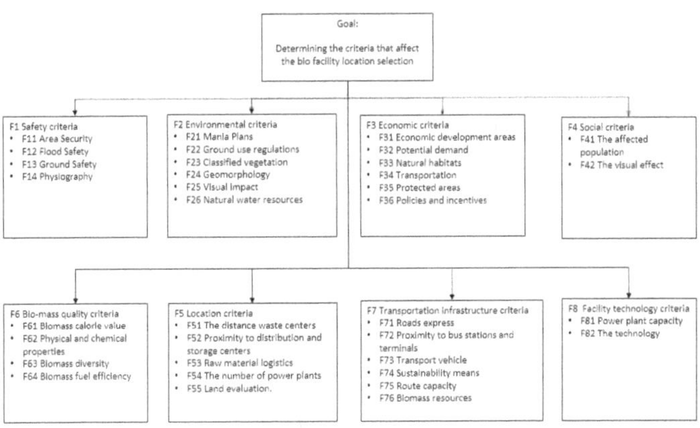

The hierarchically structured decision model is presented in Figure 2. The main criteria are the safety criteria (F1), environmental criteria (F2), economic criteria (F3), social criteria (F4), location criteria (F5), biomass quality criteria (F6), and transportation infrastructure criteria (F7), and facility technology criteria (F8) respectively.

In this study, the plant location selection for the Biopower plant installation was examined. Five alternative locations (Sivas/Merkez, Yıldızeli, Ulaş Şarkışla and Zara) have been compared for the establishment of the facility in terms of vermicompost, slaughterhouse waste and cow dung. Three different experts evaluated the criteria and sub-criteria presented in Table 2.

SOLUTIONS AND RECOMMENDATIONS

Evaluation criteria

Eight criteria and thirty-five sub-criteria for selecting biomass power plant location were deðned and prioritized. The list of these criteria and sub-criteria, presented in Table 1 and Figure 1, was determined by conducting a literature survey. The criteria are determined as safety criteria (F1), environmental criteria (F2), economic criteria (F3), social criteria (F4), location criteria (F5), biomass quality criteria (F6), transportation infrastructure criteria (F7), facility technology criteria (F8).

Safety criter (F1) is based on the four sub-criteria, e.g., area security (F11), flood safety (F12), ground safety (F13), physiography (F14). Area security (F11) means that the perimeter of the bio-power plant and storage areas is safe. Bio-power plant and storage areas should be established at a distance that will not adversely affect the urban land and industrial-scale production facility. Flood safety (F12) is another dimension of facility location safety, and it shows that the bio-plant is not affected by floods and raids that will occur. Ground safety (F13) means plant and building installation with a rock-based foundation with a topographic slope of less than 15% from the plant site. Physiography (F14) implies determining the morphological features of slopes in terms of instability and landslide hazard.

Environmental criteria (F2) are based on the six sub-criteria, e.g., mania plans (F21), ground use regulations (F22), classified vegetation (F23), geomorphology (F24), visual impact (F25), natural water resources (F26). Mania plans (F21) are city layout plans that ensure the flight safety of airports. It should be checked whether the facility location is within the flight safety zone. Ground use regulations (F22) are legal regulations that dictate that biomass power plants be built on non-urban land. Classified vegetation (F23) refers to the conservation of certain species in biomass harvesting refers to the classification of natural vegetation. Geomorphology (F24) refers to routes and docking centers threatened by erosion. The visual impact (F25) implies identifying visible areas, zero visibility of unsightly storage areas within the facility. Natural water resources (F26) indicate the proximity of the facility to underground and surface resources. The biopower plant should not pollute the nearby rivers, lakes, and pond beds.

Table 2. Criteria and sub-criteria

F1 Safety criteria	**F5 Location criteria**
F11 Area Security	F51 Distance to waste centers
F12 Flood Safety	F52 Proximity to distribution and storage centers
F13 Ground Safety	F53 Raw material logistics
F14 Physiography	F54 The number of power plants
	F55 Land evaluation.
F2 Environmental criteria	**F6 Bio-mass quality criteria**
F21 Mania Plans	F61 Biomass calorie value
F22 Ground use regulations	F62 Physical and chemical properties
F23 Classified vegetation	F63 Biomass diversity
F24 Geomorphology	F64 Biomass fuel efficiency
F25 Visual impact	
F26 Natural water resources	
F3 Economic criteria	**F7 Transportation Infrastructure criteria**
F31 Economic development areas	F71 Roads express
F32 Potential demand	F72 proximity to bus stations and terminals
F33 Natural habitats	F73 Transport vehicle
F34 Transportation	F74 Sustainability means
F35 Protected areas	F75 Route capacity
F36 Policies and incentives	F76 Biomass resources
F4 Social criteria	**F8 Facility technology criteria**
F41 Affected population	F81 Power plant capacity
F42 Visual effect	F82 The technology

Economic criteria (F3) are based on the six sub-criteria, e.g., economic development areas (F31), potential demand (F32), natural habitat (F33), transportation (F34), protected areas (F35), policies and incentives (F36). Economic development areas (F31) express whether the place of establishment is within the economic incentive areas. Potential demand (F32) determines likely demand within a residential area based on energy consumption and biomass availability. Natural habitat (F33) refers to the presence of endemic animal and plant species in a particular area. Transportation (F34) includes biomass collection and transportation costs, road length, and vehicle capacities. National laws protect natural beauties and protected areas. The foul odour and untreated water that will spread to the facility's environment adversely affect the protected areas. Protected areas (F35) are natural beauties and protected areas that will not be affected when choosing the installation sites of biopower plants and storage areas. Policies and incentives (F36) mean that the government's policy and incentives on renewable energy sources at the national level are attractive.

Social criteria (F4) are based on the two sub-criteria, e.g., affected population (F41), visual effect (F42). Affected population (F41), the distance of the facility from densely populated and residential areas, refers to the size of the affected population. The visual effect (F42) shows the effect of odour, image, pollution, and its surroundings.

Location criteria (F5) are based on the five sub-criteria, e.g., distance waste centres (F51), proximity to distribution and storage centres (F52), raw material logistics (F53), number of power plants (F54), land evaluation (F56). Distance to waste centres (F51) express the level of acceptability of the distance between the biopower plant and the waste generation production centre in the power generation facility in terms of cost. Proximity to distribution and storage centres (F52) refers to fuel tanks, water tanks, mining sites. Raw material logistics (F53) indicates that the logic of raw materials other than the biomass needed in the power plant can be provided. The number of power plants (F54) refers to the number of power plants using the same harvesting area and road route and their operating capacities. Land evaluation (F55) describes the possibility of short-term cultivation of vacant lands in the collection area and stakeholders' contribution to biowaste generation.

Biomass quality (F6) criteria are based on the four sub-criteria, e.g., biomass calorie value (F61), physical and chemical properties (F62), biomass diversity (F63), biomass fuel efficiency (F64). Biomass calorie value (F61) refers to the calorie value obtained from the unit amount harvested. Physical and chemical properties (F62) indicate the stability of the physical and chemical properties of the biomass in the collection area during the collection period. Biomass diversity (F63) shows different types of biomass such as animal wastes, logs, chips. Biomass fuel efficiency (F64) refers to converting the potential fuel biomass into kinetic energy or electrical energy.

Transportation infrastructure (F7) criteria are based on the six sub-criteria, e.g., roads express (F71), proximity to bus stations and terminals (F72), transport vehicle (F73), sustainability means (F74), route capacity (F75), biomass resources (F76). Roads (F71) express the condition of the primary and secondary roads used in the storage and transportation route. Proximity to bus stations and terminals (F72) means the proximity of the switchboard location to train bus stations and the opportunity to benefit from public transportation resources. Transport vehicle (F73) indicates types of vehicles that can be used on the storage and collection road route. Sustainability (F74) means the continuity of the supply of biomass at the collection site. Route capacity (F75) refers to the load-carrying capacity of the vehicles that can be used on the road route. Biomass resources (F76) refer to the number of supply points of biomass in the harvesting zone.

Facility technology (F8) criteria are based on the two sub-criteria, e.g., power plant capacity (F81) and technology (F82). Power plant capacity (F81) indicates the biomass power plant's daily waste processing, storage and energy production capacity. Technology (F82) shows the machine used in the plant, the production technology used, and the type of plant.

Fuzzy AHP

In decision-making problems, it is essential to consider the qualitative and quantitative evaluations of the decision-maker. Linguistic expressions of decision-makers and uncertainty are among the factors that complicate the decision-making process. Fuzzy sets can overcome such problems in modelling so that more precise results can be obtained. Fuzzy sets were proposed by Zadeh (1965) to overcome such problems (Zadeh, 1978). AHP is a method developed for solving multi-criteria decision-making problems (Büyüközkan, Havle, & Feyzioğlu, 2021; Şenyiğit & Demirel, 2018). The expert personnel determine the criteria for decision making and the purpose. Different experts create a hierarchical structure between criteria and alternatives to achieve the determined goals (Coffey & Claudio, 2021; Leśniak, Kubek, Plebankiewicz, Zima, & Belniak, 2018). The criteria are compared according to their importance (Chandna, Saini, & Kumar, 2021). Using pairwise comparison matrices, alternatives are ranked in order of importance. In the fuzzy AHP technique, indecision and uncertainty situations are modelled during the decision-making process (M. Dağdeviren, Yavuz, & Kılınç, 2009; Karam, Hussein, & Reinau, 2021). In this method, the importance of the criteria is determined to be in the value range.

There are type-1, type-2 fuzzy sets, and intuitive fuzzy sets in the literature (Güldeş et al., 2022) (Yazici et al., 2020). The best alternative is using qualitative and quantitative criteria in the multi-criteria decision-making process. The analytical hierarchy process developed by Saaty (Saaty, 1977, 1980) is a widely used method in multi-criteria decision-making probes. The uncertainty that arises with concrete and abstract concepts in the fuzzy AHP decision-making process is overcome. (Chang, 1996). In this study, Type-1 fuzzy sets were used with the AHP technique. The linguistic variables and fuzzy scale used in this study are presented in Table 3.

Table 3. Linguistic variables and fuzzy triangular numbers (Fu et al., 2020)

Linguistic Variables	Fuzzy scale		
Equally Important	(1,	1,	1)
Moderately Important	(2,	3,	4)
Important	(4,	5,	6)
Very Important	(6,	7,	8)
Extremely Important	(9,	9,	9)

Pairwise comparisons are made between the criteria, and the pairwise comparison matrices presented in Eq.1 are created. By questioning which criterion is more important, the corresponding linguistic expressions are determined. The fuzzy geometric mean is calculated by using Eq.2.

$$\tilde{a}_{ij} = \begin{bmatrix} 1 & \tilde{a}_{12} & \tilde{a}_{13} & \cdots & \tilde{a}_{1n} \\ \tilde{a}_{21} & \tilde{a}_{22} & \tilde{a}_{23} & \cdots & \tilde{a}_{2n} \\ \vdots & \vdots & \vdots & \ddots & \vdots \\ \tilde{a}_{n1} & \tilde{a}_{n2} & \tilde{a}_{n3} & \cdots & 1 \end{bmatrix} \tag{1}$$

$$\tilde{r}_i = \left(\tilde{a}_{i1} \otimes \tilde{a}_{i2} \otimes \ldots \otimes \tilde{a}_{in}\right)^{1/n} \tag{2}$$

The fuzzy weight of each criterion is calculated by using Eq.3. The vector sum of each $\tilde{r}_i$ is obtained. Take the inverse of the sum and multiply each $\tilde{r}_i$.

$$\tilde{w}_i = \tilde{r}_i \otimes \left(\tilde{r}_1 \otimes \tilde{r}_2 \otimes \ldots \otimes \tilde{r}_n\right)^{-1} \tag{3}$$

The best non-fuzzy performance value of each criterion is obtained by using Eq.4.

$$M_i = \frac{lw_i + mw_i + uw_i}{3} \tag{4}$$

Hence M is a non-fuzzy number, normalized by Eq.5

$$n_i = \frac{M_i}{\sum_{j=1}^{n} M_i} \tag{5}$$

Fuzzy sets are an essential tool and mathematical expression in modelling fuzzy logic. Fuzzy set theory helps to measure uncertainty through subjective judgments. A fuzzy set $\hat{A}$ is a function defined in the closed interval [0,1] expressed in Eq.6 (M. Dağdeviren, Akay, & Kurt, 2004).

$$\mu_{\tilde{A}} : E \rightarrow [0,1] \tag{6}$$

The interval values in the set of real numbers form a different number of fuzzy sets. Fuzzy sets are expressed with fuzzy numbers to show exact, uncertain and approximate values. Fuzzy numbers are characterized by verbal expressions such as approximately, more or less, almost.

A represents the lowest value on the left, *b* represents the best possible value, and *c* represents the highest limit; the value on the right is a fuzzy triangular set of numbers in the form of (*ab,c*). The membership function of the fuzzy number A is presented in Eq.7.

$$\mu_{\tilde{A}}(\mathrm{x}) \begin{cases} 0 & x \leq a \\ \dfrac{x-a}{b-a} & a \leq x \leq b \\ \dfrac{c-x}{c-b} & b \leq x \leq c \\ 0 & x \geq c \end{cases} \tag{7}$$

Mathematical operations can be performed on fuzzy sets as in classical sets. Let there be two positive fuzzy $\tilde{A} = (a_1, a_2, a_3)$ and $\tilde{B} = (b_1, b_2, b_3)$ fuzzy numbers, addition is presented in Eq.8, subtraction is Eq.9, multiplication is Eq.10 is division, Eq.11, multiplication by a constant number is Eq.12, the inverse is presented in Eq.13.

$$\tilde{A} + \tilde{B} = (a_1 + b_1, a_2 + b_2, a_3 + b_3) \tag{8}$$

$$\tilde{A}-\tilde{B}= (a_1-b_1, a_2-b_2, a_3-b_3) \tag{9}$$

$$\tilde{A}.\tilde{B}= (a_1.b_1, a_2.b_2, a_3.b_3) \tag{10}$$

$$\tilde{A}/\tilde{B}= (a_1/b_1, a_2/b_2, a_3/b_3) \tag{11}$$

$$k.\tilde{A}= (\mathrm{k}.a_1, \mathrm{k}.a_2, \mathrm{k}.a_3) \tag{12}$$

$$\tilde{A}^{-1}=(\frac{1}{a_1},\frac{1}{a_2},\frac{1}{a_3}) \tag{13}$$

To decide on fuzzy processing processes, it is necessary to order the fuzzy numbers. In the integral ranking method, the $\mu_{\tilde{A}}$: E®[0,1] the optimism index value is used. Triangular fuzzy number, the total integral value is calculated using Eq.14 for $\tilde{A} = (a_1, a_2, a_3)$.

$$I_T^a=(\tilde{A})=\frac{1}{2}\alpha(a_2+a_3)+\frac{1}{2}(1-\alpha)(a_1+a_2)=\frac{1}{2}[\alpha(a_2+a_3+(1-\alpha)\alpha] \tag{14}$$

An increase in the index value represents an optimistic decision-maker, and a decrease in the index value represents a pessimistic decision-maker. For fuzzy numbers $\tilde{A}$ and $\tilde{B}$ Sorting is done using Eq.15.

$$I_T^a\left(\tilde{A}\right)<I_T^a\left(\tilde{B}\right) \qquad \left(\tilde{A}\right)<\left(\tilde{B}\right) \tag{15}$$

$$I_T^a\left(\tilde{A}\right)=I_T^a\left(\tilde{B}\right) \qquad \left(\tilde{A}\right)=\left(\tilde{B}\right)$$

$$I_T^a\left(\tilde{A}\right)>I_T^a\left(\tilde{B}\right) \qquad \left(\tilde{A}\right)>\left(\tilde{B}\right)$$

Calculation example

Determining the priority among three alternatives using three criteria is prepared as an example. The criteria weights in the comparison matrix are randomly generated. Pairwise comparison value of all criteria is presented in Table 4.

Table 4. Pairwise comparison of criteria for calculation example

	C1			C2			C3		
C1	1	1	1	5	4	6	1/3	1/7	1/4
C2	1/6	1/4	1/5	1	1	1	7	8	9
C3	4	7	3	1/9	1/8	1/7	1	1	1

The geometric mean of fuzzy comparison value is calculated as follows.

$$\tilde{r}_1 = \left(1\text{x}5\text{x}\frac{1}{3}\right)^{1/3} + \left(\frac{1}{6}\text{x}1\text{x}7\right)^{1/3} + \left(4\text{x}\frac{1}{9}\text{x}1\right)^{1/3} = 1.186 + 0.830 + 1.145 = 3.002$$

$$\tilde{r}_2 = \left(1\text{x}4\text{x}\frac{1}{7}\right)^{1/3} + \left(\frac{1}{4}\text{x}1\text{x}8\right)^{1/3} + \left(7\text{x}\frac{1}{8}\text{x}1\right)^{1/3} = 1.053 + 1.260 + 1.216 = 3.046$$

$$\tilde{r}_3 = \left(1\text{x}6\text{x}\frac{1}{4}\right)^{1/3} + \left(\frac{1}{5}\text{x}1\text{x}9\right)^{1/3} + \left(3\text{x}1\text{x}1\right)^{1/3} = 0.763 + 0.956 + 0.754 = 3.115$$

$$\tilde{r}_1^{-1} = 3.002^{-1} = 0.333$$

$$\tilde{r}_2^{-1} = 3.046^{-1} = 0.328$$

$$\tilde{r}_3^{-1} = 3.115^{-1} = 0.32$$

l.m.u is ascending order of $\tilde{r}_i^{-1}$. Hence, *l.m.u* equal to 0.321, 0.328 and 0.333 respectively. The relative fuzzy weight of each criteria is calculated as follows.

$$M_1 = \frac{lw_1 + mw_2 + uw_3}{3} = \frac{0.321x1.186 + 0.328x0.830 + 0.333x1.145}{3} = 0.345$$

$$M_2 = \frac{lw_1 + mw_2 + uw_3}{3} = \frac{0.321x1.053 + 0.328x1.260 + 0.333x1.216}{3} = 0.386$$

$$M_3 = \frac{lw_1 + mw_2 + uw_3}{3} = \frac{0.321x0.763 + 0.328x0.956 + 0.333x0.754}{3} = 0.270$$

Average and normalized weight of criteria are calculated as follows.

$$n_1 = \frac{0.345}{0.345 + 0.386 + 0.270} = 0.34$$

$$n_2 = \frac{0.386}{0.345 + 0.386 + 0.270} = 0.39$$

$$n_3 = \frac{0.270}{0.345 + 0.386 + 0.270} = 0.2$$

Similarly, each alternative is compared with one another in terms of criterion. Aggregated results for each alternative according to each criterion are obtained as shown in Table 5.

Table 5. Aggregated results for each alternative

	C1	C2	C3
Weights	0.34	0.39	0.27
A1	0.06	0.06	0.22
A2	0.71	0.27	0.13
A3	0.23	0.66	0.64

Priority of the alternatives is calculated as presented below. Among the alternatives, A3 is chosen because it has thc highest score.

A1= 0.34+0.39+0.27+0.06+0.06+0.22 =0.11
A2= 0.34+0.39+.027+0.71+0.27+0.13 =0.39
A3= 0.34+0.39+0.27+0.23+0.66+0.64 =0.51

RESULTS AND FINDINGS

According to the weight values (W), the first three main criteria were environmental criteria (F2), safety criteria (F1), and economic (F3) criteria with W=0.326, W=0.182 W=0.178 weight values, respectively. The weights of the criterion are shown in Table 6.

Table 6. Ranked weights of criteria

Criteria	Weights
F2 Environmental criteria	0.292
F1 Safety criterion	0.265
F3 Economic criteria	0.156
F5 Location criteria	0.086
F7 Transportation infrastructure criteria	0.070
F4 Social criteria	0.060
F6 Bio-mass quality criteria	0.056
F8 Facility technology criteria	0.016

The most crucial criterion is the environmental criteria (F2). The power plant should be checked to see whether the facility is within the flight safety zone, built on non-urban land and routes and docking centres have covered floods, landslides, etc. The second vital criterion is the safety criterion (F1) which implies that bio-power plant and storage areas should be established at a distance that will not adversely affect the industrial-scale production, bio-plant should not be affected by floods and raids, the plant must have a rock-based foundation with a topographic slope of less than 15% from the site of the plant. The third essential criteria are the economic criteria (F3) which indicate user information, authorization-based access, ensuring the security of measurement, and evaluation system should be considered.

According to the global weight (GW) order of ten sub-criteria is flood safety (F12), area security (F11), ground use regulations (F22), mania plans (F21), natural water (F26), visual impact (F25), ground safety (F13), transportation (F34), natural habitats (F33), classified vegetation (F23), economic development areas (F31), the visual effect (F42), the affected population (F41), protected areas (F35), raw material logistics (F53) respectively. The ranked global weights (GW) of subcriteria are presented in Table 7. According to GW, the most important sub-criter is flood safety (F12) which shows that the bio-plant is not affected by floods and raids. According to global weight, the second important sub-criteria is area security (F11) which represents bio-power plants, and storage areas should be established at a fair distance between urban land and other industrial plants. According to GW, the third important sub-criter is ground use regulations (F22) which shows that location must be compatible with ground use regulations. According to global weight, the fourth important sub-criteria is mania plans (F21) that stand for power plant location should be checked in a flight safety zone. According to global weight, the fifth important sub-criteria is natural water (F26) which means that the biopower plant should not pollute the nearby river, lake, and pond beds.

Table 7. Ranked global weights of sub-criteria

Sub-Criteria	Weights	Yıldızeli	Merkez	Ulaş	Şarkışla	Zara
F12 Flood Safety	0.121	0.103	0.328	0.344	0.098	0.126
F11 Area Security	0.092	0.042	0.531	0.144	0.065	0.219
F22 Ground use regulations	0.070	0.168	0.203	0.380	0.127	0.123
F21 Mania Plans	0.067	0.124	0.346	0.340	0.132	0.057
F26 Natural water	0.061	0.144	0.336	0.351	0.111	0.057
F25 Visual impact	0.051	0.146	0.348	0.317	0.128	0.061
F13 Ground Safety	0.047	0.164	0.226	0.375	0.120	0.116
F34 Transportation	0.041	0.143	0.328	0.344	0.124	0.061
F33 Natural habitats	0.038	0.145	0.346	0.335	0.116	0.057
F23 Classified vegetation	0.034	0.154	0.290	0.371	0.117	0.067
F31 Economic development areas	0.033	0.121	0.359	0.341	0.123	0.055
F42 The visual effect	0.031	0.263	0.042	0.338	0.229	0.128
F41 The affected population	0.030	0.124	0.358	0.343	0.118	0.057
F35 Protected areas	0.028	0.140	0.340	0.335	0.125	0.061
F53 Raw material logistics	0.021	0.157	0.333	0.344	0.096	0.070
F51 The distance waste centers	0.020	0.142	0.344	0.331	0.127	0.056
F55 Land evaluation.	0.019	0.141	0.351	0.335	0.114	0.058
F54 The number of power plants	0.018	0.140	0.350	0.324	0.124	0.063
F61 Biomass calorie value	0.017	0.175	0.305	0.333	0.123	0.064

continues on following page

Table 7. Continued

Sub-Criteria	Weights	Yıldızeli	Merkez	Ulaş	Şarkışla	Zara
F64 Biomass fuel efficiency	0.015	0.150	0.345	0.327	0.116	0.062
F71 Roads express	0.014	0.141	0.339	0.314	0.141	0.066
F72 Proximity to bus stations and terminals	0.013	0.157	0.337	0.345	0.104	0.058
F62 Physical and chemical properties	0.013	0.141	0.348	0.322	0.124	0.065
F74 Sustainability means	0.012	0.145	0.328	0.351	0.118	0.059
F63 Biomass diversity	0.012	0.161	0.331	0.344	0.103	0.060
F73 Transport vehicle	0.011	0.139	0.348	0.340	0.113	0.060
F75 Route capacity	0.010	0.126	0.356	0.317	0.143	0.058
F76 Biomass resources	0.009	0.147	0.338	0.326	0.130	0.059
F81 Power plant capacity	0.009	0.121	0.380	0.331	0.111	0.057
F24 Geomorphology	0.009	0.139	0.314	0.369	0.114	0.064
F36 Policies and incentives	0.008	0.128	0.338	0.343	0.132	0.060
F52 Proximity to distribution and storage centers	0.008	0.142	0.349	0.339	0.114	0.056
F32 Potential demand	0.006	0.132	0.360	0.323	0.125	0.061
F82 The technology	0.006	0.169	0.338	0.336	0.097	0.059
F14 Physiography	0.005	0.066	0.272	0.539	0.041	0.082

FUTURE RESEARCH DIRECTIONS

In today's modern and industrialized world, studies on energy technologies show an increase to keep up with the increasing energy demand. Renewable energy sources called clean energy have an important place in these studies due to their environment-friendly nature. The literature mainly consists of facility location selection studies. In this study, a location selection was made for the biomass energy production facility.

Environmental, safety and economic criteria are the most important criteria for selecting biomass power plant location. The power plant should be checked whether the facility location is within the flight safety zone, built on non-urban land and routes. Docking centers must have covered floods and landslides Etc. Bio-power plant and storage areas should be established at a distance that will not adversely affect industrial-scale production. The potential solution must be based on environmental, safety, and economic criteria. While creating a benefit-cost balance in potential solutions, penalty weight points can be applied for flood risk and adverse effects on residential, agricultural, and industrial areas. Technical opinions of three experts were used to determine the criteria. Increasing the number of experts in determining the criteria will allow the evaluation of different perspectives. The distribution of animal numbers based on districts was accepted as homogeneous in the study. Considering the difference of village centers from city centers, it will increase the study's accuracy to take into account the village centers where the barns are located in future studies.

In future studies, facility location problems can be solved using the method presented here for different types of renewable energy. In addition, a comparison can be made by obtaining different solutions with other MCDM models. Simulation method can be used as a method to evaluate alternatives in location selection. By considering the biomass supply chain as a whole, various optimization problems can be presented by examining the problems encountered by the biomass raw material at different points of the chain.

CONCLUSION

Humanity is faced with the depletion of resources as a result of socio-economic development. The development of non-polluting renewable energy should be encouraged for resources to be long-term and sustainable. The development of energy efficiency and renewable energy sources is one of the effective solution methods in preventing climate change.

The environmental effects of fossil fuels and the depletion of resources have necessitated the development of many policies for renewable energy sources. Many criteria and criteria affect the location selecting of biomass power plants.

In this study, criteria that must be considered to select the biomass energy production facility s are determined via a comprehensive literature study. The Fuzzy Analytical Hierarchy Process was used to determine the importance levels of the criteria to select the bio-mass power plant establishment location. Eight criteria and thirty-five sub-criteria for selecting biomass power plant location were deðned and prioritized.

According to the weight values (W), the first three main criteria were F2-environmental criteria, F1- safety criteria, and F3-economic criteria with W=0.326, W=0.182 W=0.178 weight values, respectively. The findings show that the power plant should be checked whether the facility location is within the flight safety zone, built on non-urban land and routes. Docking centres must have covered floods and landslides etc. Bio-power plant and storage areas should be established at a distance that will not adversely affect industrial-scale production. Bio-plant should not be affected by floods and raids. The plant must have a rock-based foundation with a topographic slope of less than 15% from the plant site.

According to the global weight (GW), order of ten sub-criteria is determined as flood safety (F12), area security (F11), ground use regulations (F22), mania plans (F21), natural water (F26), visual impact (F25), ground safety (F13), transportation (F34), natural habitats (F33), classified vegetation (F23), economic development areas (F31), the visual effect (F42), the affected population (F41), protected areas (F35), raw material logistics (F53), respectively. The primary concern should be that the bio-plant is not affected by floods and raids. Bio-power plants and storage areas should be established reasonably far between urban land and another industrial plant. The location must be compatible with ground use regulations. The power plant location should be checked in terms of the flight safety zone. The biopower plant should not pollute the nearby rivers, lakes, and pond beds.

REFERENCES

Adua, L., Zhang, K. X., & Clark, B. (2021). Seeking a handle on climate change: Examining the comparative effectiveness of energy efficiency improvement and renewable energy production in the United States. *Global Environmental Change*, *70*, 102351. doi:10.1016/j.gloenvcha.2021.102351

Badri, M. A. (1999). Combining the analytic hierarchy process and goal programming for global facility location-allocation problem. *International Journal of Production Economics*, *62*(3), 237–248. doi:10.1016/S0925-5273(98)00249-7

Baykasoğlu, A., Kaplanoğlu, V., Durmuşoğlu, Z. D. U., & Şahin, C. (2013). Integrating fuzzy DEMATEL and fuzzy hierarchical TOPSIS methods for truck selection. *Expert Systems with Applications*, *40*(3), 899–907. doi:10.1016/j.eswa.2012.05.046

Ben Yosef, G., Navon, A., Poliak, O., Etzion, N., Gal, N., Belikov, J., & Levron, Y. (2021). Frequency stability of the Israeli power grid with high penetration of renewable sources and energy storage systems. *Energy Reports*, *7*, 6148–6161. doi:10.1016/j.egyr.2021.09.057

Böhler, L., Krail, J., Görtler, G., & Kozek, M. (2020). Fuzzy model predictive control for small-scale biomass combustion furnaces. *Applied Energy*, *276*, 115339. doi:10.1016/j.apenergy.2020.115339

Bojić, S., Đatkov, Đ., Brcanov, D., Georgijević, M., & Martinov, M. (2013). Location allocation of solid biomass power plants: Case study of Vojvodina. *Renewable & Sustainable Energy Reviews*, *26*, 769–775. doi:10.1016/j.rser.2013.06.039

Büyüközkan, G., Havle, C. A., & Feyzioğlu, O. (2021). An integrated SWOT based fuzzy AHP and fuzzy MARCOS methodology for digital transformation strategy analysis in airline industry. *Journal of Air Transport Management*, *97*, 102142. doi:10.1016/j.jairtraman.2021.102142

Cebi, S., Ilbahar, E., & Atasoy, A. (2016). A fuzzy information axiom based method to determine the optimal location for a biomass power plant: A case study in Aegean Region of Turkey. *Energy*, *116*, 894–907. doi:10.1016/j.energy.2016.10.024

Chandna, R., Saini, S., & Kumar, S. (2021). Fuzzy AHP based performance evaluation of massive online courses provider for online learners. *Materials Today: Proceedings*, *46*, 11103–11112. doi:10.1016/j.matpr.2021.02.255

Chang, D.-Y. (1996). Applications of the extent analysis method on fuzzy AHP. *European Journal of Operational Research*, *95*(3), 649–655. doi:10.1016/0377-2217(95)00300-2

Cheng, W., Zhang, Y., & Wang, P. (2020). Effect of spatial distribution and number of raw material collection locations on the transportation costs of biomass thermal power plants. *Sustainable Cities and Society*, *55*, 102040. doi:10.1016/j.scs.2020.102040

Chu, T. C. (2002). Facility location selection using fuzzy Topsis under group decisions. *International Journal of Uncertainty, Fuzziness and Knowledge-based Systems*, *10*(06), 687–701. doi:10.1142/S0218488502001739

Coffey, L., & Claudio, D. (2021). In defense of group fuzzy AHP: A comparison of group fuzzy AHP and group AHP with confidence intervals. *Expert Systems with Applications*, *178*, 114970. doi:10.1016/j.eswa.2021.114970

Dağdeviren, M., Akay, D., & Kurt, M. (2004). Analytical hierarchy process for job evaluation and application *J. Fac. Eng. Arch. Gazi Univ.*, *19*(2), 131–138.

Dağdeviren, M., Yavuz, S., & Kılınç, N. (2009). Weapon selection using the AHP and TOPSIS methods under fuzzy environment. *Expert Systems with Applications*, *36*(4), 8143–8151. doi:10.1016/j.eswa.2008.10.016

El-Halwagi, A. M., Rosas, C., Ponce-Ortega, J. M., Jiménez-Gutiérrez, A., Mannan, M. S., & El-Halwagi, M. M. (2013). Multiobjective optimization of biorefineries with economic and safety objectives. *AIChE Journal. American Institute of Chemical Engineers*, *59*(7), 2427–2434. doi:10.1002/aic.14030

Ertuğrul, İ. (2011). Fuzzy group decision making for the selection of facility location. *Group Decision and Negotiation*, *20*(6), 725–740. doi:10.100710726-010-9219-1

Fan, Y. T., Zhang, G. S., Guo, X. Y., Xing, Y., & Fan, M. H. (2006). Biohydrogen-production from beer lees biomass by cow dung compost. *Biomass and Bioenergy*, *30*(5), 493–496. doi:10.1016/j.biombioe.2005.10.009

Fehrenbach, H., Giegrich, J., Reinhardt, G., Schmitz, J., Sayer, U., Gretz, M., . . . Lanje, K. (2008). *Criteria for a sustainable use of bioenergy on a global scale*. Retrieved from Germany: https://www.osti.gov/etdeweb/servlets/purl/21240931

Fu, H.-H., Chen, Y.-Y., & Wang, G.-J. (2020). Using a Fuzzy Analytic Hierarchy Process to Formulate an Effectual Tea Assessment System. *Sustainability*, *12*(15), 6131. doi:10.3390u12156131

Güldeş, M., Atici, U., & Şahin, C. (2022). Fuzzy Resource-Constrained Project Scheduling Under Learning Considerations. Intelligent and Fuzzy Techniques for Emerging Conditions and Digital Transformation, Cham.

Guo, J. X., & Zhu, K. W. (2021). *Operation management of hybrid biomass power plant considering environmental constraints*. Sustainable Production and Consumption., doi:10.1016/j.spc.2021.09.017

Hoicka, C. E., Conroy, J., & Berka, A. L. (2021). Reconfiguring actors and infrastructure in city renewable energy transitions: A regional perspective. *Energy Policy*, *158*, 112544. doi:10.1016/j.enpol.2021.112544

Hollander, A. L., & Wright, R. E. (1980). Impact of tahanids on cattle: Blood meal size and preferred feeding sites12. *Journal of Economic Entomology*, *73*(3), 431–433. doi:10.1093/jee/73.3.431

Ilbahar, E., Kahraman, C., & Cebi, S. (2021). Location selection for waste-to-energy plants by using fuzzy linear programming. *Energy*, *234*, 121189. Advance online publication. doi:10.1016/j.energy.2021.121189

Ilbahar, E., Kahraman, C., & Cebi, S. (2022). Risk assessment of renewable energy investments: A modified failure mode and effect analysis based on prospect theory and intuitionistic fuzzy AHP. *Energy*, *239*, 121907. doi:10.1016/j.energy.2021.121907

Jeong, J. S., & Ramírez-Gómez, Á. (2018). Optimizing the location of a biomass plant with a fuzzy-DEcision-MAking Trial and Evaluation Laboratory (F-DEMATEL) and multi-criteria spatial decision assessment for renewable energy management and long-term sustainability. *Journal of Cleaner Production*, *182*, 509–520. doi:10.1016/j.jclepro.2017.12.072

Kale, R. D., Mallesh, B. C., Kubra, B., & Bagyaraj, D. J. (1992). Influence of vermicompost application on the available macronutrients and selected microbial populations in a paddy field. *Soil Biology & Biochemistry*, *24*(12), 1317–1320. doi:10.1016/0038-0717(92)90111-A

Karakuş, C. B., Demiroğlu, D., Çoban, A., & Ulutaş, A. (2020). Evaluation of GIS-based multi-criteria decision-making methods for sanitary landfill site selection: The case of Sivas city, Turkey. *Journal of Material Cycles and Waste Management*, *22*(1), 254–272. doi:10.100710163-019-00935-0

Karam, A., Hussein, M., & Reinau, K. H. (2021). Analysis of the barriers to implementing horizontal collaborative transport using a hybrid fuzzy Delphi-AHP approach. *Journal of Cleaner Production*, *321*, 128943. doi:10.1016/j.jclepro.2021.128943

Karatop, B., Taşkan, B., Adar, E., & Kubat, C. (2021). Decision analysis related to the renewable energy investments in Turkey based on a Fuzzy AHP-EDAS-Fuzzy FMEA approach. *Computers & Industrial Engineering*, *151*, 106958. doi:10.1016/j.cie.2020.106958

Lerkkasemsan, N., & Achenie, L. E. K. (2014). Pyrolysis of biomass – fuzzy modeling. *Renewable Energy*, *66*, 747–758. doi:10.1016/j.renene.2014.01.014

Leśniak, A., Kubek, D., Plebankiewicz, E., Zima, K., & Belniak, S. (2018). Fuzzy AHP application for supporting contractors' bidding decision. *Symmetry*, *10*(11), 642. doi:10.3390ym10110642

Lim, S. L., Wu, T. Y., Lim, P. N., & Shak, K. P. Y. (2015). The use of vermicompost in organic farming: Overview, effects on soil and economics. *Journal of the Science of Food and Agriculture*, *95*(6), 1143–1156. doi:10.1002/jsfa.6849

Maman Ali, M. M., & Yu, Q. (2021). Assessment of the impact of renewable energy policy on sustainable energy for all in West Africa. *Renewable Energy*, *180*, 544–551. doi:10.1016/j.renene.2021.08.084

Mourmouris, J. C., & Potolias, C. (2013). A multi-criteria methodology for energy planning and developing renewable energy sources at a regional level: A case study Thassos, Greece. *Energy Policy*, *52*, 522–530. doi:10.1016/j.enpol.2012.09.074

O'Shaughnessy, E., Heeter, J., Shah, C., & Koebrich, S. (2021). Corporate acceleration of the renewable energy transition and implications for electric grids. *Renewable & Sustainable Energy Reviews*, *146*, 111160. doi:10.1016/j.rser.2021.111160

Pattanaik, L., Duraivadivel, P., Hariprasad, P., & Naik, S. N. (2020). Utilization and re-use of solid and liquid waste generated from the natural indigo dye production process – A zero waste approach. *Bioresource Technology*, *301*, 122721. doi:10.1016/j.biortech.2019.122721

Perpiña, C., Martínez-Llario, J. C., & Pérez-Navarro, Á. (2013). Multicriteria assessment in GIS environments for siting biomass plants. *Land Use Policy*, *31*, 326–335. doi:10.1016/j.landusepol.2012.07.014

Petković, B., Agdas, A. S., Zandi, Y., Nikolić, I., Denić, N., Radenkovic, S. D., Almojil, S. F., Roco-Videla, A., Kojić, N., Zlatković, D., & Stojanović, J. (2021). Neuro fuzzy evaluation of circular economy based on waste generation, recycling, renewable energy, biomass and soil pollution. *Rhizosphere*, *19*, 100418. doi:10.1016/j.rhisph.2021.100418

Prasad, R. D., & Raturi, A. (2021). Prospects of sustainable biomass-based power generation in a small island country. *Journal of Cleaner Production*, *318*, 128519. doi:10.1016/j.jclepro.2021.128519

Ren, J., Fedele, A., Mason, M., Manzardo, A., & Scipioni, A. (2013). Fuzzy Multi-actor Multi-criteria Decision Making for sustainability assessment of biomass-based technologies for hydrogen production. *International Journal of Hydrogen Energy*, *38*(22), 9111–9120. doi:10.1016/j.ijhydene.2013.05.074

Rezk, H., Inayat, A., Abdelkareem, M. A., Olabi, A. G., & Nassef, A. M. (2022). Optimal operating parameter determination based on fuzzy logic modeling and marine predators algorithm approaches to improve the methane production via biomass gasification. *Energy*, *239*, 122072. doi:10.1016/j.energy.2021.122072

Salminen, E., & Rintala, J. (2002). Anaerobic digestion of organic solid poultry slaughterhouse waste – a review. *Bioresource Technology*, *83*(1), 13–26. doi:10.1016/S0960-8524(01)00199-7

San Cristóbal, J. R. (2011). Multi-criteria decision-making in the selection of a renewable energy project in spain: The Vikor method. *Renewable Energy*, *36*(2), 498–502. doi:10.1016/j.renene.2010.07.031

Scott, J. A., Ho, W., & Dey, P. K. (2012). A review of multi-criteria decision-making methods for bioenergy systems. *Energy*, *42*(1), 146–156. doi:10.1016/j.energy.2012.03.074

Şenyiğit, E., & Demirel, B. (2018). The selection of material in dental implant with entropy based sımple additive weighting and analytic hıerarchy process methods. *Sigma: Journal of Engineering & Natural Sciences*, *36*(3), 731–740.

Stephen, J. D., Mabee, W. E., & Saddler, J. N. (2010). Biomass logistics as a determinant of second-generation biofuel facility scale, location and technology selection. *Biofuels, Bioproducts and Biorefining, 4*(5), 503-518. doi:10.1002/bbb.239

Tsao, Y.-C., Vu, T.-L., & Lu, J.-C. (2021). Pricing, capacity and financing policies for investment of renewable energy generations. *Applied Energy, 303*, 117664. doi:10.1016/j.apenergy.2021.117664

Ulutaş, A., Karakuş, C. B., & Topal, A. (2020). Location selection for logistics center with fuzzy SWARA and CoCoSo methods. *Journal of Intelligent & Fuzzy Systems, 38*, 4693-4709. https:// doi:10.3233/JIFS-191400

Vera, D., Carabias, J., Jurado, F., & Ruiz-Reyes, N. (2010). A Honey Bee Foraging approach for optimal location of a biomass power plant. *Applied Energy, 87*(7), 2119–2127. doi:10.1016/j.apenergy.2010.01.015

Viana, H., Cohen, W. B., Lopes, D., & Aranha, J. (2010). Assessment of forest biomass for use as energy. GIS-based analysis of geographical availability and locations of wood-fired power plants in Portugal. *Applied Energy, 87*(8), 2551–2560. doi:10.1016/j.apenergy.2010.02.007

Voivontas, D., Assimacopoulos, D., & Koukios, E. G. (2001). Assessment of biomass potential for power production: A GIS based method. *Biomass and Bioenergy, 20*(2), 101–112. doi:10.1016/S0961-9534(00)00070-2

Wang, C. N., Tsai, T. T., & Huang, Y. F. (2019). A model for optimizing location selection for biomass energy power plants. *Processes (Basel, Switzerland), 7*(6), 353. doi:10.3390/pr7060353

Wang, Q., Dong, Z., Li, R., & Wang, L. (2022). Renewable energy and economic growth: New insight from country risks. *Energy, 238*, 122018. doi:10.1016/j.energy.2021.122018

Wang, S., Jena, U., & Das, K. C. (2018). Biomethane production potential of slaughterhouse waste in the United States. *Energy Conversion and Management, 173*, 143–157. doi:10.1016/j.enconman.2018.07.059

Xue, J., Ding, J., Zhao, L., Zhu, D., & Li, L. (2022). An option pricing model based on a renewable energy price index. *Energy, 239*, 122117. doi:10.1016/j.energy.2021.122117

Yazici, I., Beyca, O. F., Gurcan, O. F., Zaim, H., Delen, D., & Zaim, S. (2020). A comparative analysis of machine learning techniques and fuzzy analytic hierarchy process to determine the tacit knowledge criteria. *Annals of Operations Research*. Advance online publication. doi:10.100710479-020-03697-3

Zadeh, L. A. (1978). Fuzzy sets as a basis for a theory of possibility. *Fuzzy Sets and Systems*, *1*(1), 3–28. doi:10.1016/0165-0114(78)90029-5

Zhao, X. G., & Li, A. (2016). A multi-objective sustainable location model for biomass power plants: Case of China. *Energy*, *112*, 1184–1193. doi:10.1016/j.energy.2016.07.011

ADDITIONAL READING

Asadi, S., Nilashi, M., Iranmanesh, M., Hyun, S. S., & Rezvani, A. (2021). Effect of internet of things on manufacturing performance: A hybrid multi-criteria decision-making and neuro-fuzzy approach. *Technovation*, *102426*. Advance online publication. doi:10.1016/j.technovation.2021.102426

Chen, L., Duan, G., Wang, S., & Ma, J. (2020). A Choquet integral based fuzzy logic approach to solve uncertain multi-criteria decision making problem. *Expert Systems with Applications*, *149*, 113303. doi:10.1016/j.eswa.2020.113303

Cheraghi, M., Eslami Baladeh, A., & Khakzad, N. (2022). Optimal selection of safety recommendations: A hybrid fuzzy multi-criteria decision-making approach to HAZOP. *Journal of Loss Prevention in the Process Industries*, *74*, 104654. doi:10.1016/j.jlp.2021.104654

Dalalah, D., Hayajneh, M., & Batieha, F. (2011). A fuzzy multi-criteria decision making model for supplier selection. *Expert Systems with Applications*, *38*(7), 8384–8391. doi:10.1016/j.eswa.2011.01.031

Erbaş, M., Kabak, M., Özceylan, E., & Çetinkaya, C. (2018). Optimal siting of electric vehicle charging stations: A GIS-based fuzzy Multi-Criteria Decision Analysis. *Energy*, *163*, 1017–1031. doi:10.1016/j.energy.2018.08.140

Mohtashami, A. (2021). A novel modified fuzzy best-worst multi-criteria decision-making method. *Expert Systems with Applications*, *181*, 115196. doi:10.1016/j.eswa.2021.115196

Promentilla, M. A. B., Janairo, J. I. B., Yu, D. E. C., Pausta, C. M. J., Beltran, A. B., Huelgas-Orbecido, A. P., & Tan, R. R. (2018). A stochastic fuzzy multi-criteria decision-making model for optimal selection of clean technologies. *Journal of Cleaner Production*, *183*, 1289–1299. doi:10.1016/j.jclepro.2018.02.183

Ribeiro, R. A., Falcão, A., Mora, A., & Fonseca, J. M. (2014). FIF: A fuzzy information fusion algorithm based on multi-criteria decision making. *Knowledge-Based Systems*, *58*, 23–32. doi:10.1016/j.knosys.2013.08.032

Tavana, M., Santos Arteaga, F. J., Mohammadi, S., & Alimohammadi, M. (2017). A fuzzy multi-criteria spatial decision support system for solar farm location planning. *Energy Strategy Reviews*, *18*, 93–105. doi:10.1016/j.esr.2017.09.003

KEY TERMS AND DEFINITIONS

AHP: A method which is developed for solving multi-criteria decision-making problems.

Biomass Diversity: Different types of biomass such as animal wastes, logs, chips.

Biomass Fuel Efficiency: The converting the potential fuel biomass into kinetic energy or electrical energy.

Docking Centers: Temporary storage area where biomass is held to be collected for transport to the power plant.

Facility Location Selection: Selection of the most suitable establishment location in terms of multiple criteria.

Physical and Chemical Properties: The stability of the physical and chemical properties of the biomass in the collection area during the collection period.

Physiography: The determining the morphological features of slopes in terms of instability and landslide hazard.

Policies and Incentives: Government's policy and incentives on renewable energy sources at the national level are attractive.

Chapter 2
A Renewable Energy Assessment Method by Parametric and Non-Parametric Models' Data Analysis

Zühre Aydın Yenioğlu
https://orcid.org/0000-0002-5992-4983
Energy Market Regulatory Authority, Turkey

Vildan Ateş
Ankara Yıldırım Beyazıt University, Turkey

ABSTRACT

Consumption of renewable energy sources for countries shows a rising trend. Providing progress in the renewable energy field, countries try on related regulations and accurate investments according to renewable energy consumption and generation. European Union (EU15) countries play an essential role increasing renewable energy efficiency, which is share of Europe in total energy usage. In this chapter, deterministic and stochastic methods were used to examine whether the renewable energy efficiencies of EU15 countries and Turkey are sensitive to different data envelopment analysis and stochastic frontier analysis models using renewable energy consumption and generation parameters. The chapter presents how the renewable energy efficiency results of related countries change with different optimization models in the context of deterministic and stochastic framework, and it proposes a new method to find a common solution for the different results of different optimization models.

DOI: 10.4018/978-1-6684-2472-8.ch002

INTRODUCTION

Nowadays, if countries can be able to obtain the industrial sector energy at least cost, they can provide a competitive advantage. This situation leads countries to search alternative energy sources, namely renewable energy, due to the limited traditional fossil fuel sources such as oil, natural gas or coal. Renewable energy is an essential alternative energy source because of its environmental utilities and economical efficiencies when compared to traditional energy sources (Dağıstan, 2008). Although there is increasingly widespread use of renewable energy (hydroelectric, wind, solar, biomass, geothermal) in Turkey, a large part of the energy need is still provided from non-renewable (oil, natural gas, coal) energy sources (Koç &Kaya, 2014). On the other hand, compared with European Union countries, climatic conditions of Turkey is an advantage in renewable energy production (Dağıstan, 2008).

This chapter focuses on the stochastic approaches on renewable energy efficiency evaluation in which uncertainty has a technological structure. Decision making problems in energy efficiency field debates with uncertainty, which is affected by prices, energy demand, energy production, consumption, equipment availability, investment and expenditure. Stochastic programming provides an effective framework in which optimization problems under uncertainty are fairly formulated. In energy fields, data envelopment analysis (DEA) and stochastic frontier analysis (SFA) approaches have been used to assess energy sector markets and countries with various types of parameters such as deterministic, stochastic and fuzzy. In such cases, researchers may consider stochastic data as random indicators. By studying on random parameters and analyzing the possibility of uncertain situations, different perspectives of the available information can be obtained in energy markets. The main positive side of studying on random data is determination of accurate energy efficiency policies for future by the help of optimization problems. Parameters of stochastic models are considered random data and this seems to be a reasonable approach for future problems to account for such uncertainties while measuring such data. Analyzing of case study variables can indicates errors and noise. The noise and errors in random indicators generally leads to faulties in frontier production function and in efficiency scores.

When the energy sector researches in the literature are reviewed, it is seen that prominent methods in evaluating renewable energy efficiency are DEA and SFA. DEA was first proposed by Farrell (1957) and developed by evaluating the technical efficiency in the study of Charnes, Cooper and Rhodes. (1978). DEA does not need any assumption of the functional form and can process multiple indicators (Bal & Örkçü, 2005). The following studies can be given as examples to DEA studies in energy field. Jha and Shrestha (2006) measured the performance of hydroelectric power plants by using DEA and they used the current capacity, total number of operations, energy produced by the plant and the number of employees as inputs, total energy produced and winter and summer peak values as outputs in Nepal. In another study Chien and Hu (2007) conducted to compare renewable energy technology among 45 countries level by implementing DEA and stated that Organisation for Economic Co-operation and Development (OECD) members have a greater share of renewable energy resources than non-OECD members. In the study conducted by Barros in 2008, the efficiency of hydroelectric power plants was investigated using DEA. Sözen, Alp and Özdemir (2010) tried to evaluate the performance of thermal electric power plants in Turkey within DEA. The inputs of this chapter were the capacity utilization rate, thermal efficiency, average working time and production capacity, while the outputs were the ton amounts of carbon dioxide, sulfur dioxide, nitrogen dioxide. San Cristobal (2011), is another researcher using DEA to evaluate renewable energy development factors and efficiencies. When this study is examined, it is seen that the inputs were investment rate, implementation period, operating and maintenance costs, while the outputs were composed of power, working hours, service time and carbon dioxide tons. In another study conducted in Turkey by Sözen, Alp and Kilinc (2012), the assessment of efficiency of hydroelectric power plants were examined by DEA. The inputs were the capacity utilization factor, installed capacity, the amount of water collection in the dam, while the output variables were unit cost, operational cost and net energy production.

This chapter extended input oriented deterministic Charnes, Cooper, Rhodes (CCR) DEA model and SFA Cobb-Douglas and Translog production frontier functions to consider the stochastic variations in data and in each decision making unit's (DMU) efficiency score. CCR model compares companies that have operations in homogenous or nonhomogeneous way and creates global and local efficiency scores. In the classical DEA method, the inputs and outputs are deterministic, which leads to unpredictable measures against future inconsistencies. It can be overcome the disadvantages of classical DEA by including SFA in the analysis. By the study, it has been shown that the error in data causes stochastic inefficiency. These results suggest giving more importance to error scores to measure stochastic frontier efficiency. Technical efficiencies of countries differ between two stochastic methods. Since in deterministic DEA the randomness did not been considered in data, SFA method allows calculating noise in data. These approaches can be determined as parametric and nonparametric methods. In SFA as a parametric approach, a production frontier formulation is estimated, and in DEA as a nonparametric approach, mathematical programming techniques are used.

Renewable energy sources cannot be depleted as they constantly renew themselves. On the other hand, since it is costly to obtain energy from these sources, measuring and evaluating their efficiency has become an essential issue in terms of using the resources. The development of renewable energy facilities, equipment, component and service supply will contribute to the developed economy, high qualified employment, and improved research and development activities.

In the chapter, it is aimed to evaluate relative efficiencies within the consumption and generation of renewable energy with DEA and SFA-among Turkey and EU15 countries (Austria, Belgium, Denmark, Finland, France, Germany, Greece, Ireland, Italy, Luxembourg, Netherlands, Portugal, Spain, Sweden and United Kingdom).

Efficiency is considered as reducing energy consumption without causing a decrease in current life standards, service and production quality. In this context, primary and renewable energy consumption and generation indicators are used to provide accurate calculation of the efficiency in the consumption of renewable energy. The case study contributes to literature with how the efficiency results changes with different optimization models in the context of randomness in data by its deterministic and stochastic production functions that used for the first time in this field in Turkey. The study also compares structure between deterministic and stochastic results and different indicators within stochastic framework that will let the calculation of noisy in data. In addition, it offers a method to measure energy efficiency between stochastic and deterministic models within different results of same DMUs. According to the literature studies; input oriented nonparametric Charnes,Cooper,Rhodes (CCR) and Banker, Charnes, Cooper (BCC) models that try to minimize the input of processes in deterministic scope and parametric Cobb-Douglas and Translog production functions, which are the production functions of SFA in stochastic scope, are used.

The chapter consists of eight subchapters and in the first subchapter the aim of the study and related studies are presented. The second suchapter reveales the related literature. In third subchapter, DEA models and SFA production functions are introduced. The case study implementation to the models are studied in fourth subchapter. Results and evaluations are included in the fifth subchapter and solutions and recommendations in the sixth subchapter. In the last two subchapters, there are future research directions and conclusion sections, respectively.

BACKGROUND

This subchapter presents studies that relate to renewable energy evaluation within the DEA and SFA through the energy scope.

DEA is a nonparametric linear programming methodology that creates a frontier function for measuring efficiency by including a convex linear model of parameters. However, stochastic frontier approach includes a parametric method of frontier efficiency function and indicates error, which shows separations from the efficiency limit. The error is the accumulation of the stochastic inefficiency scores and data noisy. StoNED approach also assumes stochastic error and a nonparametric, piecewise linear frontier (Gill, Costa, Lopes & Mayrink, 2017). Lopes and Mesquita (2015) stated that SFA, DEA and StoNED models are commonly used among the European energy efficiency field for benchmarking.

Studies that were tried to make analysis of renewable energy efficiency by using DEA and SFA methods are common in the literature. When renewable energy studies are reviewed through the literature, following studies are observed. Hepbaşlı and Utlu (2004), investigated Turkey's renewable energy resources, efficient usage of sources and accurate policies. Gençoğlu (2002) studied on the status of renewable energy sources and the possibilities of efficient usage of these resources in Turkey. Külekçi (2009), investigated the importance of the geothermal energy in the field of renewable energy sources in another study. Erdal (2012), reported renewable energy investments and potential of employment creation by these investments in Turkey. Çapik, Yılmaz and Çavuşoğlu (2012) studied on the potential of Turkey's existing energy sources and stated that the current energy needs can be provided with renewable energy. In this context they reviewed Turkey's renewable energy policy. Koç and Kaya (2014) have made an overall assessment of the production of renewable energy sources. Halkos and Tzeremes (2012) used capital structure, activities and liquidity levels as three inputs, gross and operating profit margin, return of assets as four outputs to capture the profitability levels of the companies in Greece renewable energy sector by using DEA. As a result of the study, they reported that firms that operate in wind energy field tend to have better financial efficiency than firms that operate in hydroelectric energy. They also pointed out that the Greek renewable energy sector is a highly competitive industry. Menegaki (2013) takes into account of the energy inefficiencies of 31 European countries with DEA model. The variables used in the study are gross domestic product, fuel usage, carbon dioxide emissions, employment and capital. They pointed out that countries that lag behind in renewable energy are technically the most efficient countries in Europe, while countries with significant renewable energy performance have technically moderate or low efficiency. Kim et al. (2015) searched the economic aspects of renewable energy activities in Korea through DEA approach and stated that wind power is the most essential and efficient renewable energy source. Dizdarevic and Segato (2012) measured energy efficiency in the EU countries among 2000-2010. They used input oriented CCR DEA with capital, labor, energy use as input parameters and gross domestic product as the output. Li and Tao (2017) reviewed approaches and policies on performance evaluation of energy efficiency and they used DEA and SFA techniques. They stated that various models of DEA can be implemented in the energy efficiency evaluation field and is suitable for econometrics. According to this study, SFA is fundamental method for energy efficiency measurement because it allows random variables, but it is difficult to verify it through error structure. In addition, energy economic models present authorities how to make decisions and plans for future energy policies. Yenioğlu and Toklu (2021) used deterministic and stochastic DEA models for the evaluation of energy efficiency. They used multiple inputs and outputs as random variables in stochastic models. This study showed that stochastic DEA scores are more flexible than classical DEA as well as stochastic models give more results closer to the production frontier.

Umar, Girei and Yakubu (2017) studied on Cobb-Douglas and Translog frontier production functions in the analysis of agricultural technical efficiency and they showed that the flexible efficiency scores can be estimated and inefficiencies obtained from Cobb-Douglas and Translog production models have different results notably. Hence, they concluded that the determination of model for technical efficiency analysis should be based on aim of the research, standard deviation of the values and indicators. Chachuli et al. (2020) studied on systematic literature review on renewable energy performance through DEA and stated that DEA approaches, either deterministic, stochastic or fuzzy, can be deeply implemented to analyze the efficiency of renewable energy researches' based on the randomness, complexity and certainty of indicators. Hsiao et al. (2019) studied energy activities and evaluation of the Baltic Sea countries by SFA Cobb-Douglas function within statistical noise in data. In their study, they chose the capital, employment, energy consumption and CO2 emission as inputs, real gross domestic product (GDP) as output, and renewable energy usage and population as the environmental parameters. They recommended developing energy efficiency with replacing energy consumption with investments or labor and improving renewable energy with replacing the use of fossil fuels. Gökgöz and Güvercin (2018) investigated energy security and renewable energy performance evaluation by benchmarking performance of the selected EU countries by DEA. They concluded that efficiency evaluation approaches could give notable scores in studying renewable energy efficiency and energy policies of DMUs. Zeng, Guo and Zhang (2019) searched feasibility of the super efficiency approach of DEA in the scope of renewable energy and analyzed that the proposed super efficiency model has basic advantages when compared with classical DEA model. Xu et al. (2018) analyzed the production efficiency of renewable energy generation by source in 20 countries with SFA approach and concluded that renewable energy generation efficiency tends to rise up in the world and has largely developed in China. Quyang et al. (2021) applied SFA with metafrontier to evaluate industrial energy efficiency for sectors in China's 30 provinces during 1997–2016. They found that the methods for energy efficiency measurement results higher than current real results and industrial energy efficiency of China was only 0.4396, implying that there's need for energy efficiency improvement. Khan et al. (2021) studied the effects of energy efficiency on ecological footprint using the stochastic frontier analysis (SFA) of the translog production type of single output and multiple inputs. According to technically calculated energy efficiency results, urbanization is an increasing factor of ecological footprint, and investment in agriculture is also beneficial for the environment.

There are many studies on renewable energy efficiency analysis, and many further models were introduced in the context of DEA and SFA. In summarize; according to reviewed literature, the most preferred benchmarking models in energy efficiency assessment methods are DEA and SFA. In this aspect, this chapter will contribute to the literature by researching how the renewable energy efficiency results of related countries change and differ with different deterministic and stochastic optimization models and by presenting a new method to find a common solution for the different results of different optimization models.

RESEARCH METHODOLOGIES

According to literature review in previous subchapter, there are two fundamental approaches, which are parametric analysis and nonparametric methods, applied to calculation of frontiers. SFA is the parametric analysis and DEA is the nonparametric method. Implementation of nonparametric DEA on renewable energy efficiency of Turkey and EU15 countries is discussed in this part.

Data Envelopment Analysis

In this context, a performance comparison was made by input oriented CCR DEA approach in this chapter. CCR model was used to analyze the set of DMUs that were using the production function with constant returns to scale. The reason for using CCR was to provide the possibility of separately calculating technical efficiency globally. Technical efficiency calculates the DMU's overall success with related inputs. CCR model calculates the sector efficiency of a decision unit which indicates technical and scale efficiency. In this approach the assumption is that outputs rise with an increase in inputs (Li & Tao, 2017).

Mathematical equations of the input oriented CCR model is given below. It is supposed that there are n homogenous DMU*j* *(j = 1,....n)* such that all of them use m inputs *xij* *(i =1,2,.....m)* to obtain s outputs *yrj* *(r =1,2,.....s)*, and which are nonnegative and nonzero vectors. The CCR model's production possibility set suggested by CCR in 1978 is as follows (Charnes, Cooper & Rhodes, 1978):

$$TCCR=\{(X,Y) \mid \sum_{j=1}^{n}(X_j\lambda_j) \leq X, \sum_{=1}^{n}(Y_j\lambda_j) \geq Y, \lambda j\ 3\ 0, j = 1\sigma,\dots n\}$$

CCR efficiency scores can be obtained by using the envelopment input-oriented and input-oriented model (1), respectively where x_{i0} and y_{r0} represent the ith input and the rth output indicator vector of DMU_0 under calculation in models.

A DMU is named input oriented CCR efficient if its expected value in Eq. (1) is equal to unity.

$$\min \theta_0 + \varepsilon\left(\sum_{i=1}^{m} s_i^- + \sum_{r=1}^{s} s_r^+\right)$$

s.t.

$$\sum_{j=1}^{n} x_{ij}\lambda_j + s_i^- = \theta_0 x_{i0} \qquad (1)$$

$$\sum_{j=1}^{n} y_{rj}\lambda_j - y_{r0} - s_r^+$$

$$\lambda_j \geq 0,\ s_i^- \geq 0,\ s_r^+ \geq 0,\ i=1,2,\ldots,m,\ r=1,2,\ldots\ldots,s,\ j=1,2,\ldots\ldots,n$$

Here the λj represent structural variables, the s_i^-, s_r^+ represent slacks and $\varepsilon > 0$ is a ''non-Archimedean infinitesimal'' determined to be smaller than any positive real number. This means that ε is not a real number.

Stochastic Frontier Analysis

This section aims to introduce SFA, which is a parametric approach and it is used in efficiency and productivity measurement in framework of mathematical and econometric assumptions. According to literature search, an overview of SFA models has been introduced in this part.

Researchers tried to improve renewable energy efficiency by considering consumption and generation. Farrel (1957) presented SFA model to study on theoretical and empirical searches. This approach showed us that there was a parametric relation between model's input and output indicators. In 1970s, Aigner and Chu (1968) first implemented SFA in the evaluation of production function frontier. The main feature of SFA is the production of a conventional function and the determination of efficiency or inefficiency by calculating the distance of each decision unit to the curve created by this function. As it was emphasized before, the

DEA method has deterministic structure so it ignores measurement errors. In SFA, frontier emphasizes the limit of production and stochastic term implies calculated error. Literature searches showed that the random indicators may affect production. Hence, the statistical error, which has normal and one-sided distribution, included into the model. The SFA method fixes the disadvantages of measurement errors.

The Cobb-Douglas Production Function

Aigner et al. (1977) used cross-sectional data to estimate a production frontier, which was named a Cobb-Douglas production frontier. The Cobb-Douglas form of SFA model is a generally preferred functional form in SFA studies (Coelli, 1995). In SFA studies, Cobb-Douglas production functional form is used because of its various advantages such as its understandable structure. The most discussed disadvantages of Cobb-Douglas production function is its inflexible functional structure. But Lau (1986) has presented that, "this model makes computations easy and has the features of explicit representability, uniformity, parsimony and flexibility" (Lau, 1986).

Cobb-Douglas production frontier function's form is given as below:

$$\ln(y_i) = \beta_0 + \sum_{n=1}^{N} \beta_n lnx_{ni} - u_i,\ i=1,2,\ldots,N \qquad (2)$$

Where y_i is the vector size of output that is produced by *i*th DMU, x_{ni} is the vector size of nth input that is used by *i*th DMU, β is unknown parameter and u_i is positive random variable that indicates technical inefficiency. But the Eq. (2) is a deterministic frontier as $exp(\beta_0 + \sum_{n=1}^{N} \beta_n lnx_{ni})$.

This deterministic frontier counts out the mathematical possibility of measurement error and statistical noises, and remarks all deviations from the frontier as only technical inefficiency (Aigner & Chu, 1968). Hence, stochastic production frontier function was presented for specifying random indicators represented by statistical noise.

Aigner et al. (1977) defined the model's production function in deterministic and stochastic ways. Broeck and Meeusen (1977) introduced a new stochastic production function (Aigner & Chu, 1968). They included symmetric random error for calculation of statistical noise. The formula is as follows.

$$y_i = f\left(x_i, \beta_i\right) + v_i - u_i = f\left(x_i, \beta_i\right) + \varepsilon_i,\ i=1,2,\ldots,N \qquad (3)$$

where v_i independent random variable showing the $N\left(0,\sigma_i^2\right)$ distribution and $\varepsilon_i \leq 0$. In the Eq. (4) it is assumed that ε_i is a composite error parameter consisting of two independent parameters, v_i and u_i. v_i includes noisy, errors that occur in determining the production function and also omissions caused by the independent parameter x. If a DMU provides expected production with full efficiency, the technical efficiency is "1", but if it produces expected outputs under the optimal capacity its efficiency measure is less than 1, that is this DMU is inefficient. Another point to be considered is how the efficiency is calculated.

If the problem is output maximization (production maximization), then the composite error term calculation is valid and it is $\varepsilon_i = v_i - u_i$.

If the efficiency problem is input minimization (cost function), then the equation $\varepsilon_i = v_i + u_i$ is valid (Aigner & Chu, 1968).

In this study, FRONTIER version 4.1 software was used for SFA that transforms formulation (3) to a logarithmic function as below Eq. (4) as in Broeck and Meeusen (1977).

$$\ln(y_i) = \beta_0 + \sum_{n=1}^{N} \beta_n lnx_{ni} + v_i - u_i ,\ i=1,2,\ldots,N \tag{4}$$

Eq. (4) represents the logarithm of inputs and outputs.

According to Coelli et.al. (2005), Cobb-Douglas stochastic frontier function is as below in Eq. (5, 6) (Coelli et al., 2005).

$$\ln(y_i) = \beta_0 + \beta_1 lnx_i + v_i - u_i ,\ i=1,2,\ldots,N \tag{5}$$

$$y_i = \exp(\beta_0 + \beta_1 lnx_i + v_i - u_i) ,\ i=1,2,\ldots,N \tag{6}$$

The Cobb-Douglas production frontier as in Eq.(7) will be used for in this study's renewable energy efficiency including cross-sectional data and supposing a half normal distribution.

Eq.(7):

$$\ln(y_i) = \ln(x_i, \beta) + v_i - u_i ,\ i=1,2,\ldots,N \tag{7}$$

The SFA form of Cobb-Douglas production function is implemented between 2017 and 2019 and is modeled as follows in Eq. (8) for related inputs and output.

$$ln(y_i) = \beta_0 + \beta_1 (lnx_1) + \beta_2 (lnx_2) + \beta_3 (lnx_3) + v_i - u_i \tag{8}$$

In Eq. (7), y_i is the ith country's "renewable energy generation" output's log, x_i is the ith metropolitan's inputs' log, u_i is the ith metropolitan's inefficiency level. In model (8) x_1, x_2, x_3 implies respectively; primary energy consumption, renewable energy consumption and share of electricity from renewables.

The Translog Production Function

Baltagi and Griffin (1988) developed the Translog production frontier form that provides flexible approximation including Cobb-Douglas as an essential case. Both production functions have the advantage that they can be linearized by natural logarithms. Since Cobb-Douglas form is a situation in general set of production estimations with constant flexibility of substitution, the Translog functional form relaxes the constant flexibility of substitution and reduces to the Cobb-Douglas production frontier when there is constant elasticity of substitution (Prasad, 2015). The Translog form does not need constraints and it allows more flexible general determination of the frontier form since it includes any arbitrary form of production. Thus, it can be said that it has been broadly used in efficiency analysis studies. It has been operated to examine input substitution, technical change and productive efficiency.

In this chapter, the Translog production frontier has the following form as in Eq. (9) for related inputs and output.

$$\begin{aligned} ln(y_i) = {} & \beta_0 + \beta_1 (lnx_{1i}) + \beta_2 (lnx_{2i}) + \beta_3 (lnx_{3i}) + 0,5.\beta_{11} (lnx_{1i})^2 + 0,5.\beta_{22} (lnx_{2i})^2 \\ & + 0,5.\beta_{33} (lnx_{3i})^2 + \beta_{12} (lnx_{1i}.lnx_{2i}) + \beta_{13} (lnx_{1i}.lnx_{3i}) + \beta_{23} (lnx_{2i}.lnx_{3i}) + v_i - u_i \end{aligned} \tag{9}$$

The Translog production frontier allows to make a transition from a linear relationship to a nonlinear one between the output indicator and the production inputs. Due to its properties, the Translog production function can be used for the total factor productivity estimation of a linear and homogenous production.

CASE STUDY

As understood from the literature review, SFA and DEA are widely used methods for measuring efficiency of countries in the field of renewable energy. This chapter compares and shows renewable energy consumption and generation situations of

Turkey and EU15 countries within parameters such as; primary energy consumption, renewable energy consumption, share of electricity from renewables and renewable energy generation.

In this review, primary energy consumption implies commercially traded fuels, including renewables used to produce electricity. Renewable energy consumption based on gross electric output, is the total amount of electrical energy generation by transforming other forms of energy, for example nuclear or solar power. Renewable energy generation by source based on gross electric generation is the indicator for output parameter. Share of electricity production from renewables includes electricity generation from solar, wind, biomass and waste, hydropower, geothermal, tidal and wave sources. Countries' renewable energy economy and policies must be planned based on primary energy and renewable energy consumed by source. Heating and transport are more dependent on oil and gas, hence these sectors tend to be harder to decarbonize. So it can be stated that renewables have a higher share in the electricity for the total energy mix. These indicators contribute to analysis of this study in the context of renewable energy supply-demand balance, energy economy and energy policies.

It is also aimed to evaluate efficiency of EU15 counties and Turkey, whose efficiencies are measured with the nonparametric DEA and parametric SFA methods. For this purpose, the use of error components model among stochastic boundary models was preferred, and the stochastic production frontiers was prepared by taking into account of the full logarithmic production functions. Study wants to compare and analyze the renewable energy efficiency of Turkey and EU15 countries by common methods SFA and VZA. The data sets were added from the bulletins published by the World Bank and British Petroleum (BP) between 2017 and 2019 (BP, 2021; The World Bank, 2021). Stochastic methods are used, because decision makers deal with uncertain and imprecise conditions. The noise factor and randomness in data often cause errors in production frontier function and technical efficiency scores. In these cases, data analysts can consider noisy data as random indicator. By working with random inputs and outputs and realizing the possibility of uncertain cases, different perspectives of the usable information can be detected in energy efficiency researches to carry out energy policy accurately. The main utility of random data in SFA models is the prediction of efficiencies in future optimization problems.

DATA STATISTICS AND EFFICIENCY RESULTS

In this subsection, dataset is implemented from 2017 to 2019 to evaluate renewable energy efficiency of DMUs. The inputs are primary energy consumption, renewable energy consumption, share of electricity from renewables and output is renewable energy generation. Table 1 shows the average statistics of data that were added to models.

The results of deterministic DEA models, Cobb-Douglas Production Function and Translog Production Function SFA models were implemented on mathematical programming and optimization systems, General Algebraic Modeling System (GAMS) and FRONTIER 4.1.

Table 1. Descriptive statistics of the indicators

Country	Primary Energy Consumption			Renewable Energy Consumption			Share of Electricity From Renewables			Renewable Energy Generation by Source		
DMU	Mean	StdD	Var	Mean	StdD	Var	Mean	StdD	Var	Mean	StdD	Var
Austria	1.468	0.001	0.031	0.136	0.000	0.006	76.553	0.5208	0.722	51.135	20.161	4.4901
Belgium	2.653	0.004	0.062	0.172	0.000	0.013	20.903	5.4170	2.327	17.306	2.3136	1.5210
Denmark	0.702	0.000	0.006	0.207	0.000	0.012	71.526	15.586	3.948	22.136	1.8174	1.3481
Finland	1.129	0.001	0.027	0.171	0.000	0.010	46.320	0.2547	0.505	30.910	0.0967	0.3109
France	9.750	0.011	0.105	0.548	0.004	0.061	18.746	3.2756	1.801	104.99	154.80	12.442
Germany	13.45	0.104	0.322	1.989	0.014	0.111	36.280	12.039	3.461	228.46	205.58	14.338
Greece	1.158	0.000	0.011	0.101	0.000	0.007	28.983	12.250	3.500	15.170	1.4947	1.2225
Ireland	0.659	0.000	0.008	0.091	0.000	0.011	32.796	14.056	3.741	10.190	1.7161	1.3100
Italy	6.466	0.007	0.082	0.648	0.000	0.015	38.526	7.3114	2.704	109.17	37.879	6.1546
Luxembourg	0.165	0.000	0.006	0.010	0.000	0.001	69.953	7.4576	2.731	0.6833	0.0072	0.0850
Netherlands	3.523	0,000	0,010	0.199	0.001	0.028	16.696	3.6361	1.907	19,553	6,3916	2.5281
Portugal	1.065	0.000	0.019	0.167	0.000	0.009	47.693	57.816	7.604	26.423	12.238	3.4983
Spain	5.751	0.003	0.052	0.707	0.001	0.034	35.790	9.9619	3.156	98.443	82.837	9.1015
Sweden	2.206	0.001	0.038	0.337	0.000	0.022	57.386	1.9781	1.406	95.056	17.370	4.1678
Turkey	6.384	0.010	0.101	0.340	0.005	0.072	36.226	57.403	7.577	106.50	614.91	24.797
United Kingdom	7.932	0.007	0.081	0.982	0.010	0.102	32.926	14.440	3.800	109.37	105.34	10.263

Cobb-Douglas production function and Translog production function SFA models' analysis results are presented in Table 2 and Table 3. According to Table 3 the parameter γ=0.95 is represents statistical significance at the 1% level. This means that most sources of inefficiency, in combined error term ($\mathcal{E}$) are caused due to almost 95% of technical inefficiency and 5% of random errors. In this context, it could be stated that the technical inefficiency has a high rate within the combined error term, and the existence of random errors are very low. The likelihood ratio test shows that inefficiency scores are statistically significant according to renewable energy efficiency among countries. In Table 2 it can be seen that parameter coefficients of independent variables are also significant and positive. This implies that, parameters have positive size of elasticity and can be used in other models certainly. The one-sided error LR test is taken into consideration to evaluate the technical efficiency of DMUs. LR ratio was found to be approximately 5.931 and this value should be compared with the table value of 2.706 in the Kodde-Palm with a restriction of 1 at 0.05 significance level (Kodde & Palm, 1986).

Table 2. Analysis results of Cobb-Douglas production function model

Parameter	Coefficient	Standard Error	T Ratio
β_0	1.857	0.413	4.498
β_1(Primary Energy Consumption)	0.482	0.102	4.688
β_2(Renewable Energy Consumption)	0.680	0.184	3.684
β_3(Share of Electricity from Renewables)	0.766	0.033	22.91
σ_2	0.281	0.123	2.286
Γ	0.95	0.090	11.11
Log-likelihood	-3.819		
LR test of the one-sided error	5.931		

Table 3. Analysis results of Translog production function model

Parameter	Coefficient	Standard Error	T Ratio
β_0	-28.27	0.992	-28.4
β_1(Primary Energy Consumption)	12.26	0.947	12.94
β_2(Renewable Energy Consumption)	-14.25	0.958	-14.87
β_3 (Share of Electricity from β_0 Renewables)	7.434	0.87	8.543
$\beta_4\,(\beta_1{*}\beta_1)$	-3.26	0.919	-3.54
$\beta_5(\beta_1{*}\beta_2)$	-0.542	0.908	-5.976
$\beta_6(\beta_1{*}\beta_3)$	-0.818	0.59	-1.385
$\beta_7\,(\beta_2{*}\beta_2)$	2.978	0.388	7.673
$\beta_8(\beta_2{*}\beta_3)$	- 1.408	4.641	-0.303
$\beta_9(\beta_3{*}\beta_3)$	1.838	0.433	4.241
σ_2	0.045	0.186	0.242
Γ	0.982	0.923	1.063
μ	-0.0014	0.991	0.0015
Log-likelihood	12.216		
LR test of the one-sided error	5.0083		

$H_0 : \gamma = 0$
$H_1 : \gamma^1 0$

Since 5.931 score is bigger than the table value of 2.706 according to Kodde-Palm, the null hypothesis (H0) is rejected. This situation implies that, there is a statistically significant technical inefficiency in the model.

In Table 3 the parameter γ=0.982 represents statistical significance at the 1% level. That also means most sources of inefficiency, in combined error term (Ɛ) caused with almost 98% of technical inefficiency and 2% of random errors. It means that the existence of random errors is very little. In Table 3 it is seen that parameter coefficients of independent variables are also significant except renewable energy consumption. Renewable energy consumption parameter coefficient has negative and non-significant value due to the relationship between chosen input indicators in this model. The one sided error LR ratio was found to be approximately 5.0083 and when this value is compared the table value of 2.706 in the Kodde-Palm with a restriction of 1 at 0.05 significance level, null hypothesis is rejected. The one sided error LR ratio was found to be approximately 5.0083 and comparing the table value of 2.706 in the Kodde-Palm with a restriction of 1 at 0.05 significance level, we can say that $\mathbf{H_0}$ hypothesis is rejected. In addition, there is a statistically significant technical inefficiency in the Translog model the analyzed coefficients are also consistent within both models results' from literature studies. Table 4 shows average estimated 2017, 2018 and 2019 efficiency results of the DEA and SFA models. Figure1 presents models' scores separation on y axis for each DMU on x axis.

Figure 1. Separation of average efficiency estimated results

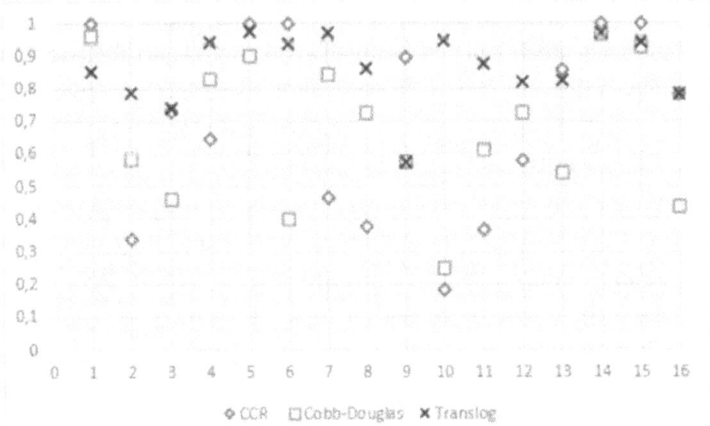

Table 4. Average efficiency estimated results of models

Country	CCR DEA	Cobb-Douglas	Translog
Austria	1.000	0.959	0.851
Belgium	0.340	0.583	0.787
Denmark	0.730	0.460	0.737
Finland	0.643	0.830	0.944
France	1.000	0.897	0.974
Germany	1.000	0.397	0.939
Greece	0.464	0.841	0.972
Ireland	0.377	0.723	0.862
Italy	0.896	0.572	0.574
Luxembourg	0.184	0.246	0.947
Netherlands	0.364	0.612	0.874
Portugal	0.578	0.723	0.820
Spain	0.856	0.542	0.825
Sweden	1.000	0.966	0.970
Turkey	1.000	0.927	0.944
United Kingdom	0.782	0.437	0.782

It can be seen from Figure 1 and Table 4 that mean efficiency results of DEA CCR is 0,701, Translog production frontier is 0,862 and Cobb Douglas production function is 0,670. Cobb Douglas production function has the lowest mean efficiency. According to Table 4, Cobb Douglass and Translog models' results differ significantly. Austria, France, Germany, Sweden and Turkey are pure technically/locally and globally efficient countries. These countries have constant returns to scale and their scale efficiencies are 1. But according to DEA model and Translog production function Germany is the only high efficient country that has very low efficiency in Cobb Douglas model. Also, in Figure 1, Translog and Cobb Douglas models tend to converge to CCR results, hence SFA production functions show same efficiency tendency through globally overall technical efficiency. As seen from Table 4, Sweden, Turkey, Austria and France are the reference DMUs that they can be chosen as samples for renewable energy policy implications. Since Cobb Douglas production function results has the lowest value in this model, Luxembourg, United Kingdom, Denmark and Germany could be revised in renewable energy policies and practices. Translog production function has more elasticity in frontier model and its scores are high and parallel to DEA model. Italy has the lowest technical efficiency score in

Translog model. Belgium, Ireland, Greece, Luxembourg and Netherlands have the lowest globally technical efficiency according to CCR DEA model.

Examining the Figure 1, it can be stated that Luxembourg (DMU10) has the technical inefficiency under 50% in both CCR DEA and Cobb Douglas models. Sweden (DMU14) has the highest efficiency scores within CCR DEA, Translog and Cobb Douglass models.

Table 5. Consistency testing of technical efficiency scores of models

Test/Model	Cobb_Douglas vs Translog	CCR_DEA vs Cobb_ Douglas	CCR_DEA vs Translog
Spearman	0.040	0.137	0.639
Mann-Whitney U	0.010	0.539	0.323

Spearman and Mann-Whitney U tests' results between models' efficiency scores are presented in Table 5. According to Table 5, CCR DEA and SFA models have consistency through related p values. Cobb Douglas and Translog models do not have consistency according to their efficiency scores as seen in Table 4. This implies the elasticity of each indicator is affected by the other indicators due to interactive relationship between the parameters in Translog frontier function. The highest compatibility is between CCR DEA and Translog models. As can be seen from the results, it is not possible to make a precise renewable energy efficiency measurement with different results of different models. Therefore, the energy efficiency results of different stochastic and deterministic approaches need to be investigated for energy policy studies.

Consequently, this chapter proposes a new method to measure renewable energy efficiency of different stochastic and deterministic models. This new proposed method takes mean of CCR DEA and Translog models' efficiency results that have highest consistency in consistency testing. Afterward, the efficient and inefficient decision making units are decided according their ranks.

According to Table 6, France, Sweden and Turkey are the first three DMUs that they can be chosen as references for renewable energy policy implications. France, which has low efficiency in the Cobb- Douglas model, has become the most effective decision making unit with the removal of Cobb-Douglas results among the consistent models. In other words, the efficiency of some units has increased by removing the Cobb-Douglas from the efficiency analysis. France, Sweden and Turkey are the first three DMUs that they can be chosen as references for renewable energy policy implications. Belgium, Luxembourg and Netherlands are the last three DMUs that have the lowest technical efficiency according to the most consistent models. These DMUs' can be revised for renewable energy policies and practices. With consistency testing it can be easily seen that Translog production function has more elasticity in frontier model and its scores are high and parallel to CCR DEA model.

Table 6. Mean of efficiency scores of the most consistent models and related new ranks

Country	Mean of Efficiency Scores	Ranks
Austria	0.9255	5
Belgium	0.5635	16
Denmark	0.7335	10
Finland	0.7935	7
France	0.987	1
Germany	0.9695	4
Greece	0.718	11
Ireland	0.6195	13
Italy	0.735	9
Luxembourg	0.5655	15
Netherlands	0.619	14
Portugal	0.699	12
Spain	0.8405	6
Sweden	0.985	2
Turkey	0.972	3
United Kingdom	0.782	8

In summary, inefficient countries should focus on increasing the capacity they have to generate in return for the inputs they consume. The differences between models are due to randomness of data in stochastic models, variable returns to scale and constant returns to scale approaches show different scores, since there are various energy implications of each countries that differ within energy policies.

SOLUTIONS AND RECOMMENDATIONS

This chapter highlighted the notable stochastic and deterministic optimization methodologies for renewable energy efficiency measurement through generation and consumption parameters. The significancy and unsignificancy of stochastic and deterministic optimization methods are presented by a novel measurement suggestion. The evaluation of relative efficiencies within the consumption and generation of renewable energy with DEA and SFA is done among Turkey and EU15 countries, which leads renewable energy field in Europe.

Through the methodology and objectives of the chapter we can conclude the solutions and recommendations as below for deterministic and stochastic implementations in the field of renewable energy.

- SFA Translog model efficiency scores can give higher results than the expected results. SFA provides best results because it could determine the relationship between the dependent and independent variables. In addition; the elasticity of each parameter is affected by the others since there is interactive relationship between the parameters in Translog frontier function.
- It can be seen that Translog production function has more elasticity in frontier and its results are high and similar to CCR DEA model. Therefore; among different stochastic and deterministic models, the most consistent results are between CCR DEA and Translog SFA models.
- Authorities imply the balance among electricity demand, generation and consumption and less environmental pollution within renewable energy efficiency. Hence; for measuring renewable energy efficiency with both parametric and nonparametric methods, researchers can use these indicators since they are the most important energy efficiency factors.
- From the analysis results, it is seen that the effective DMUs use their inputs and output at optimum level and these countries operate their energy policies at optimum scale.
- As CCR scores increase, efficiency scale of DMUs increases. That is, there is a positive relationship between efficiency scale and CCR scores. Therefore, investment decisions should be planned according to energy demand, consumption and generation rates.
- In order to ensure overall technical efficiency for the future and to improve inefficient countries; increasing the capacity of generation for requested consumption is essential, hence there should be true policies and investment strategies.
- The differences between used models could be specified as randomness of data in stochastic models, variable returns to scale and constant returns to scale approaches and the energy implications of each countries that differs in energy policies around used indicators of the chapter.
- According to the results, countries are effective in scale in the field of renewable energy. SFA and DEA methods, can be used to evaluate renewable energy efficiency accurately.

- This chapter shows that; the determination of methodology for measuring renewable energy efficiency depends on the scope of research, related input and output parameters and randomness of data.
- Consistency testing of various technical efficiency scores of models are essential to show significancy and insignificancy between parametric and nonparametric benchmarking approaches.
- The highest compatibility is between CCR DEA and Translog models. As can be seen from the results, it is not possible to make a precise renewable energy efficiency measurement with different results of different models. Therefore, the energy efficiency results of different stochastic and deterministic approaches need to be investigate for energy policy studies.

FUTURE RESEARCH DIRECTIONS

In future studies, performance comparisons can be implemented by different input and output parameters with high relationship status, different but homogeneous DMUs and different parametric and nonparametric methodologies involving quantitative and qualitative approaches. The authors suggest investigating new indicators that can affect the production and consumption capacities of renewable energy to enhance the resolution of the empirical results. The methods that were presented by the paper can also be considered to investigate the capabilities of different countries in other aspects of energy. Implementing SFA and DEA methods can provide calculation of noise in data and can fit the criterias to see readiness and efficiency of utilizing needed energy production and low cost energy consumption. Future research regarding renewable energies can be helpful for many countries' governments and policymakers to assess their current performance in terms of their use of renewable resources within accurate and available data. This chapter is also evidence that benchmarking methods such as SFA and DEA can be a more effective evaluation model in the context of deterministic and stochastic data analysis framework. Their results can be compared together to obtain highly accurate results. In addition; in the concept of stochastic programming, robust optimization techniques can be implemented within the scope of renewable energy under uncertainty.

CONCLUSION

In the chapter, deterministic and stochastic methods were used to examine whether the renewable energy efficiencies of EU15 countries and Turkey are sensitive to different DEA and SFA models. Four different models were implemented in deterministic and stochastic framework. Since there is difference between models, consistency tests are performed and a new method is proposed to measure renewable energy efficiency of different stochastic and deterministic models through the most consistent CCR DEA and Translog SFA models. This is the first study measuring renewable energy efficiency with both parametric and nonparametric methods using primary energy consumption, renewable energy consumption, share of electricity from renewables and renewable energy generation indicators.

According to the fact that, there are not very high differences between CCR technical efficiency scores and this shows that countries are effective in scale in the field of renewable energy. From the analysis, it is understood that the effective countries use their inputs and output at optimum level compared to other countries. In other words, these countries operate their energy policies at optimum scale. As CCR scores increase, efficiency scale of countries increases. In other words, there is a positive relationship between efficiency scale and CCR scores. Thus, investment decisions should be implemented with the renewable energy regulations and energy plans should be organized according to energy consumption and generation rates. These situations should be applied especially in inefficient countries in order to ensure overall technical efficiency for the future.

In stochastic models, higher efficiency scores are expected and in this study's results are higher than the expected especially in Translog model. In addition, results of SFA models showed that inefficiency scores, elasticities and efficicency scores of Translog and Cobb Douglas frontier functions differ significantly. Hence, it can be stated that the determination of functional form for measuring renewable energy efficiency should be based on the scope of research, chosen indicators and randomness of data.

According to primary energy consumption, renewable energy consumption, share of electricity from renewables and renewable energy generation parameters showed that elasticity sizes and coefficients of them are positive and significiant in both SFA models except renewable energy consumption in Translog functional form. This implies that this indicator can be implemented into Translog model with different input parameters. Stochastic operational researches in the field of renewable energy can produce reliable studies due to rising of big data concept, complexity and uncertainty of renewable problems.

In this chapter, a new method is proposed to measure renewable energy efficiency of different stochastic and deterministic models and it is mentioned that countries' renewable energy plans, economies and policies should be devised according to primary energy consumption, renewable energy consumption and renewable energy generation. Renewables have higher share in the electricity for the total energy mix; hence, share of electricity from renewables is an essential indicator in deterministic and stochastic optimization methodologies. These parameters are significant in models and contributed to optimization analysis of this study in the context of renewable energy supply-demand balance, energy economy and energy policies. If the renewable energy sustainability is desired, then there should be efficient energy usage to decrease the negative effects of energy environmentally. Balanced energy consumption of renewable sources by contribution of all renewable sources should be stimulated by utilization of renewables, connecting renewable energy to the electricity grid, increasing incentives provided by the government and providing low cost of energy generation.

The study highlighted significancy and insignificancy cases of stochastic and deterministic approaches in renewable energy applications. This review also determined that the SFA and DEA methods, either stochastic or deterministic, might be comprehensively implemented to analyze renewable energy efficiency researches taking into account the availability and accuracy of data and significance of parameters.

ACKNOWLEDGMENT

This research received no specific grant from any funding agency in the public, commercial, or not-for-profit sectors.

REFERENCES

Aigner, D., Lovell, C. K., & Schmidt, P. (1977). Formulation and estimation of stochastic frontier production function models. *Journal of Econometrics*, *6*(1), 21–37. doi:10.1016/0304-4076(77)90052-5

Aigner, D. J., & Chu, S. F. (n.d.). On estimating the industry production function. *The American Economic Review*, *58*(4), 826–839.

Bal, H., & Örkçü, H. H. (2005). Combining the discriminant analysis and the data envelopment analysis in view of multiple criteria decision making: A new model. *Gazi University Journal of Science*, *18*(3), 355–364.

Baltagi, B. H., & Griffin, J. M. (1988). A general index of technical change. *Journal of Political Economy*, *96*(1), 20–41. doi:10.1086/261522

Barros, C. (2008). Efficiency analysis of hydroelectric generating plants: A case study for Portugal. *Energy Economics*, *1*(1), 59–75. doi:10.1016/j.eneco.2006.10.008

BP. (2021). *BP Statistical Review of World Energy*. https://www.bp.com/content/dam/bp/business-sites/en/global/corporate/pdfs/energy-economics/statistical-review/bp-stats-review-2020-renewable-energy.pdf

Broeck, V., & Meeusen, W. (1977). Efficiency estimation from Cobb-Douglas production functions with composed error. *International Economic Review*, *18*(2), 435–444. doi:10.2307/2525757

Çapik, M., Yılmaz, A. O., & Çavuşoğlu, İ. (2012). Present situation and potential role of renewable energy in Turkey. *Renewable Energy*, *46*, 1–13. doi:10.1016/j.renene.2012.02.031

Chachuli, F. S. M., Ludin, N. A., Mat, S., & Sopian, K. (2020). Renewable energy performance evaluation studies using the data envelopment analysis (DEA): A systematic review. *Journal of Renewable and Sustainable Energy*, *12*(6), 062701. doi:10.1063/5.0024750

Charnes, A., Cooper, W. W., & Rhodes, E. (1978). Measuring the efficiency of decision making units. *European Journal of Operational Research*, *2*(6), 429–444. doi:10.1016/0377-2217(78)90138-8

Chien, T., & Hu, J. (2007). Renewable energy and macro economic efficiency of OECD and non-OECD economies. *Energy Policy*, *35*(7), 3606–3615. doi:10.1016/j.enpol.2006.12.033

Coelli, T. J. (1995). Estimators and hypothesis tests for a stochastic frontier function: A Monte Carlo analysis. *Journal of Productivity Analysis*, *6*(3), 247–268. doi:10.1007/BF01076978

Coelli, T. J., Rao, D. S. P., O'Donnell, C. J., & Battese, G. E. (2005). *An Introduction to Efficiency and Productivity Analysis*. Springer.

Dağıstan, H. (2008). Our renewable energy and geothermal resources. In *Proceedings 5th World Water Forum*. Sözkesen Publishing.

Erdal, L. (2012). Renewable energy investments and potential for green jobs in Turkey. *Journal of Social and Human Sciences*, *4*(1), 171–181.

Farrell, M. J. (1957). The measurement of productive efficiency. *Journal of the Royal Statistical Society. Series A (General)*, *120*(3), 253–290. doi:10.2307/2343100

Gençoğlu, M. T. (2002). Importance of renewable energy resources for Turkey. *Fırat University Journal of Science and Engineering Sciences*, *14*(2), 57–64.

Gil, D. R. G., Costa, M. A., Lopes, A. L. M., & Mayrink, V. D. (2017). Spatial statistical methods applied to the 2015 Brazilian energy distribution benchmarking model: Accounting for unobserved determinants of inefficiencies. *Energy Economics*, *64*, 373–383. doi:10.1016/j.eneco.2017.04.009

Gökgöz, F., & Güvercin, M. T. (2018). Energy security and renewable energy efficiency in EU. *Renewable & Sustainable Energy Reviews*, *96*, 226–239. doi:10.1016/j.rser.2018.07.046

Halkos, G. E., & Tzeremes, N. G. (2012). Analyzing the Greek renewable energy sector: A data envelopment analysis approach. *Renewable & Sustainable Energy Reviews*, *16*(5), 2884–2893. doi:10.1016/j.rser.2012.02.003

Hepbaşlı, A., & Utlu, Z. (2004). Evaluating the energy utilization efficiency of Turkey's renewable energy sources during 2001. *Renewable & Sustainable Energy Reviews*, *8*(3), 237–255. doi:10.1016/j.rser.2003.11.001

Hsiao, W., Hu, J., Hsiao, C., & Chang, M. (2019). Energy efficiency of the Baltic Sea countries: An application of stochastic frontier analysis. *Energies*, *12*(1), 104. doi:10.3390/en12010104

Jha, D. K., & Shrestha, R. (2006). Measuring efficiency of hydropower plants in Nepal using data envelopment analysis. *IEEE Transactions on Power Systems*, *21*(4), 1502–1511. doi:10.1109/TPWRS.2006.881152

Khan, D., Nouman, M., Popp, J., Khan, M., Ur Rehman, F., & Olah, J. (2021). Link between technically derived energy efficiency and ecological footprint: Empirical evidence from the Asean region. *Energies*, *14*(13), 13. doi:10.3390/en14133923

Kim, K. T., Lee, D. J., Park, S. J., Zhang, Y., & Sultanov, A. (2015). Measuring the efficiency of the investment for renewable energy in Korea using data envelopment analysis. *Renewable & Sustainable Energy Reviews*, *47*, 694–702. doi:10.1016/j.rser.2015.03.034

Koç, E., & Kaya, K. (2014). Energy resources–state of renewable energy. *Engineer and Machine*, *56*(668), 36–47.

Kodde, D. A., & Palm, F. (1986). Wald criteria for jointly testing equality and inequality restrictions. *Econometrica*, *54*(5), 1243–1248. doi:10.2307/1912331

Külekçi, Ö. C. (2009). Place of geothermal energy in the content of renewable energy sources and it's importance for Turkey. *Ankara University Journal of Environmental Sciences*, *2*(2), 83–91.

Lau, L. J. (1986). Functional forms in econometric model building. Handbook of Econometrics, 3, 1515-1566.

Li, M., & Tao, W. (2017). Review of methodologies and polices for evaluation of energy efficiency in high energy-consuming industry. *Applied Energy*, *187*, 203–215. doi:10.1016/j.apenergy.2016.11.039

Lopes, A. L. M., & Mesquita, R. B. (2015). *Tariff regulation of electricity distribution: A comparative analysis of regulatory benchmarking models*. In The 14th European Workshop on Efficiency and Productivity Analysis 2015, Helsinki, Finland.

Menegaki, A. N. (2013). Growth and renewable energy in Europe: Benchmarking with data envelopment analysis. *Renewable Energy*, *60*, 363–369. doi:10.1016/j.renene.2013.05.042

Ouyang, X., Chen, J., & Du, K. (2021). Energy efficiency performance of the industrial sector: From the perspective of technological gap in different regions in China. *Energy*, *214*, 118865. doi:10.1016/j.energy.2020.118865

Prasad, R. (2015). The production function methodology for estimating the value of spectrum. *Telecommunications Policy*, *39*(1), 77–88. doi:10.1016/j.telpol.2014.12.007

San Cristóbal, J. R. A. (2011). Multi criteria data envelopment analysis model to evaluate the efficiency of the renewable energy technologies. *Renewable Energy*, *36*(10), 2742–2746. doi:10.1016/j.renene.2011.03.008

Sözen, A., Alp, İ., & Kilinc, C. (2012). Efficiency assessment of the hydro-power plants in Turkey by using data envelopment analysis. *Renewable Energy*, *46*, 192–202. doi:10.1016/j.renene.2012.03.021

Sözen, A., Alp, İ., & Özdemir, A. (2010). Assessment of operational and environmental performance of thermal power plants in Turkey by using data envelopment analysis. *Energy Policy*, *38*(10), 6194–6203. doi:10.1016/j.enpol.2010.06.005

The World Bank. (2021). *World development indicators*. https://databank.worldbank.org/data/reports.aspx?source=world-development-indicators&Type=TABLE&preview=on

Umar, H. S., Girei, A. A., & Yakubu, D. (2017). Comparison of cobb-douglas and Translog frontier models in the analysis of technical efficiency in dry-season tomato production. *Agrosearch*, *17*(2), 67. doi:10.4314/agrosh.v17i2.6

Vlahinić-Dizdarević, N., & Šegota, A. (2021). Total-factor energy efficiency in the EU countries. *Zbornik Radova Ekonomskog Fakulteta U Rijeci: Časopis Za Ekonomsku Teoriju I Praksu, 30*(2), 247-265.

Xu, X., Chen, H. H., Feng, Y., & Tang, J. (2018). The production efficiency of renewable energy generation and its influencing factors: Evidence from 20 countries. *Journal of Renewable and Sustainable Energy, 10*(2), 025901. doi:10.1063/1.5006844

Yenioğlu, Z. A., & Toklu, B. (2021). Performance measurement with stochastic data envelopment analysis: Comparative analysis of Turkish electricity distribution companies. *Journal of Polytechnic of Gazi University, 24*, 87–101.

Zeng, Y., Guo, W., & Zhang, F. (2019). Comprehensive evaluation of renewable energy technical plans based on data envelopment analysis. *Energy Procedia, 158*, 3583–3588. doi:10.1016/j.egypro.2019.01.907

ADDITIONAL READING

Banker, R. D., Charnes, A., & Cooper, W. W. (1984). Some models for estimating technical and scale inefficiencies in data envelopment analysis. *Management Science, 30*(9), 1078–1092. doi:10.1287/mnsc.30.9.1078

Ezici, B., Eğilmez, G., & Gedik, R. (2020). Assessing the eco-efficiency of US manufacturing industries with a focus on renewable vs. non-renewable energy use: An integrated time series MRIO and DEA approach. *Journal of Cleaner Production, 253*, 119630. doi:10.1016/j.jclepro.2019.119630

Han, Q., Meng, F., Hu, T., & Chu, F. (2017). Non-parametric hybrid models for wind speed forecasting. *Energy Conversion and Management, 148*, 554–568. doi:10.1016/j.enconman.2017.06.021

Jebali, E., Essid, H., & Khraief, N. (2017). The analysis of energy efficiency of the Mediterranean countries: A two-stage double bootstrap DEA approach. *Energy, 134*, 991–1000. doi:10.1016/j.energy.2017.06.063

Maradin, D., & Cerovi, L. (2014). Possibilities of applying the DEA method in the assessment of efficiency of companies in the electric power industry: Review of wind energy companies. *International Journal of Energy Economics and Policy, 4*(3), 320.

Mohd Chachuli, F. S., Ahmad Ludin, N., Mat, S., & Sopian, K. (2020). Renewable energy performance evaluation studies using the data envelopment analysis (DEA): A systematic review. *Journal of Renewable and Sustainable Energy*, *12*(6), 062701. doi:10.1063/5.0024750

Rodrigues, M., Montañés, C., & Fueyo, N. (2010). A method for the assessment of the visual impact caused by the large-scale deployment of renewable-energy facilities. *Environmental Impact Assessment Review*, *30*(4), 240–246. doi:10.1016/j.eiar.2009.10.004

Ueasin, N., Liao, S. Y., & Wongchai, A. (2015). The technical efficiency of rice husk power generation in Thailand: Comparing data envelopment analysis and stochastic frontier analysis. *Energy Procedia*, *75*, 2757–2763. doi:10.1016/j.egypro.2015.07.518

Wang, C. N., Dang, T. T., Tibo, H., & Duong, D. H. (2021). Assessing renewable energy production capabilities using DEA window and fuzzy TOPSIS model. *Symmetry*, *13*(2), 334. doi:10.3390ym13020334

Xin-gang, Z., & Zhen, W. (2019). The technical efficiency of China's wind power list enterprises: An estimation based on DEA method and micro-data. *Renewable Energy*, *133*, 470–479. doi:10.1016/j.renene.2018.10.049

Zhang, C., Cui, C., Zhang, Y., Yuan, J., Luo, Y., & Gang, W. (2019). A review of renewable energy assessment methods in green building and green neighborhood rating systems. *Energy and Building*, *195*, 68–81. doi:10.1016/j.enbuild.2019.04.040

KEY TERMS AND DEFINITIONS

Data Analysis: The process that collects raw data and turns it into meaningful and useful information using statistical methods.

Data Envelope Analysis: A non-parametric method used in operations research and economics for estimating production limits.

Decision-Making Process: The process that in case of a need, choosing the most suitable one from the available options in order to meet this need.

Efficiency: A performance dimension determining the degree of achievement of the objectives as a result of the activities.

Non-Parametric Model: The tests used for data series that are not suitable for normal distribution in statistics.

Parametric Model: The statistical model accepting that the data comply with the random distribution principle and makes inferences according to the probability distribution parameters.

Renewable Energy: The energy obtained from the existing energy flow in continuous natural processes.

Renewable Energy Efficiency: The balance of electricity demand, generation and consumption and less environmental pollution.

Stochastic Frontier Analysis: A parametric method used to measure the effectiveness of decision-making units.

Chapter 3
Analysis of the Criteria Used to Evaluate Renewable Energy Sources in Turkey With Fuzzy SWARA

Murat Kemal Keleş
Keçiborlu Vocational School, Isparta University of Applied Sciences, Turkey

Aşkın Özdağoğlu
Faculty of Business, Dokuz Eylul University, Turkey

Melik Ziya Yakut
Faculty of Technology, Isparta University of Applied Sciences, Turkey

ABSTRACT

The aim of this study is to determine the criteria used in the evaluation of renewable energy resources with the extremely high strategic importance that Turkey has and to find their degree of importance. For this purpose, an application has been made to find which criteria come to the fore and the weights of these criteria in order to evaluate Turkey's renewable energy resources. Five main criteria (technical, economic, environmental, social, and political) and a total of 22 sub-criteria related to these criteria were included in the scope of the study. Fuzzy SWARA method, one of the multi-criteria decision-making methods, has been used in the study. According to the results of the analysis, the most important criterion among the main criteria was "environmental criteria." As a result, it has shown the importance of the priority criteria for Turkey and environmental evaluation criteria, which are important for the common future of humanity in parallel with the results in the world.

DOI: 10.4018/978-1-6684-2472-8.ch003

INTRODUCTION

The world population continues to increase day by day. Unplanned use of limited life resources and wastage will negatively affect the quality of life of future generations. The increasing energy demand of the increasing population needs new opportunities and new contributions in terms of energy policies of countries. Today, the global competition between countries continues to be fierce. Countries that use the existing resources of the country in the most effective and efficient way in global competition, that make their own production with these resources, that attach importance to R&D, and produce products with high added value by transforming scientific knowledge into innovative products will gain an advantage over their competitors. A foreign trade deficit occurs in a country that produces with imported resources rather than its own resources. Therefore, production should be done by increasing the use of domestic resources and foreign trade deficit should be prevented. One of the most important production factors required for production is energy.

With the first industrialization revolution, directing the current economic potential of the countries to the main areas of growth such as production, technology, research and development activities brought along many advantages and disadvantages. The innovation brought by technology in every field in the world, which has entered a rapid transformation process, has revealed an intense energy demand that will support the sustainability of these innovations. The demand for the consumption of energy and natural resources is increased as a result of population growth, urbanization, industrialization and globalization in which commercial gains are constantly increased. Therefore, it is necessary with a sustainable policy to present energy resources to consumers in a sufficient, high quality, low cost, safe manner and with the utmost sensitivity to the environment. Unfortunately, although the damage to the environment and human health caused by the use of fossil fuels has become evident over time, countries have preferred their policies primarily according to their economic growth targets. Countries have delayed the development of technology and its use as an attractive resource by avoiding the necessary investments that are costly in the development of clean and sustainable renewable energy sources. Although the importance of renewable energy sources is now supported by international agreements by policymakers in countries, environmental quality is not currently positively affected by these developments.

Countries can meet their energy needs with fossil fuels, nuclear energy and renewable energy sources. All three energy sources have advantages and disadvantages. Although fossil fuels have low production costs, it is not environmentally friendly due to toxic gas emissions. High energy can be obtained from nuclear energy sources. However, the high risk of accidents and the intense radioactivity emitted should be considered. The advantages of renewable energy sources can be expressed as follows. Today, the production cost of renewable energy sources has decreased compared to the past. Renewable energy sources prevent energy import and do not cause foreign trade deficit. Renewable energy sources are more environmentally friendly than other energy sources.

While the rapid steps to be taken due to climate change caused by global warming are trying to keep countries away from fossil fuels, the concerns of countries about the sustainability and security of their energy policies are increasing day by day with the emergence of problems such as the transmission costs of natural gas and the increasing energy prices due to these. Renewable energy sources, on the other hand, show a reassuring development for the future in creating more independent and sustainable policies and creating serious employment opportunities in energy, with their increasing efficiency thanks to the ever-decreasing energy production costs and continuous development technologies.

Turkey is a developing country in its region and has a growing economy. Turkey's ever-increasing young population and high-tech daily life demand more energy. The increasing refugee population, the problems caused by the pandemic process and the need for protection from the negative effects of the climate crisis have created imbalances in the planning of energy production and consumption. While Turkey is signing to continue production at the production site of the 2021 Paris climate, it also invests in the sustainable and clean energy it needs. While completing the largest projects in the country in renewable energy resources, it also plans to extract and operate the natural gas resources it finds in its own field in the Black Sea until 2023.

The aim of this study is to determine the criteria used in the evaluation of renewable energy resources with extremely high strategic importance that Turkey has, and to find their degree of importance. For this purpose, an application has been made to find out which criteria come to the fore and the weights of these criteria in order to evaluate Turkey's renewable energy resources. First of all, the studies on renewable energy sources in the literature were examined and the criteria used in the evaluation of renewable energy sources were determined. Among the determined criteria, the criteria to be used in this study were selected and grouped by three academicians who are experts in renewable energy sources. The opinions of the experts in question are taken in determining the degree of importance of the criteria determined within the scope of the study. As evaluation criteria; 5 main criteria as technical, economic, environmental, social and political and a total of 22 sub-criteria related to these criteria were included in the scope of the study. Fuzzy SWARA method, one of the multi-criteria decision making methods, is used in the study.

In the literature research, it is possible to come across studies in which multi-criteria decision-making methods are used on different issues related to renewable energy sources. However, a study using the Fuzzy SWARA method to find the weights of the criteria used in the evaluation, covering all renewable energy sources as a whole, has not been found in the literature. In this respect, it is thought that the study will contribute to the literature.

The fuzzy SWARA method has emerged by integrating the fuzzy logic approach and the SWARA method, which is a subjective weighting method. In other words, it is a method based on fuzzy logic. It provides the opportunity to make calculations with the verbal expressions that decision makers use while evaluating. In this way, the opinions of the decision makers are easily included in the process. It is used to find the importance of evaluation criteria in decision problems. Having many criteria in decision-making problems complicates the evaluation process. However, the presence of many different criteria in the decision making process is not a problem for Fuzzy SWARA.

Within the scope of the study, the criteria to be used in the evaluation of renewable energy resources in Turkey are analyzed by ranking and scoring by three academicians who are experts in their fields according to the Fuzzy SWARA algorithm.

BACKGROUND

In the literature research, it is possible to come across studies in which multi-criteria decision-making methods are used on different issues related to renewable energy sources. However, a study using the Fuzzy SWARA method to find the weights of the criteria used in the evaluation, covering all renewable energy sources as a whole, has not been found in the literature. In this respect, it is thought that the study will contribute to the literature. Considering the studies on renewable energy sources in which multi-criteria decision-making methods are used, the following information can be given.

Yakıcı Ayan and Pabuçcu (2013) evaluated the renewable energy resources investment projects that are planned to be implemented in Turkey with the Analytical Hierarchy Process Approach.

Vafaeipour et al. (2014) evaluated the priority of regions for the implementation of solar energy projects in Iran with SWARA and WASPAS methods.

Büyüközkan and Güleryüz (2016) evaluated the existing renewable energy resources in Turkey from the perspective of investors by using the DEMATEL and ANP method.

Damgaci et al. (2017) evaluated the renewable energy sources in Turkey using the Intuitive Fuzzy Topsis method.

Karaca et al., (2017) evaluated the most suitable renewable energy source alternative in Turkey and the employment-increasing effect of renewable energy investments with the COPRAS method.

Maghsoodi et al. (2018) applied the Renewable Energy Technology Selection Problem in Iran using SWARA and MULTIMOORA methods.

Alkan and Albayrak (2020) made the selection and ranking of renewable energy sources in Turkey with Fuzzy Entropi, Fuzzy COPRAS and Fuzzy MULTIMOORA.

Rani et al. (2020) used Pythagorean Fuzzy SWARA–VIKOR methods for solar panel selection.

Ghenai et. al. (2020), the sustainability of renewable energy sources has been evaluated by SWARA and ARAS methods.

Mishra et. get. (2020) evaluated the bioenergy production process with Fuzzy SWARA and COPRAS.

Karaaslan and Aydın (2020) determined the most suitable renewable energy source alternative for Turkey by using AHP, COPRAS and MULTIMOORA methods.

Karaca and Ulutaş (2018) used Entropy and WASPAS methods to determine the most suitable renewable energy source to meet the energy need in Turkey.

Derse and Yontar (2020) conducted a study to determine the most suitable renewable energy source for Turkey by using the SWARA-TOPSIS method.

Albayrak (2020) examined the Multi-Criteria Decision Making techniques used when evaluating alternatives for renewable energy sources within the scope of studies between the years 2017-2020, the criteria used in the evaluation and the methods used when determining the weights of different evaluation criteria.

Karakul (2020) prioritized among renewable energy sources in Turkey using the Fuzzy AHP method.

Solan meat. get. (2019) evaluated renewable energy resources in Pakistan using Integrated Delphi, AHP and Fuzzy TOPSIS methods.

Yücenur and İpekçi (2021) used SWARA and WASPAS methods to determine the location of the power plant that is planned to be established in order to generate energy from the sea current.

Ecer (2021) analyzed the criteria affecting the wind farm establishment location, which is planned to be established, with the FUCOM method.

Wang et. get. (2021) made the selection of renewable energy sources required for Pakistan with SWOT analysis and Fuzzy AHP.

Bilgiç et al., (2021) selected the most suitable renewable energy source for a private energy company that wants to invest in the Central Anatolia Region.

Examples of study in which the Fuzzy SWARA method applied work are as follows:

Ghorabaee et al., (2017) evaluated construction equipment with Fuzzy SWARA, Fuzzy CRITIC and Fuzzy EDAS methods to ensure sustainability in the construction industry.

Zarbakhshnia et al., (2018) selected the most suitable sustainable third party reverse logistics provider using Fuzzy SWARA and the improved Fuzzy COPRAS.

Ulutaş et.al. (2020) used Fuzzy SWARA and CoCoSo methods to select the most suitable logistics center location for the city of Sivas in Turkey.

Agarwal et. Al., (2020) evaluated the solutions to supply chain management problems arising from personnel with Fuzzy SWARA and Fuzzy WASPAS methods.

Sahebi et al., (2020), analyzed the barriers to organizational transformation with Fuzzy SWARA.

Özdağoğlu et al., (2021) used the Fuzzy SWARA and Fuzzy MARCOS methods in an integrated way to select cabin attendants in civil aviation.

Arsu and Ayçin (2021), evaluated the third party reverse logistics service provider evaluation criteria with Fuzzy SWARA.

Sengul et al., (2021), made the valuation of the works in a newly established assembly line in a company, using Fuzzy SWARA and Intuitive Fuzzy AHP methods.

Poyraz (2021), in his master's thesis, first analyzed the errors in a demand planning process in the air conditioning and heating sector with the classical Failure Mode and Effect Analysis (FMEA) method, then analyzed them by using Fuzzy SWARA and Fuzzy Copras methods in an integrated way, and compared the results.

Keleş et al., (2021) made a general evaluation from the perspective of passengers of Süleyman Demirel, Denizli Çardak and Uşak Airports operating in Turkey using the Fuzzy SWARA, CODAS, ARAS, Fuzzy CODAS, Fuzzy ARAS methods.

METHODOLOGY: FUZZY SWARA

Fuzzy SWARA (Fuzzy Stepwise Weight Assessment Ratio) is a new method for assessing the weights of the criteria in the problem. The Fuzzy SWARA calculation process can be seen in Table 1 (Percin, 2019, 534). First, the criteria are ranked from the most important criterion to the least important criterion by the decision-makers independently. In the second step of Fuzzy SWARA method, decision-makers evaluate the criteria by using the fuzzy scale in Table 2 independently.

Table 1. Fuzzy SWARA steps

Step	Equation
Ranking the criteria	$\begin{cases} j = 1 \Rightarrow \text{the most important criterion} \\ j = n \Rightarrow \text{the least important criterion} \end{cases}$ (1)

continues on following page

Table 1. Continued

Step	Equation
Calculation of coefficient value	$\begin{cases} j=1 \Rightarrow k_{jdl}=1 \\ j>1 \Rightarrow k_{jdl}=1+s_{jdl} \end{cases}$ (2) $\begin{cases} j=1 \Rightarrow k_{jdm}=1 \\ j>1 \Rightarrow k_{jdm}=1+s_{jdm} \end{cases}$ (3) $\begin{cases} j=1 \Rightarrow k_{jdu}=1 \\ j>1 \Rightarrow k_{jdu}=1+s_{jdu} \end{cases}$ (4)
Finding the fuzzy recalculated weights	$\begin{cases} j=1 \Rightarrow q_{jdl}=1 \\ j>1 \Rightarrow q_{jdl}=\dfrac{q_{\{j-1\}dl}}{k_{jdu}} \end{cases}$ (5) $\begin{cases} j=1 \Rightarrow q_{jdm}=1 \\ j>1 \Rightarrow q_{jdm}=\dfrac{q_{\{j-1\}dm}}{k_{jdm}} \end{cases}$ (6) $\begin{cases} j=1 \Rightarrow q_{jdu}=1 \\ j>1 \Rightarrow q_{jdu}=\dfrac{q_{\{j-1\}du}}{k_{jdl}} \end{cases}$ (7)
Calculation of fuzzy relative weights	$w_{jdl}=\dfrac{q_{jdl}}{\sum_{j=1}^{n} q_{jdu}}$ (8) $w_{jdm}=\dfrac{q_{jdm}}{\sum_{j=1}^{n} q_{jdm}}$ (9) $w_{jdu}=\dfrac{q_{jdu}}{\sum_{j=1}^{n} q_{jdl}}$ (10)

continues on following page

Table 1. Continued

Step	Equation
Integration of the decision-makers` opinions	$w_{jl} = \frac{\sum_{d=1}^{D} w_{jdl}}{D}$ (11) $w_{jm} = \frac{\sum_{d=1}^{D} w_{jdm}}{D}$ (12) $w_{ju} = \frac{\sum_{d=1}^{D} w_{jdu}}{D}$ (13)
Defuzzification of the weights	$w_j = \frac{w_{jl} + w_{jm} + w_{ju}}{3}$ (14)
Normalization of the defuzzified weights	$w_j^* = w_j / (\sum_{j=1}^{n} w_j)$ (15)

Reference: Percin, 2019, 534-535

Where

j: criterion; $j=1,2,3,\ldots,n$
d: decision maker; $d=1,2,3,\ldots,D$
l: triangular fuzzy number lower limit value
m: triangular fuzzy number the most promising value
u: triangular fuzzy number upper limit value
$\tilde{s}_{jd}$: fuzzy evaluation value for criterion j according to decision maker d
s_{jdl}: fuzzy evaluation lower limit value
s_{jdm}: fuzzy evaluation the most promising value
s_{jdu}: fuzzy evaluation upper limit value
$\tilde{k}_{jd}$: coefficient value for decision maker d
k_{jdl}: coefficient lower limit value
k_{jdm}: coefficient the most promising value
k_{jdu}: coefficient upper limit value
$\tilde{q}_{jd}$: fuzzy recalculated weight for decision maker d
q_{jdl}: recalculated weight lower limit value
q_{jdm}: recalculated weight the most promising value
q_{jdu}: recalculated weight upper limit value
$\tilde{w}_{jd}$: fuzzy relative weight for decision maker d

w_{jdl}: relative weight lower limit value
w_{jdm}: relative weight the most promising value
w_{jdu}: relative weight upper limit value
$\tilde{w}_j$: aggregated fuzzy relative weight
w_{jl}: aggregated relative weight lower limit value
w_{jm}: aggregated relative weight the most promising value
w_{ju}: aggregated relative weight upper limit value
w_j: aggregated defuzzified relative weight
w_j^* : normalized weight of criterion j

Table 2. Evaluation scale for Fuzzy SWARA

Linguistic Term	s_{jdl}	s_{jdm}	s_{jdu}
Very low	0.00	0.00	0.30
Low	0.00	0.25	0.50
Medium	0.30	0.50	0.70
High	0.50	0.75	1.00
Very high	0.70	1.00	1.00

ANALYSES AND FINDINGS

Evaluating the Criteria Set

In the first phase of the study, the interviews were conducted with the experts about renewable energy resources. Main criteria and sub criteria can be seen in Table 3.

Table 3. Main criteria and sub-criteria

Criterion code	Criterion name
T	**Technical**
T1	Energy efficiency (%)
T2	Operating life (year)
T3	Land use
T4	Energy resource potential
T5	Sustainability
T6	Facility construction time

continues on following page

Table 1. Continued

Criterion code	Criterion name
EC	**Economic**
EC1	Investment cost ($/kW)
EC2	Facility O&M cost
EC3	Cost of electricity ($/kW-hour)
EC4	Return on investment
EC5	Profitability
EC6	Impact on local economy
EC7	External dependency rate
EN	**Environmental**
EN1	Greenhouse gas emissions
EN2	Air pollution emissions
EN3	Land requirement m^2/kWs
EN4	Ecological Impact
S	**Social**
S1	Post-Plant Employment rate (per MW)
S2	Pre-Plant Employment rate (per MW)
S3	Social Acceptance
P	**Politics**
P1	Government Incentive, support
P2	Permission Process

Four main criteria were determined and each was completed with sub-criteria. Explanations of the selected sub-criteria are given below:

(T1) Energy efficiency (%): The amount of energy obtained from an energy source ratio to the amount of energy used.

(T2) Operating life (year): Operating life as the year duration of the energy plant.

(T3) Land use: Amount of space the energy plant will use as land.

(T4) Energy resource potential: The potential amount of the energy source for the energy plant where it will be installed.

(T5) Sustainability: The possibility of continuously being present at the place where the energy source for the energy plant will be established.

(T6) Facility construction time: Construction time for the power plant up to energy production.

(EC1) Investment cost ($/kW): The initial investment cost of an energy plant until it is able to produce energy.

(EC2) Facility O&M cost : Maintenance and operating costs during the working period of an energy plant.

(EC3) Cost of electricity ($/kW-hour): 1 kW-hour of the plant, the amount of energy generation cost in dollars.

(EC4) Return on investment: Return on investment in the energy plant with the electricity it produces.

(EC5) Profitability: The profitability rate of investment in the energy plant, which can be revealed by considering the risks.

(EC6) Impact on local economy: Local economy contribution to the energy plant as a result of all processes.

(EC7) External dependency rate: Inability to be self-sufficient with local resources in all processes of the energy plant.

(EN1) Greenhouse gas emissions : Carbon dioxide (CO2), nitrous oxide (N2O), methane, three groups of fluorinated gases (sulfur hexafluoride (SF6), hydrofluorocarbons (HFCs) and perfluorocarbons (PFCs)) are the major anthropogenic greenhouse gases. It is the emissions of these gases that pollute the atmosphere in the installation and operation of the energy plant.

(EN2) Air pollution emissions: It is the release of gases that harm the lives of man and living things. Many gases can be counted as examples. There are many different types of air pollutants, such as gases (including ammonia, carbon monoxide, sulfur dioxide, nitrous oxides, methane, carbon dioxide and chlorofluorocarbons), particulates (both organic and inorganic), and molecule biologicals.

(EN3) Land requirement m^2/kWs : Specifies how many m^2 of space is needed for 1 kW/s of electricity generation of an energy plant.

(EN4) Ecological Impact: Effects of environmental damage during the installation and operation of the energy plant.

(S1) Post-Plant Employment rate (per MW): The rate of employment during the operation of the energy plant.

(S2) Pre-Plant Employment rate (per MW): The rate at which the energy plant is employed during construction.

(S3) Social Acceptance: Social acceptability of the energy source's preference.

(P1) Government Incentive, support: The electability and supportability of the source of energy in government policies.

(P2) Permission Process: Defines the process of obtaining permission from the relevant management units for the installation of the energy source.

In the first phase of the study, the experts evaluate the criteria. The evaluations of the expert 1 for main criteria can be seen in Table 4.

Table 4. Evaluations of expert 1 (main criteria)

	s_{j1l}	s_{j1m}	s_{j1u}
P			
EC	0,0000	0,0000	0,3000
EN	0,0000	0,0000	0,3000
T	0,3000	0,5000	0,7000
S	0,0000	0,0000	0,3000

Calculations of the coefficient values according to the answers of expert 1 for main criteria can be seen in Table 5.

Table 5. Coefficients (expert 1-main criteria)

	k_{j1l}	k_{j1m}	k_{j1u}
P	1,0000	1,0000	1,0000
EC	1,0000	1,0000	1,3000
EN	1,0000	1,0000	1,3000
T	1,3000	1,5000	1,7000
S	1,0000	1,0000	1,3000

Fuzzy recalculated weights according to the answers of expert 1 for main criteria can be seen in Table 6.

Table 6. Fuzzy recalculated weights (expert 1-main criteria)

	q_{j1l}	q_{j1m}	q_{j1u}
P	1,0000	1,0000	1,0000
EC	0,7692	1,0000	1,0000
EN	0,5917	1,0000	1,0000
T	0,3481	0,6667	0,7692
S	0,2677	0,6667	0,7692

Fuzzy relative weights according to the answers of expert 1 for main criteria can be seen in Table 7.

Table 7. Fuzzy relative weights (expert 1-main criteria)

	w_{j1l}	w_{j1m}	w_{j1u}
P	1,0000	1,0000	1,0000
EC	0,7692	1,0000	1,0000
EN	0,5917	1,0000	1,0000
T	0,3481	0,6667	0,7692
S	0,2677	0,6667	0,7692

The evaluations of the expert 1 for technical sub criteria can be seen in Table 8.

Table 8. Evaluations of expert 1 (technical sub criteria)

	s_{j1l}	s_{j1m}	s_{j1u}
T1			
T5	0,3000	0,5000	0,7000
T4	0,0000	0,0000	0,3000
T2	0,3000	0,5000	0,7000
T3	0,0000	0,2500	0,5000
T6	0,0000	0,0000	0,3000

Calculations of the coefficient values according to the answers of expert 1 for technical sub criteria can be seen in Table 9.

Table 9. Coefficients (expert 1- technical sub criteria)

	k_{j1l}	k_{j1m}	k_{j1u}
T1	1,0000	1,0000	1,0000
T5	1,3000	1,5000	1,7000
T4	1,0000	1,0000	1,3000
T2	1,3000	1,5000	1,7000
T3	1,0000	1,2500	1,5000
T6	1,0000	1,0000	1,3000

Fuzzy recalculated weights according to the answers of expert 1 for technical sub criteria can be seen in Table 10.

Table 10. Fuzzy recalculated weights (expert 1- technical sub criteria)

	q_{j1l}	q_{j1m}	q_{j1u}
T1	1,0000	1,0000	1,0000
T5	0,5882	0,6667	0,7692
T4	0,4525	0,6667	0,7692
T2	0,2662	0,4444	0,5917
T3	0,1774	0,3556	0,5917
T6	0,1365	0,3556	0,5917

Fuzzy relative weights according to the answers of expert 1 for technical sub criteria can be seen in Table 11.

Table 11. Fuzzy relative weights (expert 1- technical sub criteria)

	w_{j1l}	w_{j1m}	w_{j1u}
T1	0,2318	0,2866	0,3816
T5	0,1364	0,1911	0,2935
T4	0,1049	0,1911	0,2935
T2	0,0617	0,1274	0,2258
T3	0,0411	0,1019	0,2258
T6	0,0316	0,1019	0,2258

The evaluations of the expert 1 for economical sub criteria can be seen in Table 12.

Calculations of the coefficient values according to the answers of expert 1 for economical sub criteria can be seen in Table 13.

Fuzzy recalculated weights according to the answers of expert 1 for economical sub criteria can be seen in Table 14.

Fuzzy relative weights according to the answers of expert 1 for economical sub criteria can be seen in Table 15.

The evaluations of the expert 1 for environmental sub criteria can be seen in Table 16.

Table 12. Evaluations of expert 1 (economical sub criteria)

	s_{j1l}	s_{j1m}	s_{j1u}
EC5			
EC1	0,3000	0,5000	0,7000
EC6	0,0000	0,0000	0,3000
EC7	0,0000	0,0000	0,3000
EC3	0,3000	0,5000	0,7000
EC4	0,3000	0,5000	0,7000
EC2	0,3000	0,5000	0,7000

Table 13. Coefficients (expert 1- economical sub criteria)

	k_{j1l}	k_{j1m}	k_{j1u}
EC5	1,0000	1,0000	1,0000
EC1	1,3000	1,5000	1,7000
EC6	1,0000	1,0000	1,3000
EC7	1,0000	1,0000	1,3000
EC3	1,3000	1,5000	1,7000
EC4	1,3000	1,5000	1,7000
EC2	1,3000	1,5000	1,7000

Table 14. Fuzzy recalculated weights (expert 1- economical sub criteria)

	q_{j1l}	q_{j1m}	q_{j1u}
EC5	1,0000	1,0000	1,0000
EC1	0,5882	0,6667	0,7692
EC6	0,4525	0,6667	0,7692
EC7	0,3481	0,6667	0,7692
EC3	0,2047	0,4444	0,5917
EC4	0,1204	0,2963	0,4552
EC2	0,0708	0,1975	0,3501

Calculations of the coefficient values according to the answers of expert 1 for environmental sub criteria can be seen in Table 17.

Fuzzy recalculated weights according to the answers of expert 1 for environmental sub criteria can be seen in Table 18.

Table 15. Fuzzy relative weights (expert 1- economical sub criteria)

	w_{j1l}	w_{j1m}	w_{j1u}
EC5	0,2126	0,2539	0,3591
EC1	0,1250	0,1693	0,2762
EC6	0,0962	0,1693	0,2762
EC7	0,0740	0,1693	0,2762
EC3	0,0435	0,1129	0,2125
EC4	0,0256	0,0752	0,1634
EC2	0,0151	0,0502	0,1257

Table 16. Evaluations of expert 1 (environmental sub criteria)

	s_{j1l}	s_{j1m}	s_{j1u}
EN4			
EN2	0,3000	0,5000	0,7000
EN1	0,3000	0,5000	0,7000
EN3	0,0000	0,2500	0,5000

Table 17. Coefficients (expert 1- environmental sub criteria)

	k_{j1l}	k_{j1m}	k_{j1u}
EN4	1,0000	1,0000	1,0000
EN2	1,3000	1,5000	1,7000
EN1	1,3000	1,5000	1,7000
EN3	1,0000	1,2500	1,5000

Table 18. Fuzzy recalculated weights (expert 1- environmental sub criteria)

	q_{j1l}	q_{j1m}	q_{j1u}
EN4	1,0000	1,0000	1,0000
EN2	0,5882	0,6667	0,7692
EN1	0,3460	0,4444	0,5917
EN3	0,2307	0,3556	0,5917

Fuzzy relative weights according to the answers of expert 1 for environmental sub criteria can be seen in Table 19.

Table 19. Fuzzy relative weights (expert 1- environmental sub criteria)

	w_{j1l}	w_{j1m}	w_{j1u}
EN4	0,3387	0,4054	0,4619
EN2	0,1992	0,2703	0,3553
EN1	0,1172	0,1802	0,2733
EN3	0,0781	0,1441	0,2733

The evaluations of the expert 1 for social sub criteria can be seen in Table 20.

Table 20. Evaluations of expert 1 (social sub criteria)

	s_{j1l}	s_{j1m}	s_{j1u}
S3			
S2	0,5000	0,7500	1,0000
S1	0,3000	0,5000	0,7000

Calculations of the coefficient values according to the answers of expert 1 for social sub criteria can be seen in Table 21.

Table 21. Coefficients (expert 1- social sub criteria)

	k_{j1l}	k_{j1m}	k_{j1u}
S3	1,0000	1,0000	1,0000
S2	1,5000	1,7500	2,0000
S1	1,3000	1,5000	1,7000

Fuzzy recalculated weights according to the answers of expert 1 for social sub criteria can be seen in Table 22.

Fuzzy relative weights according to the answers of expert 1 for social sub criteria can be seen in Table 23.

The evaluations of the expert 1 for political sub criteria can be seen in Table 24.

Calculations of the coefficient values according to the answers of expert 1 for political sub criteria can be seen in Table 25.

Fuzzy recalculated weights according to the answers of expert 1 for political sub criteria can be seen in Table 26.

Table 22. Fuzzy recalculated weights (expert 1- social sub criteria)

	q_{j1l}	q_{j1m}	q_{j1u}
S3	1,0000	1,0000	1,0000
S2	0,5000	0,5714	0,6667
S1	0,2941	0,3810	0,5128

Table 23. Fuzzy relative weights (expert 1- social sub criteria)

	w_{j1l}	w_{j1m}	w_{j1u}
S3	0,4588	0,5122	0,5574
S2	0,2294	0,2927	0,3716
S1	0,1349	0,1951	0,2858

Table 24. Evaluations of expert 1 (political sub criteria)

	s_{j1l}	s_{j1m}	s_{j1u}
P2			
P1	0,3000	0,5000	0,7000

Table 25. Coefficients (expert 1- political sub criteria)

	k_{j1l}	k_{j1m}	k_{j1u}
P2	1,0000	1,0000	1,0000
P1	1,3000	1,5000	1,7000

Table 26. Fuzzy recalculated weights (expert 1- political sub criteria)

	q_{j1l}	q_{j1m}	q_{j1u}
P2	1,0000	1,0000	1,0000
P1	0,5882	0,6667	0,7692

Fuzzy relative weights according to the answers of expert 1 for political sub criteria can be seen in Table 27.

The procedure repeats for all experts. As a holistic approach, all results derived from the answers of Expert 1 can be seen in Table 28.

Table 27. Fuzzy relative weights (expert 1- political sub criteria)

	w_{j1l}	w_{j1m}	w_{j1u}
P2	0,5652	0,6000	0,6296
P1	0,3325	0,4000	0,4843

Table 28. All results (expert 1)

Criteria	w_{j1l}	w_{j1m}	w_{j1u}
T	0,0767	0,1538	0,2584
EC	0,1695	0,2308	0,3359
EN	0,1304	0,2308	0,3359
S	0,0590	0,1538	0,2584
P	0,2203	0,2308	0,3359
T1	0,2318	0,2866	0,3816
T2	0,0617	0,1274	0,2258
T3	0,0411	0,1019	0,2258
T4	0,1049	0,1911	0,2935
T5	0,1364	0,1911	0,2935
T6	0,0316	0,1019	0,2258
EC1	0,1250	0,1693	0,2762
EC2	0,0151	0,0502	0,1257
EC3	0,0435	0,1129	0,2125
EC4	0,0256	0,0752	0,1634
EC5	0,2126	0,2539	0,3591
EC6	0,0962	0,1693	0,2762
EC7	0,0740	0,1693	0,2762
EN1	0,1172	0,1802	0,2733
EN2	0,1992	0,2703	0,3553
EN3	0,0781	0,1441	0,2733
EN4	0,3387	0,4054	0,4619
S1	0,1349	0,1951	0,2858
S2	0,2294	0,2927	0,3716
S3	0,4588	0,5122	0,5574
P1	0,3325	0,4000	0,4843
P2	0,5652	0,6000	0,6296

The results of Expert 2, 3 and 4 can be seen in Table 29, 30 and 31 respectively.

Table 29. All results (expert 2)

Criteria	w_{j2l}	w_{j2m}	w_{j2u}
T	0,0889	0,1434	0,2300
EC	0,0593	0,1147	0,2300
EN	0,1778	0,2509	0,3451
S	0,0456	0,1147	0,2300
P	0,3023	0,3763	0,4486
T1	0,0474	0,1075	0,1832
T2	0,0806	0,1613	0,2381
T3	0,1363	0,1613	0,2381
T4	0,2725	0,3226	0,4048
T5	0,1048	0,1613	0,2381
T6	0,0316	0,0860	0,1832
EC1	0,0991	0,1838	0,3218
EC2	0,0127	0,0420	0,1262
EC3	0,0165	0,0420	0,1262
EC4	0,2528	0,3447	0,4184
EC5	0,1487	0,2298	0,3218
EC6	0,0248	0,0525	0,1262
EC7	0,0496	0,1051	0,2146
EN1	0,2099	0,3175	0,4500
EN2	0,0700	0,1270	0,2647
EN3	0,1049	0,1587	0,2647
EN4	0,3148	0,3968	0,4500
S1	0,1429	0,2254	0,3636
S2	0,2143	0,2817	0,3636
S3	0,4286	0,4930	0,5455
P1	0,3148	0,3333	0,3922
P2	0,6296	0,6667	0,6667

The opinions of the experts are integrated by using Equation 11, 12 and 13. These weights are defuzzified by using Equation 14. The results can be seen in Table 32.

Table 30. All results (expert 3)

Criteria	w_{j3l}	w_{j3m}	w_{j3u}
T	0,0684	0,1782	0,3730
EC	0,0889	0,1782	0,3730
EN	0,2000	0,2784	0,3730
S	0,0456	0,1425	0,3730
P	0,1333	0,2227	0,3730
T1	0,1435	0,1957	0,3085
T2	0,0499	0,1304	0,2373
T3	0,0849	0,1957	0,3085
T4	0,1104	0,1957	0,3085
T5	0,1865	0,1957	0,3085
T6	0,0294	0,0870	0,1826
EC1	0,0845	0,1515	0,2794
EC2	0,0500	0,1515	0,2794
EC3	0,1099	0,1515	0,2794
EC4	0,0650	0,1515	0,2794
EC5	0,1429	0,1515	0,2794
EC6	0,0333	0,1212	0,2794
EC7	0,0257	0,1212	0,2794
EN1	0,1923	0,2632	0,3629
EN2	0,1479	0,2632	0,3629
EN3	0,0986	0,2105	0,3629
EN4	0,2500	0,2632	0,3629
S1	0,2317	0,3030	0,3884
S2	0,1545	0,2424	0,3884
S3	0,3939	0,4545	0,5050
P1	0,5000	0,5556	0,6000
P2	0,3333	0,4444	0,6000

Table 32 stated that the main priority criterion begins with environmental and political decisions. In particular, the energy potential of the region where the facility is intended to be established stands out, while economic criteria stand out in the profitability sub-criteria in priority decision making. Environmental criteria also highlight the importance of ecological impacts, while social criteria have received the highest response according to social acceptance. The importance of permission-taking processes has emerged in political criteria. The table shows the local weights of the sub criteria. The weight values of main criteria are multiplied with the related sub criteria. Then the global weights of the sub criteria can be found. The last step is to normalize these global weights by using Equation 15. These results can be seen in Table 33.

Table 31. All results (expert 4)

Criteria	w_{j4l}	w_{j4m}	w_{j4u}
T	0,1357	0,2432	0,3432
EC	0,2293	0,2432	0,3432
EN	0,1764	0,2432	0,3432
S	0,0798	0,1622	0,2640
P	0,0469	0,1081	0,2031
T1	0,1499	0,2195	0,3249
T2	0,0678	0,1463	0,2499
T3	0,0452	0,1171	0,2499
T4	0,1949	0,2195	0,3249
T5	0,1153	0,2195	0,3249
T6	0,0266	0,0780	0,1923
EC1	0,1429	0,1471	0,2766
EC2	0,1099	0,1471	0,2766
EC3	0,0845	0,1471	0,2766
EC4	0,0500	0,1471	0,2766
EC5	0,0650	0,1471	0,2766
EC6	0,0385	0,1471	0,2766
EC7	0,0257	0,1176	0,2766
EN1	0,2727	0,2800	0,3764
EN2	0,2098	0,2800	0,3764
EN3	0,0807	0,1600	0,2509
EN4	0,1614	0,2800	0,3764
S1	0,2500	0,3544	0,5000
S2	0,3750	0,4430	0,5000
S3	0,1250	0,2025	0,3333
P1	0,5652	0,6000	0,6296
P2	0,3325	0,4000	0,4843

Table 32. Integration

Criterion	w_{jl}	w_{jm}	w_{ju}	w_j
T	0,0924	0,1797	0,3012	0,1911
EC	0,1367	0,1917	0,3205	0,2163
EN	0,1712	0,2508	0,3493	0,2571
S	0,0575	0,1433	0,2814	0,1607
P	0,1757	0,2345	0,3401	0,2501
T1	0,1432	0,2023	0,2995	0,2150
T2	0,0650	0,1414	0,2378	0,1481
T3	0,0769	0,1440	0,2556	0,1588
T4	0,1707	0,2322	0,3329	0,2453
T5	0,1358	0,1919	0,2913	0,2063
T6	0,0298	0,0882	0,1959	0,1047
EC1	0,1129	0,1629	0,2885	0,1881
EC2	0,0469	0,0977	0,2020	0,1155
EC3	0,0636	0,1134	0,2237	0,1336
EC4	0,0984	0,1796	0,2845	0,1875
EC5	0,1423	0,1956	0,3092	0,2157
EC6	0,0482	0,1225	0,2396	0,1368
EC7	0,0437	0,1283	0,2617	0,1446
EN1	0,1980	0,2602	0,3657	0,2746
EN2	0,1567	0,2351	0,3390	0,2439
EN3	0,0906	0,1684	0,2880	0,1823
EN4	0,2662	0,3363	0,4128	0,3385
S1	0,1899	0,2695	0,3845	0,2813
S2	0,2433	0,3150	0,4059	0,3214
S3	0,3516	0,4156	0,4853	0,4175
P1	0,4281	0,4722	0,5265	0,4756
P2	0,4652	0,5278	0,5952	0,5294

When the results in Table 33 are examined, it is seen that political sub-criteria and ecological impact are more important than other sub-criteria. In this study, the experts state that the priority choice should now be the establishment of a viable and environmentally compatible energy plant and energy source for Turkey. In particular, the change that took place when the use of coal, which was the biggest cause of air pollution in the past, was replaced by natural gas, the fact that agriculture is still very valid in rural areas, the reflection of the climate crisis as a major problem, has better shown the importance of choosing the right energy source in the country. Here, as a result, the effects of the climate crisis in the country become palpable, and the future anxiety caused by climate problems comes before the choice of energy source over other criteria.

Table 33. Global weights of sub criteria and normalization

Criterion	w_j	w_j^*
T1	0,0411	0,0363
T2	0,0283	0,0250
T3	0,0303	0,0268
T4	0,0469	0,0414
T5	0,0394	0,0348
T6	0,0200	0,0177
EC1	0,0407	0,0360
EC2	0,0250	0,0221
EC3	0,0289	0,0255
EC4	0,0406	0,0359
EC5	0,0467	0,0413
EC6	0,0296	0,0262
EC7	0,0313	0,0276
EN1	0,0706	0,0624
EN2	0,0627	0,0554
EN3	0,0469	0,0414
EN4	0,0870	0,0769
S1	0,0452	0,0400
S2	0,0517	0,0457
S3	0,0671	0,0593
P1	0,1190	0,1052
P2	0,1324	0,1171

FUTURE RESEARCH DIRECTIONS

In this study, Fuzzy SWARA method was used for the renewable energy resources sector. In another study, the weights of the evaluation criteria for a different sector can be calculated with Fuzzy SWARA or another MCDM method. Renewable energy source alternatives were not evaluated in this study. In another study, renewable energy source alternatives can be included in the study and listed using different MCDM methods. Also, the criteria to be used in evaluating the renewable energy resources of a different country can be determined and analyzed. Renewable energy sources in different regions can be compared with MCDM methods.

In future studies, the weights of the criteria to be used in the evaluation of renewable energy sources can be found by a different multi-criteria decision-making method. This study was carried out in Turkey. In future studies, the importance of criteria to evaluate renewable energy sources on a global scale or in a different country can be found. In addition, renewable energy source alternatives can be evaluated with new multi-criteria decision making methods used in recent studies in the literature.

CONCLUSION

As a result, the study was tried to be customized for Turkey, but achieved similar results due to similar problems in the world. Since the negative effects of the climate crisis have been seen, scientists and social consciences have tried to protect the environment, and according to the similar point of conclusion, the study has taken importance according to international values.

Environmental problems and the climate crisis are turning into key social problems and are destroying the most basic rights of life. Turkey, which lives in the most regional sense, is trying to make a choice for a clean future with its additional problems. With this study, it has been revealed that the most fundamental problem for the country should be started with this criterion and it will be the best solution to the problems of the developing and growing population of Turkey.

Due to the impact of global problems in Turkey, the energy source compatible with the environment and its political support stand out and other criteria determined afterwards can only be important. Supporting environmentally compatible production systems in energy policies in Turkey will allow future economic and technical criteria to become more talkable. Unless renewable energy sources supported by policies are sufficiently prominent, it will not be possible to talk about models that are beneficial to the environment and humanity.

REFERENCES

Agarwal, S., Kant, R., & Shankar, R. (2020). Evaluating solutions to overcome humanitarian supply chain management barriers: A hybrid fuzzy Swara–fuzzy Waspas approach. *International Journal of Disaster Risk Reduction, 101838*. Advance online publication. doi:10.1016/j.ijdrr.2020.101838

Albayrak, Ö. K. (2020). Multi criteria decision making techniques used in evaluation of renewable energy resources and analysis of evaluation criteria: 2017-2020. *Ataturk University Journal of Economics and Administrative Sciences*, *34*(4), 1287–1310. doi:10.16951/atauniiibd.717808

Alkan, Ö., & Albayrak, Ö. K. (2020). Ranking of renewable energy sources for regions in Turkey by fuzzy entropy based fuzzy Copras and fuzzy Multimoora. *Renewable Energy*, *162*, 712–726. doi:10.1016/j.renene.2020.08.062

Arsu, T., & Ayçin, E. (2021). Evaluation of third-party reverse logistics service provider selection criteria with fuzzy Swara method. *Journal of Yasar University*, *16*(63), 1282–1300.

Bilgiç, S., Torğul, B., & Paksoy, T. (2021). Evaluation of renewable energy resources with Bwm for sustainable energy management. *Journal of Productivity*, (2), 95–110.

Büyüközkan, G., & Güleryüz, S. (2016). An integrated Dematel-Anp approach for renewable energy resources selection in Turkey. *International Journal of Production Economics*, *182*, 435–448. doi:10.1016/j.ijpe.2016.09.015

Damgacı, E., Boran, K., & Boran, F. E. (2017). Evaluation of Turkey's renewable energy using intuitionistic fuzzy Topsis method. *Journal of Polytechnic*, *20*(3), 628–637.

Derse, O., & Yontar, E. (2020). Determination of the most appropriate renewable energy source by Swara-Topsis method. *Journal of Industrial Engineering*, *31*(3), 389–419.

Ecer, F. (2021). An analysis of the factors affecting wind farm site selection through Fucom subjective weighting method. *Pamukkale University Journal of Engineering Sciences*, *27*(1), 24–34. doi:10.5505/pajes.2020.93271

Ghenai, C., Albawab, M., & Bettayeb, M. (2020). Sustainability indicators for renewable energy systems using multi-criteria decision-making model and extended Swara/Aras hybrid method. *Renewable Energy*, *146*, 580–597. doi:10.1016/j.renene.2019.06.157

Ghorabaee, M. K., Amiri, M., Zavadskas, E. K., & Antucheviciene, J. (2018). A new hybrid fuzzy MCDM approach for evaluation of construction equipment with sustainability considerations. *Archives of Civil and Mechanical Engineering*, *18*(1), 32–49. doi:10.1016/j.acme.2017.04.011

Karaaslan, A., & Aydın, S. (2020). Evaluation of renewable energy resources with multi criteria decision making techniques: Evidence from Turkey. *Ataturk University Journal of Economics and Administrative Sciences*, *34*(4), 1351–1375. doi:10.16951/atauniiibd.749466

Karaca, C., & Ulutaş, A. (2018). The selection of appropriate renewable energy source for Turkey by using Entropy and Waspas methods. *Ege Academic Review*, *18*(3), 483–494. doi:10.21121/eab.2018341150

Karaca, C., Ulutaş, A., & Eşgünoğlu, M. (2017). Determination of renewable energy source in turkey by Copras and analysis of the employment-enhancing effect of renewable energy investments. *Maliye Dergisi*, *172*, 111–132.

Karakul, A. K. (2020). Selection of renewable energy source using fuzzy Ahp method. *Bingöl University Journal of Social Sciences Institute*, *10*(19), 127–150. doi:10.29029/busbed.640162

Keleş, M. K., Özdağoğlu, A., & Işıldak, B. (2021). An Application with Multi-Criteria Decision-Making Methods for the Evaluation of Airports from Passengers' View. *Ankara Hacı Bayram Veli University Journal of the Faculty of Economics and Administrative Sciences*, *23*(2), 419–456.

Maghsoodi, A. I., Maghsoodi, A. I., Mosavi, A., Rabczuk, T., & Zavadskas, E. K. (2018). Renewable energy technology selection problem using integrated H-Swara-Multimoora approach. *Sustainability*, *10*(12), 4481. doi:10.3390u10124481

Mishra, A. R., Rani, P., Pandey, K., Mardani, A., Streimikis, J., Streimikiene, D., & Alrasheedi, M. (2020). Novel multi-criteria intuitionistic fuzzy Swara–Copras approach for sustainability evaluation of the bioenergy production process. *Sustainability*, *12*(10), 4155. doi:10.3390u12104155

Özdağoğlu, A., Keleş, M. K., & Işıldak, B. (2021). Cabin crew selection in civil aviation with fuzzy Swara and fuzzy Marcos methods. Gümüşhane University Journal of Social Sciences Institute, 12(2), 284-302.

Perçin, S. (2019). An integrated fuzzy Swara and fuzzy AD approach for outsourcing provider selection. *Journal of Manufacturing Technology Management*, *30*(2), 531–552. doi:10.1108/JMTM-08-2018-0247

Poyraz, P. (2021). *Fault analysis with process phase fuzzy multiccriterial decision making methods in supply chain risk managament* [M.Sc. Thesis]. Karabük University Institute of Graduate Programs Department of Industrial Engineering, Karabük, Turkey.

Rani, P., Mishra, A. R., Mardani, A., Cavallaro, F., Štreimikienė, D., & Khan, S. A. R. (2020). Pythagorean fuzzy Swara–Vikor framework for performance evaluation of solar panel selection. *Sustainability*, *12*(10), 4278. doi:10.3390u12104278

Sahebi, I. G., Arab, A., & Toufighi, S. P. (2020). Analyzing the barriers of organizational transformation by using fuzzy Swara. *Journal of Fuzzy Extension & Applications*, *1*(2), 88–103. doi:10.22105/jfea.2020.249191.1010

Şengül, D., Çağıl, G., & Ardalı, Z. (2021). Fuzzy Swara and interval-valued intuitionistic fuzzy analytic hierarchy process application in job evaluation process. *Journal of Management and Economics*, *28*(2), 243–263. doi:10.18657/yonveek.731727

Solangi, Y. A., Tan, Q., Mirjat, N. H., Valasai, G. D., Khan, M. W. A., & Ikram, M. (2019). An integrated Delphi-AHP and fuzzy TOPSIS approach toward ranking and selection of renewable energy resources in Pakistan. *Processes (Basel, Switzerland)*, *7*(2), 118. doi:10.3390/pr7020118

Ulutaş, A., Karakuş, C. B., & Topal, A. (2020). Location selection for logistics center with fuzzy Swara and Cocoso methods. *Journal of Intelligent & Fuzzy Systems*, *38*(4), 4693–4709. doi:10.3233/JIFS-191400

Vafaeipour, M., Zolfani, S. H., Varzandeh, M. H. M., Derakhti, A., & Eshkalag, M. K. (2014). Assessment of regions priority for implementation of solar projects in Iran: New application of a hybrid multi-criteria decision making approach. *Energy Conversion and Management*, *86*, 653–663. doi:10.1016/j.enconman.2014.05.083

Wang, Y., Xu, L., & Solangi, Y. A. (2020). Strategic renewable energy resources selection for Pakistan: Based on swot-fuzzy Ahp approach. *Sustainable Cities and Society*, *52*(101861), 1–14. doi:10.1016/j.scs.2019.101861

Yakıcı Ayan, T., & Pabuçcu, H. (2013). Evaluation of the renewable energy investment project with analytic hierarchy process method. *Suleyman Demirel University. The Journal of Faculty of Economics and Administrative Sciences*, *18*(3), 89–110.

Yücenur, G. N., & Ipekçi, A. (2021). Swara/Waspas methods for a marine current energy plant location selection problem. *Renewable Energy*, *163*, 1287–1298. doi:10.1016/j.renene.2020.08.131

Zarbakhshnia, N., Soleimani, H., & Ghaderi, H. (2018). Sustainable third-party reverse logistics provider evaluation and selection using fuzzy Swara and developed fuzzy Copras in the presence of risk criteria. *Applied Soft Computing*, *65*, 307–319. doi:10.1016/j.asoc.2018.01.023

ADDITIONAL READING

Ali, S., Taweekun, J., Techato, K., Waewsak, J., & Gyawali, S. (2019). GIS based site suitability assessment for wind and solar farms in Songkhla, Thailand. *Renewable Energy*, *132*, 1360–1372. doi:10.1016/j.renene.2018.09.035

Ayçin, E., & Arsu, T. (2019). CODAS ve ENTROPİ yöntemleri ile yenilenebilir enerji kaynaklarının düzey 1 bölgelerine göre incelenmesi. *Avrasya Uluslararası Araştırmalar Dergisi*, *7*(18), 425–447. doi:10.33692/avrasyad.595695

El-Bayeh, C. Z., Alzaareer, K., Brahmi, B., Zellagui, M., & Eicker, U. (2021). An original multi-criteria decision-making algorithm for solar panels selection in buildings. *Energy*, *217*, 119396. doi:10.1016/j.energy.2020.119396

Kamali Saraji, M., Streimikiene, D., & Ciegis, R. (2022). A novel Pythagorean fuzzy-SWARA-TOPSIS framework for evaluating the EU progress towards sustainable energy development. *Environmental Monitoring and Assessment*, *194*(1), 1–19. doi:10.100710661-021-09685-9 PMID:34939168

Karatop, B., Taşkan, B., Adar, E., & Kubat, C. (2021). Decision analysis related to the renewable energy investments in Turkey based on a Fuzzy AHP-EDAS-Fuzzy FMEA approach. *Computers & Industrial Engineering*, *151*, 106958. doi:10.1016/j.cie.2020.106958

Lee, H. C., & Chang, C. T. (2018). Comparative analysis of MCDM methods for ranking renewable energy sources in Taiwan. *Renewable & Sustainable Energy Reviews*, *92*, 883–896. doi:10.1016/j.rser.2018.05.007

Lo, H. W., Hsu, C. C., Chen, B. C., & Liou, J. J. (2021). Building a grey-based multi-criteria decision-making model for offshore wind farm site selection. *Sustainable Energy Technologies and Assessments*, *43*, 100935. doi:10.1016/j.seta.2020.100935

Mojaver, M., Hasanzadeh, R., Azdast, T., & Park, C. B. (2022). Comparative study on air gasification of plastic waste and conventional biomass based on coupling of AHP/TOPSIS multi-criteria decision analysis. *Chemosphere*, *286*, 131867. doi:10.1016/j.chemosphere.2021.131867 PMID:34411931

KEY TERMS AND DEFINITIONS

Economic Criteria: Economic criterion is the evaluation of the energy facility and energy resource in terms of its economic worthiness.

Environmental Criteria: Environmental criterion defines the impact on the environment from the energy source to the energy production of the energy plant.

Fuzzy SWARA Method: Fuzzy Stepwise Weight Assessment Ratio is a new method for assessing the weights of the criteria in the problem.

Multi-Criteria Decision Making: Evaluating all alternatives by taking into account many different criteria together.

Political Criteria: The political criterion defines the support for the installation and production of the energy source in the country and the procedures for the construction of the energy plant.

Renewable Energy Resources: Renewable energy is an inexhaustible source of energy that is continuously available from the natural environment. It is obtained from natural sources including solar energy, wind power, biomass energy, geothermal energy, hydraulic energy, wave energy.

Social Criteria: Social criterion defines the social worthiness of the energy plant and its contribution to society.

Technical Criteria: Technical criterion is the computational evaluation of the energy plant and the energy source.

Chapter 4
Evaluating the Most Effective Solar Panel by ENTROPY and COPRAS Methods

Ezgi Demir
Sumer Robotics Engineering and Consultancy Ltd., UK

Figen Balo
Fırat University, Turkey

ABSTRACT

The terms "sustainable development" and "sustainability" have become more popular because of the critical troubles faced through mankind such as growing humankind effect on ecology and the risk of energy source depletion. Solar energy is one of the most rapidly developing sources of sustainable energy available today. The solar panel is the foundation of a photovoltaic system. In this study, the COPRAS and ENTROPY methods have been applied to select the most effective solar panel (100W) for a solar farm design. The five various solar panel brands have been evaluated, and professionals' choices have been dependent on the most important characteristics of the solar panels. From the top panel firms worldwide, the solar panel data utilized in this chapter is obtained. The most effective panel selection has been analyzed to affect the solar panel property potency employing in scales, in conjunction with several competing solar panels from which corporations must choose the top requirements, using COPRAS and ENTROPY techniques.

DOI: 10.4018/978-1-6684-2472-8.ch004

INTRODUCTION

Due to continued industrial expansion, the decline of fossil fuels, and increased environmental awareness, the need for renewable energy sources has risen steadily this century. The sun is the renewable source of all life on Earth. Through fusing H_2 into He at its core, this inexhaustible source radiates energy and supplies with both light and heat. This known as solar irradiation. Only approximately half of the sun's energy reaches the surface of Earth. The rest is either reflected or absorbed through the atmosphere and clouds. Even so, the sun provides enough energy to meet the needs of the whole human race — millions of times over. Solar energy is one of the options to petroleum-based fuels for producing power in tandem with today's quickly expanding industries. It is the cleanest and the most significant power resource that it can be utilized in almost every area. The first of these areas is lighting systems. Solar energy or sunlight, can be utilized directly for lighting and heating businesses and homes, for hot water heating, and for producing electricity, cooling with sun, and a range of other industrial and commercial uses. The fact that electricity production by sun is an excellent option to electricity generated from fossil fuels, with no water and air pollution, no global pollution by warming, threats to public health, and no electricity price spikes' risk, is especially significant. The amount of solar energy available is immense. The quantity of sunshine that touches the earth's surface in an hour and a half is adequate to power the entire world's energy usage for a year, according to the United States Department of Energy. The amount of energy contained in all of the world's natural gas, coal, and oil reserves is equal to sunshine's eighteen days on Earth.

Once a system is in location to use the sun source and convert it into beneficial power is free. Since 2010, the mean price of photovoltaic panels has decreased by more than 60 percent, while the solar electricity system's cost has decreased through roughly 50 percent. The solar electric is regarded to be cost-competitive with other forms of energy now. Without considering the given installation' specifics, there are solar energy industries' 2 primary kinds: concentrating and photovoltaic solar energy. When the solar irradiances are come on one of photovoltaic panels, photons from the sun are adsorbed through the solar cells, causing an electricity area to form across the sheets and energy to flow. Photovoltaic panels can be installed on the ground, on the roof, or on the wall. They can be permanently oriented to maximize output and value. In the same way, they can be mounted on trackers that follow the sun across the sky. In the future, our capacity to use solar and other sustainable energy resources is clearly dependent on our ability to do so. The tax incentives, expanding technologies, and utility firms adjusting to solar users are all positive signs in the solar energy business. The most significant thing to remember is that one m^2 collects 4.2 kWh of energy from the sun every day, which is about equivalent to nearly a barrel of oil per year when averaged over the entire surface of the earth. The key conditions for operating and developing an efficient solar energy station are: Right site (favorable climate and available land), right equipment choice (such as panels, inverters), and solar power availability (preferably direct sunlight unobstructed). Photovoltaic panels are one of the most expensive pieces of equipment due to their manufacture and installation expenses among used equipment.

In this reason, a photovoltaic panel is photovoltaic system's significant component, and much research has been done around the world to minimize material prices while boosting energy efficiency. The solar panel's total cost depends on, the brand, size (in W), the durability/longevity, the physical size, and panel's any certifications. During the last years, photovoltaic panels have been utilized for fewer scale energy production, especially for residential or commercial use in individual or complex buildings with performance ranges between 18 percent to 12 percent. Generally, solar panels have a whole life cycle of twenty-five years. Solar panel selection for a photovoltaic system is a multi criteria decision-making problem involving both quantitative and qualitative characteristics. Multi criteria decision-making is a combined decision-making mechanism as it includes both qualitative and quantitative criterions.

Photovoltaic panels are made up of bonded solar cells that produce electricity when exposed to sunlight. The existence of solar rays has a huge impact on their efficiency. The more sun rays that the photovoltaic panel receives, the more electrical energy it may produce. However, as the sun's rays increase, so does the temperature of the photovoltaic panel, resulting in a reduction in efficiency.

The photovoltaic panels are usually evaluated based on the specific characteristics: technical characteristics, which includes efficiency (the quantity of energy generated through collecting generated power and exposing it to solar radiation), reliability, financial results, safety, operation and maintenance ease, long service time and profitability, material and kind of photovoltaic panel (Setiawan, Kurniawan & Setiawan, 2015). There are photovoltaic panels' three main kinds; mono-crystalline, thin-film (amorphous), and poly-crystalline. The mono-crystalline photovoltaic panels are the most efficient. This is, they take the least amount of potential area to set up. But they are also costly. Although poly-crystalline types are fewer efficient than mono-crystalline types, they are far more affordable. Thin-film types are the least expensive, but they require the most space to set up – almost twice as much as mono-crystalline types. In photovoltaic design systems, the crystalline types (poly or mono crystalline) are the most common type of photovoltaic panel used. As a result, the solar panel selection is one of the most important factors in photovoltaic design systems.

During the last years, academic researcheres have begun to concentrate on the renewable energy technologies' evaluation. Criteria that are frequently contradictory must be compared throughout multiple assessments.

In the renewable energy business, the multi-criteria decision-making approach offers a scientific-technical decision-assistance apparatus that can demonstrate its selections consistently and clearly. In Korean, Brownson and Suh performed Geographic Information System Fuzzy and Analytic Hierarchy Process methods to determine the most appropriate solar energy stations through analyzing economic and social circumstances like mean system costs and power production; such as meteorology, local climate, topography, and economics (Suh, Brownson,2016). Through utilizing Fuzzy Analytic Hierarchy Process and Geographic Information System, Asakereh and his co-workers assessed the terrain convenience for solar energy station builds in Iran (Shodirwan region) (Asakereh, Omid, Alimardani & Sarmadian,2014). Balo and his co-workers suggested an Analytic Hierarchy Process methodology to determine the most appropriate photovoltaic panel for the solar energy-based production facility plan considering environmental, financial, electrical, and mechanical in addition to customer satisfaction effectiveness of each of the panels (Balo, Sagbansua, 2016). Kengpol and his co-workers assessed the solar energy station placements through implementing Technique for Order of Preference by Similarity to Ideal Solution and Fuzzy Analytic Hierarchy Process methodologies (Kengpol, Rontlaong & Tuominen, 2013). Kaa and his co-workers examined 5 photovoltaic industries to acquire the best photovoltaic industry through implementing the fuzzy (logarithmic fuzzy preference programming) and Analytic Hierarchy Process with crisp(Kaa, Rezaei, Kamp, Winter, 2014). Sahin and Turk utilized PV Analytic Hierarchy Process and Geographical Information System to assess option places and to choose the most appropriate location for solar energy-based production facility with minimal the total expense and maximal performance for electrical efficiency (Turk, Sahin,2018). In the Mediterranean region, Mandalaki and Stamatakis used a multi-criteria decision-making method to examine solar panels mounted to typical south-facing shading areas of houses (Stamatakis, Mandalaki & T., 2016). Zeyuan compared and assessed various types of solar cells using technique for order of preference by similarity to ideal solution. The findings revealed that monocrystalline-silicon photovoltaic cells are more effective, less expensive, and more socially beneficial for the enterprise than thin-film and polycrystalline silicon photovoltaic cells (Zeyuan, 2013). Beltran and his co-workers used ANP model to choose solar energy projects. Utilizing the two diverse ANP approaches, the impacts between the net's parameters (such as risks and alternatives) were defined and analyzed (Aragonés-Beltrán, Chaparro-González, Pastor-Ferrando & Rodríguez-Pozo, 2010). Utilizing both the Analytic Hierarchy Process and the solar home system model, Sahin and his co-workers assessed the selection of the most proper solar panel (monocrystalline) for the 3 cities of Turkey (Sinop, Adana, and Yozgat) (Şahin, Alakoç & Keçeci, 2010). Raina and his co-workers compared an archetype of 180 W peak crystalline photovoltaic panels from various companies to the optimal fill factor with actual fill factor for the photovoltaic panels' real-time monitoring ability (Raina, Mumbai & Hedau, 2013). In India, Sindhu and his co-workers suggested a hybrid methodology through using fuzzy technique for order of preference by similarity to ideal solution and Analytic Hierarchy Process to determine the best solar energy-based production facility in the solar energy efficiency thinking varied chief options: technical, social, economic, political, and environmental ways (Sindhu, Nehra & Luthra, 2017).Using the geographic information system and an analytic hierarchy process approach in China, Xiao and his co-workers suggested a solar farm place choice methodology for desert energy stations that took into account land cover forms, climate conditions, and terrain geographical considerations(Xiao, Yao, Qu, Sun, 2013). Bruce researched the efficiency of photovoltaic panels in terms of total cost and payback time. He used an experimental research to find differences in photovoltaic panel production owing to aging, temperature, shade, inclination angle, and direction. Data is assessed to compare performance against total cost, according to the firms' expertise (Bruce, 2011). Under same circumstances, Saetre and Midtgard studied on the performance of three various photovoltaic panels, each connected to an electrical load, sustaining in parallel to provide the 3 diverse V–I characteristic properties (Midtgård & Sætre, 2006). Giurca and his co-workers suggested using the PROMETHEE approach to choose technic solutions in the case of multi-junction photovoltaic panels (Giurca, Aşchilean, Safirescu & Mureşan, 2014). Poli-Si, Thin film, and M-Si panels were explored in three solar system configurations by Yılmaz and his co-workers (Yilmaz, Ozcalik, Kesler, Dincer, Yelmen, 2015).

To investigate firm-level data gathered from solar panel firms in the Crystalline-Silicon solar panel. Industry, the Analytic Hierarchy Process and technique for order of preference by similarity to ideal solution for multi criterion decision making methodologies were used by Chen and Yang (Chen & Yang, 2014). Gupta and his co-workers applied the Analytic Hierarchy Process in the Indian solar panel industry to assess production sustainability through multiple manufacturing uses (Gupta, Dangayach, Singh, Rao, 2015.). Guenounou and his co-workers examined the output of solar panels from several companies over the course of a year in natural environmental circumstances in Algeria. They investigated four types of solar panels: micromorph silicon (l-Si), amorphous silicon (a-Si), monocrystalline-silicon, and polycrystalline-silicon (Guenounou, Malek & Aillerie, 2016). Badea and his co-workers employed the Onicescu approach to identify the best multi junction solar panel for their photovoltaic energy system (Badea, Naghiu, Safirescu, Mureşan, Badea, & Megyesi, 2014). Salah and his co-workers (2008) designed a multi criteria fuzzy-algorithm for energy management in order to link native devices on photovoltaic panels (Salah, Chaabenea, Ammara, 2008). Khorasaninejad and his co-workers employed a multi-criteria hybrid approach sourced on fuzzy-DEMATEL, fuzzy-PROMETHEE, and fuzzy-ANP to choose the best alternative among fuel cells, solar panels, gas engines, gas turbines, and diesel engines (Khorasaninejad, Fetanat & Hajabdollahi, 2016). Based on literature and professional interviews, Kaa and his co-workers used LFPP and fuzzy-AHP among multi criteria decision-making methodologies to assess the twelve parameters of the five PV designs (Van de Kaa, Rezaei, Kamp, De Winter, 2014). Using the Electre-Boldur Methodology, Naghiu and his co-workers investigated the best resolution for the solar panels' concentration ratio (Naghiu, Giurca, Achilean & Badea, 2016). Cavallaro used an outranking approach to investigate the thin film solar panel industry's production procedures (Cavallaro, 2010). Amin and his co-workers conducted a field investigation on the performance of various solar panels (Amin, Lung & Sopian, 2009).

In the photovoltaic cell energy business, Lee and his colleagues created a conceptual approach for produce policy, combining "the benefits, opportunities, costs and risks", "fuzzy analytical network process", and "interpretative structural model" the idea to assess viable strategic (Lee, Chen, Kang, 2011).

The photovoltaic panel choice for solar industry is a multiple criteria decision-making question. The purpose of this paper is to identify the most optimal photovoltaic panel through assessing the multi directional features of diverse photovoltaic panel trademarks performed into the real solar farms. In this study, the COPRAS and ENTROPY methods for photovoltaic panel analysis is performed through utilizing existent characteristics of photovoltaic panel trademarks for 100W. Among the most commonly utilized photovoltaic panel trademarks, the most effective photovoltaic panel selection is obtained through the most important criteria' categories (environment, electrical, financial, customer, and mechanic) for photovoltaic panels.

ENTROPY AND COPRAS IN PHOTOVOLTAIC PANELS SELECTION

Entropy Method

Step 1: Applying a positive transform to data containing negative values

In this method, Z-score standardization is applied to the criteria data Xij values. It is expressed by the following mathematical expression.

$$Z_{if} = \frac{X_{ij} - \bar{X}_j}{\sigma_j} \tag{1}$$

Here $\bar{X}_j$ and σj respectively, j. are the mean and standard deviations of the criterion. Then the data is made positive by making coordinate transformation:

$$Z'_{ij}=Z_{ij} + A,\ A > |\min Z_{ij}| \tag{2}$$

In the decision matrix, Z'ij values are now written instead of Xij criterion values. For Z'ij values, it shows the correspondence of i = 1,2,3,.....,m alternative values to j = 1,2,3,...., n criterion values.

Step 2. Conversion of criteria into benefit or cost analysis

$$R_{ij} = \frac{X_{ij}}{\max X_{ij}} \text{(benefit criteria), (i =1,2,3,..,m number of alternatives)} \tag{3}$$

$$R_{ij} = \frac{\min X_{ij}}{X_{ij}} \text{(cost criteria). (j=1,2,3,....,n number of alternatives)} \tag{4}$$

Step 3. Normalizing the decision matrix

$$P_{if} = \frac{R_{ij}}{\sum_{i=1}^{m} R_{ij}},\ \forall j \tag{5}$$

i = alternatives

j = criteria,

P_{ij} = normalized values,

R_{ij} = Converted values by benefit or cost status.

Step 4. Calculation of Entropy Values

$$E_j = -k.\ \sum_{i=1}^{m} P_{ij}.In(P_{ij}),\forall j \tag{6}$$

k (entropy) $0 \leq E_j \leq 1$ is the entropy value that provides the expression

$$k = \frac{1}{In(m)}$$; m, the number of alternatives (7)

P_{ij} = stands for normalized values

Step 5. Calculation of degrees of difference

The D_j value, which represents the degree of difference of the information for each criterion, is calculated as follows.

$D_j = 1 - E_j$, j=1,2,…,n index of criteria (8)

Step 6. Calculation of weights

The significance weights (Wj) of the kits are calculated as follows by normalizing the degree of difference (Dj)

$$W_j = \frac{D_j}{\sum_{j=1}^{n} D_j}$$, j=1,2,…,n index of criteria (9)

Copras Method

Step 1: Creating the decision matrix

$$X = \begin{bmatrix} X_{11} & X_{12} & \cdots & X_{1N} \\ & \vdots & \ddots & \vdots \\ X_{M1} & X_{M2} & \cdots & X_{MN} \end{bmatrix} \quad (10)$$

i = 1,2,3,…,m number of alternatives j = 1,2,3,…., n number of criteria of X_{ij}

Step 2: Generating the Normalized Decision Matrix

$$X_{ij}^{*} = \frac{X_{ij}}{\sum_{i=1}^{m} X_{ij}} \quad \forall\, j = 1,2,3,\ldots,n \quad (11)$$

Step 3: Creating a Weighted Decision Matrix

The weighted decision matrix is obtained by multiplying the normalized decision matrix by the weight of each criterion value, W_j

$$D = d_{ij} = X_{ij} * W_j \tag{12}$$

Step 4. Calculation of Useful and Useless Metrics

$$S_i^+ = \sum_{j=1}^{k} d_{ij}, \text{ j = 1,2,3,...,k Useful Metrics} \tag{13}$$

$$S_i^- = \sum_{j=k+1}^{n} d_{ij}, \text{ j = k+1, k+2, k+3,...,n Useless Metrics} \tag{14}$$

Step 5. Q_i Calculation of Relative Significance Degree

$$Q_i = S_i^+ + \frac{\sum_{i=1}^{m} S_i^-}{S_i^- . \sum_{i=1}^{m} \frac{1}{S_i^-}} \tag{15}$$

Step 6. Calculation of Peak Relative Significance Values

$$Q_{maksimum} = \text{maksimum}\{Q_i\} \ \forall_i = 1,2,3,\ldots,m \tag{16}$$

Step 7. Calculation of Performance Index Pi Values for Alternatives

$$P_i - \frac{Q_i}{Q_{maksimum}} . \%100 \tag{17}$$

In this study, it has been aimed to apply a multi-criteria decision-making methodology for photovoltaic panel selection. For this purpose, criteria affecting panel selection have been determined. There are 23 criteria that affect the panel selection. The Entropy method, which allows to determine the weights of the criteria, has been chosen from the multi-criteria decision-making methods. Then, the data of 5 photovoltaic panels corresponding to the criteria have been taken and sorted. Another name for photovoltaic panel is solar panel. Sorting has been done with the COPRAS method, which is one of the ranking methods.

In the first stage, the initial decision matrix was created with the entropy method. The initial decision matrix has been shown in Table 1.

Table 1. The important parameters of solar panel (Solar panel datasheet catalogue, Wikipedia, 2021)

Alternatives/Criteria	STC Power per unit of area (W/m^2)	Power Tolerances (%)	Imp (A)	Isc (A)	NOCT (^0C)	Temp. Coefficient of Voltage (V/K)	Maximum System Voltage (V)	PTC power rating (W)	Peak Efficiency (%)	Number of Cells	Vmp (V)	Voc (V)	Temp. Coefficient of Power (%K)	Series Fuse Rating (A)	Lower energy density(W/m^2)	Length* Width* Depth (mm)	Weight (kg)	Price ($)	Support of government	Cost per Watt ($)	Spare part	Service support	Reliability
SP1	138.9	5	1.44	1.57	45	-0.245	1000	92.7	13.89	216	69.4	87.6	-0.29	4	8.22	4896000	12	135	0.27	0.43	2	1	1
SP2	137.3	2	5.75	6	47	-0.074	600	89.14	13.73	36	17.5	21.9	-0.46	15	7.98	25486125	9	120	0.25	0.38	3	2	2
SP3	150.7	5	7.96	8.72	55	-0.051	600	87.2	12.74	24	12.56	15.32	-0.42	15	7.52	14918374,02	8.2	115	0.25	0.37	1	3	3
SP4	66.4	5	1.31	1.6	47	-0.288	600	75.15	6.64	42	74.7	96	-0.21	3	5.45	60918000	26.4	95	0.22	0.30	4	4	5
SP5	64.3	5	1.3	1.67	45	-0.312	1000	74.23	6.43	32	71	94.5	-0.32	3	5.48	47190000	25	89	0.22	0.28	5	5	4

Table 2. Normalized decision matrix

Alternatives/Criteria	SP1	SP2	SP3	SP4	SP5
STC Power per unit of area (W/m^2)	0,249	0,246	0,270	0,119	0,115
Power Tolerances (%)	0,227	0,091	0,227	0,227	0,227
Imp (A)	0,081	0,324	0,448	0,074	0,0732
Isc (A)	0,080	0,307	0,446	0,082	0,085
NOCT (^{0}C)	0,188	0,197	0,230	0,197	0,188
Temp. Coefficient of Voltage (V/K)	0,253	0,076	0,053	0,297	0,322
Maximum System Voltage (V)	0,263	0,158	0,158	0,158	0,263
PTC power rating (W)	0,222	80,213	0,208	0,180	0,177
Peak Efficiency (%)	0,260	0,257	0,238	0,124	0,120
Number of Cells	0,617	0,103	0,069	0,12	0,091
Vmp (V)	0,283	0,071	0,051	0,305	0,290
Voc (V)	0,277	0,069	0,049	0,304	0,300
Temp. Coefficient of Power (%K)	0,171	0,271	0,247	0,124	0,188
Series Fuse Rating (A)	0,1	0,375	0,375	0,075	0,075
Lower energy density(W/m^2)	0,237	0,230	0,217	0,157	0,158
Length* Width* Depth (mm)	0,032	0,166	0,097	0,397	0,308
Weight (kg)	0,149	0,112	0,102	0,328	0,310
Price ($)	0,244	0,217	0,208	0,171	0,161
Support of government	0,223	0,207	0,207	0,182	0,182
Cost per Watt ($)	0,244	0,216	0,210	0,170	0,159
Spare part	0,133	0,2	0,067	0,267	0,333
Service support	0,067	0,133	0,2	0,267	0,333
Reliability	0,067	0,133	0,2	0,333	0,267

The negative values and unit values are different in the decision matrices in Table 1, the matrix should be normalized. Therefore, the variables have been normalized as the second step of the Entropy method. The normalized decision matrix has been shown in Table 2.

In the third step, entropy values have been found according to each criterion. Then, the weighted entropy values and the weights of the criteria have been found in Table 3.

After the weights of the criteria have been found, the problem of choosing the best solar panel has started.

Values found in criterion weighting have been used in the selection problem of solar panels. The initial decision matrix of the COPRAS method has been discussed as in Table 4.

The criteria have been divided into two as benefit-oriented and cost-oriented. It is expected that the benefit-oriented criteria will be maximum. And also cost-oriented criteria have been expected to be minimal. Therefore, the transformed decision matrix has been given in Table 5.

According to the next step of the COPRAS method, the transformed decision matrix should also be normalized due to the unit difference. For this reason, the normalization process has been performed in Table 6.

In the next step, the normalized matrix has been weighted. The weighted matrix has been given in Table 7.

In the last step of the COPRAS method, the optimum values have been obtained and ranked. These values have been shown in Table 8.

CONCLUSION

Because of growing energy requisition, investment accomplished on imported power resources raise petroleum-based oil prices and greenhouse gas emissions. As a sustainable power resource, sun power is safe to the human health and environment compared to petroleum-based oils, meets power requisition with lesser costs and gives mostly business potential in uncivilized fields. Thus, sun power investments have been getting more carefulness through policy-makers during the last years because of the great sun power potential in the World. A solar farm is utilized to obtain solar power through solar panels. The solar panel is one of a solar farm's critical components and there has been lots of studies recorded with the purpose of decreasing its material prices with improved energy performance. To choose the ideal photovoltaic panel, it must strike a balance between intangible and tangible characteristics that are at odds with one another.

Alternatives/Criteria	SP1	SP2	SP3	SP4	SP5		Wi values	Optimum
Reliability	1	2	3	5	4	Max	0,042	5
Service support	1	2	3	4	5	Max	0,042	5
Spare part	2	3	1	4	5	Max	0,042	5
Cost per Watt ($)	0.43	0.38	0.37	0.3	0.28	Min	0,004	0.28
Support of government	0.27	0.25	0.25	0.22	0.22	Max	0,001	0.27
Price ($)	135	120	115	95	89	Min	0,004	89
Weight (kg)	12	9	8.2	26.4	25	Min	0,043	8.2
Length* Width* Depth (mm)	4896000	25486125	14918374.02	60918000	47190000	Min	0,087	4896000
Lower energy density(W/m²)	8.22	7.98	7.52	5.45	5.48	Max	0,006	8.22
Series Fuse Rating (A)	4	15	15	3	3	Max	0,090	15
Temp. Coefficient of Power (%K)	-0.29	-0.46	-0.42	-0.21	-0.32	Max	0,013	-0.21
Voc (V)	87.6	21.9	15.32	96	94.5	Max	0,070	96
Vmp (V)	69.4	17.5	12.56	74.7	71	Max	0,068	74.7
Number of Cells	216	36	24	42	32	Max	0,149	216
Peak Efficiency (%)	13.89	13.73	12.74	6.64	6.43	Max	0,019	13.89
PTC power rating (W)	92.7	89.14	87.2	75.15	74.23	Max	0,001	92.7
Maximum System Voltage (V)	1000	600	600	600	1000	Max	0,012	1000
Temp. Coefficient of Voltage (V/K)	-0.245	-0.074	-0.051	-0.288	-0.312	Max	0,066	-0.051
NOCT (°C)	45	47	55	47	45	Max	0,001	55
Isc (A)	1.57	6	8.72	1.6	1.67	Max	0,096	8.72
Imp (A)	1.44	5.75	7.96	1.31	1.3	Max	0,105	7.96
Power Tolerances (%)	5	2	5	5	5	Max	0,016	5
STC Power per unit of area (W/m²)	138.9	137.3	150.7	66.4	64.3	Max	0,022	150.7

Table 4. Initial decision matrix

Table 5. Transformed decision matrix

Alternatives/Criteria	SP1	SP2	SP3	SP4	SP5		Optimum
STC Power per unit of area (W/m²)	138.9	137.3	150.7	66.4	64.3	Max	150.7
Power Tolerances (%)	5	2	5	5	5	Max	5
Imp (A)	1.44	5.75	7.96	1.31	1.3	Max	7.96
Isc (A)	1.57	6	8.72	1.6	1.67	Max	8.72
NOCT (°C)	45	47	55	47	45	Max	55
Temp. Coefficient of Voltage (V/K)	-0.245	-0.074	-0.051	-0.288	-0.312	Max	-0.051
Maximum System Voltage (V)	1000	600	600	600	1000	Max	1000
PTC power rating (W)	92.7	89.14	87.2	75.15	74.23	Max	92.7
Peak Efficiency (%)	13.89	13.73	12.74	6.64	6.43	Max	13.89
Number of Cells	216	36	24	42	32	Max	216
Vmp (V)	69.4	17.5	12.56	74.7	71	Max	74.7
Voc (V)	87.6	21.9	15.32	96	94.5	Max	96
Temp. Coefficient of Power (%K)	-0.29	-0.46	-0.42	-0.21	-0.32	Max	-0.21
Series Fuse Rating (A)	4	15	15	3	3	Max	15
Lower energy density(W/m²)	8.22	7.98	7.52	5.45	5.48	Max	8.22
Length* Width* Depth (mm)	2,04248E-07	3,9237E-08	6,70314E-08	1,64155E-08	2,11909E-08	Min	2,04248E-07
Weight (kg)	0.083	0.111	0.122	0.038	0.04	Min	0.122
Price ($)	0.007	0.008	0.009	0.011	0.011	Min	0.011
Support of government	0.27	0.25	0.25	0.22	0.22	Max	0.27
Cost per Watt ($)	2.326	2.632	2.703	3.333	3.571	Min	3.571
Spare part	2	3	1	4	5	Max	5
Service support	1	2	3	4	5	Max	5
Reliability	1	2	3	5	4	Max	5

Table 6. Normalized matrix

Alternatives/Criteria	SP1	SP2	SP3	SP4	SP5		Optimum
Reliability	0.050	0.100	0.150	0.250	0.200	Max	0.250
Service support	0.050	0.100	0.150	0.200	0.250	Max	0.250
Spare part	0.100	0.150	0.050	0.200	0.250	Max	0.250
Cost per Watt ($)	0.128	0.145	0.149	0.184	0.197	Min	0.197
Support of government	0.182	0.169	0.169	0.149	0.149	Max	0.182
Price ($)	0.129	0.145	0.151	0.183	0.196	Min	0.196
Weight (kg)	0.161	0.215	0.236	0.073	0.077	Min	0.236
Length* Width* Depth (mm)	0.370	0.071	0.121	0.030	0.038	Min	0.370
Lower energy density(W/m^2)	0.192	0.186	0.175	0.127	0.128	Max	0.192
Series Fuse Rating (A)	0.073	0.273	0.273	0.055	0.055	Max	0.273
Temp. Coefficient of Power (%K)	0.152	0.241	0.220	0.110	0.168	Max	0.110
Voc (V)	0.213	0.053	0.037	0.233	0.230	Max	0.233
Vmp (V)	0.217	0.055	0.039	0.234	0.222	Max	0.234
Number of Cells	0.382	0.064	0.042	0.074	0.057	Max	0.382
Peak Efficiency (%)	0.206	0.204	0.189	0.099	0.096	Max	0.206
PTC power rating (W)	0.181	0.174	0.171	0.147	0.145	Max	0.181
Maximum System Voltage (V)	0.208	0.125	0.125	0.125	0.208	Max	0.208
Temp. Coefficient of Voltage (V/K)	0.240	0.072	0.050	0.282	0.306	Max	0.050
NOCT (°C)	0.153	0.160	0.187	0.160	0.153	Max	0.187
Isc (A)	0.056	0.212	0.308	0.057	0.059	Max	0.308
Imp (A)	0.056	0.224	0.309	0.051	0.051	Max	0.309
Power Tolerances (%)	0.185	0.074	0.185	0.185	0.185	Max	0.185
STC Power per unit of area (W/m^2)	0.196	0.194	0.213	0.094	0.091	Max	0.213

Table 7. Weighted matrix

Alternatives/Criteria	SP1	SP2	SP3	SP4	SP5		Optimum
Reliability	0.002	0.004	0.006	0.011	0.008	Max	0.011
Service support	0.002	0.004	0.006	0.008	0.011	Max	0.011
Spare part	0.004	0.006	0.002	0.008	0.011	Max	0.011
Cost per Watt ($)	0.0006	0.0006	0.0006	0.0008	0.0009	Min	0.0009
Support of government	0.0002	0.0002	0.0002	0.0002	0.0002	Max	0.0002
Price ($)	0.001	0.0006	0.0006	0.0008	0.0008	Min	0.0008
Weight (kg)	0.007	0.009	0.010	0.003	0.003	Min	0.010
Length* Width* Depth (mm)	0.032	0.006	0.011	0.003	0.003	Min	0.032
Lower energy density(W/m²)	0.001	0.001	0.001	0.0007	0.0007	Max	0.001
Series Fuse Rating (A)	0.007	0.025	0.025	0.005	0.005	Max	0.025
Temp. Coefficient of Power (%K)	0.002	0.003	0.003	0.001	0.002	Max	0.001
Voc (V)	0.015	0.004	0.003	0.016	0.016	Max	0.016
Vmp (V)	0.015	0.004	0.003	0.016	0.015	Max	0.016
Number of Cells	0.057	0.009	0.006	0.011	0.008	Max	0.057
Peak Efficiency (%)	0.004	0.004	0.004	0.002	0.002	Max	0.004
PTC power rating (W)	0.0003	0.0003	0.0002	0.0002	0.0002	Max	0.0003
Maximum System Voltage (V)	0.002	0.001	0.001	0.001	0.002	Max	0.002
Temp. Coefficient of Voltage (V/K)	0.016	0.005	0.003	0.019	0.020	Max	0.003
NOCT (°C)	0.0002	0.0002	0.0002	0.0002	0.0002	Max	0.0002
Isc (A)	0.005	0.020	0.029	0.005	0.006	Max	0.029
Imp (A)	0.006	0.024	0.033	0.005	0.005	Max	0.033
Power Tolerances (%)	0.003	0.001	0.003	0.003	0.003	Max	0.003
STC Power per unit of area (W/m²)	0.004	0.004	0.005	0.002	0.002	Max	0.005

Table 8. Optimization function and sorting

	Si	Ki	Sorting
Optimum	0.272	1	
SP1	0.186	0.684	1
SP2	0.1371	0.504	3
SP3	0.1555	0.572	2
SP4	0.1233	0.453	5
SP5	0.1261	0.464	4

This study is based on information from current literature, photovoltaic technology investigations, and professional comments from the technology, companies, and manufacturers of solar panel. In this study Entropy-based Aras method has been used for the selection of solar panels. For this purpose, the criteria in the selection of solar panels have been weighted by the Entropy method. In this context, it has been seen that the most important criterion weight is the "Number of cells". The least important criterion weights have been found to be "NOCT", "PTC power rating" and "Support of government". Then, in order to find the best solar panel, sorting has been done with the COPRAS method. The most suitable alternative order can be listed as SP1, SP3, SP2, SP5 and SP4. The study is quite comprehensive within the scope of criterion weighting of the solar panel. Therefore, it is important. Objective data have been used in the study. The Entropy-based Copras method has been applied to the solar panel system for the first time. For further studies, it has been planning to benefit from fuzzy decision making methods by inviting leading names in the sector as decision makers.

REFERENCES

Amin, N., Lung, C., & Sopian, K. A. (2009). Practical field study of various solar cells on their performance in Malaysia. *Renewable Energy Journal*, *34*(8), 1939–1946.

Aragonés-Beltrán, P., Chaparro-González, F., Pastor-Ferrando, J. P., & Rodríguez-Pozo, F. (2010). An ANP-based approach for the selection of photovoltaic solar power plant investment projects. *Renewable & Sustainable Energy Reviews*, *14*(1), 249–264. doi:10.1016/j.rser.2009.07.012

Asakereh, A., Omid, M., Alimardani, R., & Sarmadian, F. (2014). Developing a GIS-based fuzzy AHP model for selecting solar energy sites in shodirwan region in Iran. *Int J Adv Sci Technol.*, *68*, 37–48. doi:10.14257/ijast.2014.68.04

Badea, G., Naghiu, G.S., Safirescu, C., Mureşan, D., Badea, F. & Megyesi, E. (2014). Choosing The Optimal Multi-Junctions Photovoltaic Cells for Application in The Field of Concentrated Photovoltaic. *Computer Applications in Environmental Sciences and Renewable Energy*, 144-150.

Balo, F., & Sagbansua, L. (2016). The selection of the best solar panel for the photovoltaic system design by using AHP. *Energy Procedia*, *100*, 50–53. doi:10.1016/j.egypro.2016.10.151

Bruce, T. (2011). *Investigation of Cost and Performance Characteristic of Photovoltaic Panels*. University of Southern Queensland Faculty of Engineering and Surveying.

Cavallaro, F. A. (2010). Comparative assessment of thin-film photovoltaic production processes using the ELECTRE III method. *Energy Policy*, *38*, 463–474.

Chen, H. C., & Yang, C. H. (2014). A Multi-Criterion Analysis of Cross-Strait Co-Opetitive Strategy in the Crystalline Silicon Solar Cell Industry. Mathematical Problems in Engineering.

Giurca, I., Aşchilean, I., Safirescu, C. O., & Mureşan, D. (2014). Choosing Photovoltaic Panels Using The Promethee Method. *Proceedings of the 8th International Management Conference "Management Challenges For Sustainable Development"*.

Guenounou, A., Malek, A., & Aillerie, M. (2016). Comparative performance of PV panels of different technologies over one year of exposure: Application to a coastal Mediterranean region of Algeria. *Energy Conversion and Management*, *114*, 356–363.

Gupta, S., Dangayach, G. S., Singh, A. K., & Rao, P. N. (2015). Analytic Hierarchy Process (AHP) Model for Evaluating Sustainable Manufacturing Practices in Indian Electrical Panel Industries. *Procedia: Social and Behavioral Sciences*, *189*, 208–216.

Kaa, G., Rezaei, J., Kamp, L., & Winter, A. (2014). Photovoltaic technology selection: A fuzzy MCDM approach. *Renewable & Sustainable Energy Reviews*, *32*, 662–670. doi:10.1016/j.rser.2014.01.044

Kengpol, A., Rontlaong, P., & Tuominen, M. (2013). A decision support system for selection of solar power plant locations by applying fuzzy AHP and TOPSIS: An empirical study. *J Software Eng Appl*, *6*(9), 470–481. doi:10.4236/jsea.2013.69057

Khorasaninejad, E., Fetanat, A., & Hajabdollahi, H. (2016). Prime mover selection in thermal power plant integrated with organic Rankine cycle for waste heat recovery using a novel multi criteria decision making approach. *Applied Thermal Engineering*, *102*, 1262–1279.

Lee, A. H. I., Chen, H. H., & Kang, H. Y. (2011). A model to analyze strategic products for photovoltaic silicon thin-film solar cell power industry. *Renewable & Sustainable Energy Reviews*, *15*, 1271–1283.

Midtgård, O. M., & Sætre, T. O. (2006). Seasonal variations in yield for different types of PV modules measured under real life conditions in northern Europe. *21st European photovoltaic solar energy conference,* 2383–6.

Naghiu, G. S., Giurca, I., Achilean, I., & Badea, G. (2016). Multicriterial Analysis on Selecting Solar Radiation Concentration Ration for Photovoltaic Panels Using Electre-Boldur Method. *Procedia Technology, 22*, 773 – 780.

Raina, G., Mumbai, N., & Hedau, R. (2013). A Novel Technique for PV Panel Performance Prediction. *IJCA Proc. Int. Conf. Work. Emerg. Trends Technol, 4*, 19–24.

Şahin, S., Alakoç, N. P., & Keçeci, B. A. (2010). DSS Based Selectıon of Solar Panels for Dıfferent Regıons Of Turkey. *10th International Conference On Clean Energy (Icce-2010).*

Salah, C. B., Chaabenea, M., & Ammara, M. B. (2008). Multi-criteria fuzzy algorithm for energy management of a domestic photovoltaic panel. *Renewable Energy*, *33*, 993–1001.

Setiawan, E. A., Kurniawan, K., & Setiawan, A. (2015). Gaussıan approach to compare crystallıne solar panel performance. *International Journal of Technology*, *3*(3), 336–344. doi:10.14716/ijtech.v6i3.1474

Sindhu, S., Nehra, V., & Luthra, S. (2017). Investigation of feasibility study of solar plants deployment using hybrid AHP-TOPSIS analysis, Case study of India. *Renew Sustain Energy,* (73), 496–511.

Stamatakis, M., Mandalaki, T. T., & Tsoutsos, T. (2016). Multi-criteria analysis for PV integrated in shading devices for Mediterranean region. *Energy and Building*, *117*, 128–137. doi:10.1016/j.enbuild.2016.02.007

Suh, J., & Brownson, J. (2016). Solar farm suitability using geographic information system fuzzy sets and analytic hierarchy processes: Case study of ulleung island, korea. *Energies*, *9*(8), 648. doi:10.3390/en9080648

Turk, S., & Sahin, G. (2018). Multi-criteria decision-making in the location selection for a solar PV power plant using AHP. *Measurement.*, *129*, 218–226. doi:10.1016/j.measurement.2018.07.020

Van de Kaa, G., Rezaei, J., Kamp, L., & De Winter, A. (2014). Photovoltaic technology selection: A fuzzy MCDM approach. *Renewable & Sustainable Energy Reviews*, *32*, 662–670.

Wikipedia. (2021). https://en.wikipedia.org/wiki/Growth_of_photovoltaics#Forecast

Xiao, J., Yao, Z., Qu, J., & Sun, J. (2013). Research on an optimal site selection model for desert photovoltaic power plants based on analytic hierarchy process and geographic information system. *Journal of Renewable and Sustainable Energy*, *5*(2), 1–15. doi:10.1063/1.4801451

Yilmaz, S., Ozcalik, H. R., Kesler, S., Dincer, F., & Yelmen, B. (2015). The analysis of different PV power systems for the determination of optimal PV panels and system installation— A case study in Kahramanmaras, Turkey. *Renewable & Sustainable Energy Reviews*, *52*, 1015–1024.

Zeyuan, Y. (2013). Selection of Solar Cell based on TOPSIS Method. *International Conference on Advanced Information Engineering and Education Science (ICAIEES)*, 151-154.

Chapter 5
Evaluation of the Criteria Used in the Selection of Renewable Energy Sources With the Plithogenic PIPRECIA Method

Alptekin Ulutaş
https://orcid.org/0000-0002-8130-1301
Sivas Cumhuriyet University, Turkey

Ayşe Topal
Nigde Omer Halisdemir, Turkey

ABSTRACT

The use of fossil fuels has decreased compared to the past due to the gradual depletion of fossil fuels and the greenhouse gases emerging in their use. As a result of this decrease, many countries have turned to alternative energy sources instead of fossil fuels. The most popular energy sources among these alternatives are renewable energy sources. Compared to fossil fuels, renewable energy sources cause little or no harm to nature. While renewable energy sources differ according to the countries, it is necessary to determine the most optimal renewable energy source for countries. In the literature, the most optimal renewable energy source selection has been made many times, and while this selection is made, multi-criteria decision-making (MCDM) methods are generally used. In these studies, the most optimal renewable energy source was selected by considering multiple criteria. In this chapter, criteria that are frequently used in the literature will be evaluated with the Plithogenic PIPRECIA method. In addition, the most important criterion will be determined.

DOI: 10.4018/978-1-6684-2472-8.ch005

INTRODUCTION

Energy has been used substantially by the economies to maintain development for centuries as it is a significant input for industries. The best option for energy use was fossil fuel initially as fossil fuels are easily to accessible and cheaper options compared to other resources. Therefore, most of the countries around the globe depend on fossil fuels to meet their energy needs (Sugiawan and Managi, 2019). However, using fossil fuels as a major source of energy has had a variety of negative environmental implications, such as atmospheric pollution and climate change.

Pollution causes a host of health problems as well as social and economic ramifications. Energy use leads to pollution on soil, water, and air. Air pollution is a serious public health issue, and one of the most significant issues created by energy use is air pollution. Poor air quality is responsible for around 6.5 million fatalities per year, which makes it fourth leading cause of mortality in the world after high blood pressure, nutritional diseases, and smoking (IEA, 2016). The devastating toll of air pollution on living beings is expected to increase unless the world changes how it generates and consumes energy.

Climate change has been a devastating problem recently due to the greenhouse gases with the increasing use of fossil fuels. To mitigate climate change, countries have decided to reduce the fossil fuel use in comparison to the past. Global efforts are also being undertaken to avert the effects of climate change by establishing international accords that lead to local laws tailored to each signing nation's development. Furthermore, fossil fuels are depleting, and new generations may find them limited or extinct. As a result, several countries have shifted toward alternative energy resources.

Renewable energy sources (RES) are less harmful to the environment in comparison to fossil fuels therefore they have been presented as a target in the Sustainable Development Goals (SDGs). SDGs present a compelling framework for international collaboration to create a sustainable future for the planet. They aim for eradicating extreme poverty, combating inequality and injustice, and safeguarding the planet's ecosystem such as dealing with climate change. SDG 7 has three major goals: guaranteeing affordable, reliable, and accessible energy; considerably increasing the proportion of RES in the global energy supply; and doubling the rate of energy efficiency improvement globally (McCollum et al., 2017).

Increased reliance on fossil fuels has exacerbated environmental problems, particularly in developing countries such as Turkey (Sinha et al., 2017). Turkey with its geographical location has high potential in RES, specifically solar energy. However, while RES are better options compared to fossil fuels, they vary by country therefore it is vital to decide which RES is best for each country. MCDM approaches are thought to be extremely beneficial in studying versatile energy challenges and offering paths for a sustainable future (Martín-Gamboa et al., 2017). Therefore, MCDM methods such as AHP, ANP, PROMETHEE, ELECTRE, DEA, TOPSIS, and VIKOR were commonly used in RES selection (Kumar et al., 2017). In this chapter, RES selection problem in Turkey has been studied with a novel MCDM method called Plithogenic PIPRECIA method. Plithogenic PIPRECIA method is a multi-criteria decision making (MCDM) which are used to determine the optimal one among several options. This method is useful to aggregated decision makers' opinion. Therefore, Plithogenic PIPRECIA method is used to evaluate criteria used in the selection of RES.

The rest of this chapter is organized as the following. A literature review about RES selection studies in Turkey is conducted in section 2. In section 3, the methodology is explained. Section 4 presents findings of Plithogenic PIPRECIA method. Section 5 of the chapter finishes with recommendations for the future.

BACKGROUND

Literature about RES selection decision problem has been reviewed. In the first paragraph, studies globally focusing on RES selection has been reviewed. In the second one, RES selection studies undertaken in Turkey has been assessed.

Al Garni et al. (2016) employed an AHP-based MCDM technique to evaluate RES options to rank Saudi Arabia's RES. It was found that solar PV is the most optimal option, followed by solar thermal and wind. Algarin et al. (2017) employed the AHP to assess RES in Colombia. They came to the conclusion that solar energy was the best RES option. Ijadi Maghsoodi et al. (2018) applied SWARA and MULTIMOORA methods to the RES selection problem in Iran. They presented that the best RES option is solar PV, followed by solar thermal and wind energy. Lee and Chang (2018) used WSM, VIKOR, TOPSIS and ELECTRE to compare RES options for Taiwan energy supply. Hydraulic energy was shown to be the most optimal choice. Wu et al. (2018) compared RES for China with triangular fuzzy number, AHP and cumulative prospect theory. They found that solar PV has been confirmed to be the best in China. Solangi et al. (2019) examined RES in Pakistan by using Delphi-AHP and fuzzy TOPSIS methods. It was found that wind energy emerges as the best RES option for generating electricity in Pakistan. Zhang et al. (2019) used extended TODIM to assess RES for China and found out the most significant criterion is greenhouse gas (GHG) emission, with the wind power project being the best RES option for China. Abdel-Basset et al. (2020) evaluated RES options under uncertainty in Egypt with an integrated method including triangular neutrosophic numbers, AHP, VIKOR and TOPSIS. They determined that concentrated solar energy is the best RES option for Egypt.

Uysal (2011) has assessed RES options for Turkey with graph theory and matrix approach. There were five criteria in total; technology, environment, socio-politic, economic and energy potential to examine alternatives. The findings have suggested that solar energy is the most optimal option for Turkey. Kuleli et al. (2015) used ANP and TOPSIS techniques to model the energy selection problem. They identified that hydro is the most optimal RES option in Turkey. Erdogan and Kaya (2015) defined the ranking of RES for Turkey. They employed type-2 fuzzy AHP to weight the parameters before ranking RES choices with type-2 fuzzy TOPSIS. The findings revealed that wind is the best RES. To assess the most acceptable RES option in Turkey from investment perspective, Büyüközkan and Güleryüz (2016) created a hybrid MCDM method incorporating DEMATEL and ANP approaches. Wind has been chosen as the most optimal RES in Turkey. Celikbilek and Tuysuz (2016) introduced a grey-based MCDM approach that combines DEMATEL, ANP, and VIKOR methodologies for choosing best RES option for Turkey. It was found that solar power is the most optimal option. Büyüközkan and Güleryüz (2017) used linguistic fuzzy hybrid method including DEMATEL, ANP and TOPSIS to rank RES options in Turkey. The most optimal RES option for Turkey has been found as geothermal and biogas. Karakaş and Yildiran (2019) utilized fuzzy AHP to compare RES and select best option for Turkey. They discovered that solar energy is the most optimal RES option and wind is the second in Turkey. Alkan and Albayrak (2020) compared RES in twenty-six regions of Turkey by using the fuzzy Entropy approach for weight calculation and fuzzy COPRAS and fuzzy MULTIMOORA methods for alternative ranking. Both fuzzy COPRAS and fuzzy MULTIMOORA methods have found that hydroelectric is a best option for most of the regions in Turkey. Derse and Yontar (2020) utilized an integrated MCDM method including SWARA and TOPSIS for choosing an optimal RES in Turkey. According to the findings of the study, hydroelectric plants should be given higher priority in Turkey. Yilan et al. (2020) assessed energy resources from sustainability perspective in Turkey with MAUT and WSM methods. They indicated that hydroelectricity (dam) is the most optimal choice to be used in Turkey.

Table 1 presents methodologies and criteria used in the literature.

METHODOLOGY

The Plithogenic PIPRECIA method is utilised in this study to evaluate criteria used in RES selection.

Table 1. Details of studies in the literature

Studies	Method	Economic Criteria	Environmental Criteria	Technical Criteria	Social/Political Criteria
Kuleli et al. (2015)	ANP and TOPSIS	- Dependency on import	- Energy use - Land pollution - Water pollution - Impact on climate change	- Efficiency	- Fatalities
Erdogan and Kaya (2015)	Fuzzy AHP and fuzzy TOPSIS	- Capital cost - Cost of operation and maintenance - Financial value - Economic risk - Operation life - Economic development of region - Supply security - Resource sustainability - Durability	- Air pollution - Land use - Disruption of land - Pollution of water - Waste	- Efficiency - Exergy efficiency - Maturity - Reliability - Feasibility - Risk - Knowhow - Performance sustainability and predictability - Implementation time - Preparation time - Technological and operational reliability - Accessibility	- Job creation - Acceptance of public and politics - Impact on labour - Compatibility with energy policy
Al Garni et al. (2016)	AHP	- Capital cost - Cost of operation and maintenance - Energy cost - Economic development of country	- CO2 emissions - Land use	- Efficiency - Maturity - Safety - Availability - Easy decentralisation	- Job creation - Acceptance of public and politics - Preserving leadership position
Büyüközkan and Güleryüz (2016)	DEMATEL and ANP	- Capital cost - Cost of operation and maintenance - Research & Development cost - Rate of return - Cost of production	- CO2 emissions - Land use/requirement - Impact on ecosystem	- Efficiency - Maturity - Reliability - Availability of resource - Innovation in technology - Investment capacity	- Job creation - Acceptance of public - Benefits for society - Compatibility with the legislation - Compatibility with energy policy - Government policy and financial assistance

Studies	Method	Economic Criteria	Environmental Criteria	Technical Criteria	Social/Political Criteria
Celikbilek and Tuysuz (2016)	DEMATEL, ANP, and VIKOR	- Capital cost	- Accessibility and sustainability of energy resource - Impact on environment	- Efficiency - Usability - Storability - Conveyance efficiency - Facility's simplicity - Technological requirement - Maintenance need	- Risk of accident
Algarin et al. (2017)	AHP	- Capital cost - Cost of operation and maintenance - Payback period - Operation life - Investment risk	- CO_2 emissions - Land use - Water use - Visual impact - Hazardous waste - Natural disaster risk	- Efficiency - Maturity - Availability of spare parts - Infrastructure - Reliability - Technical risk	- Acceptability of public - Job creation - The use of energy in rural health and education - The usage of indigenous land - Armed conflict risk
Ijadi Maghsoodi et al. (2018)	SWARA and MULTIMOORA	- Capital cost - Cost of operation and maintenance - Potential revenue	NA	- Efficiency - Amount Energy Generated - Level of Sophistication - Dependability - Air Pollution	- Job creation - Risk for long term - Global and local support
Lee and Chang (2018)	WSM, VIKOR, TOPSIS and ELECTRE	- Capital cost - Cost of operation and maintenance - Electricity cost	- GHG - Land use	- Efficiency - Capacity factor - Maturity	- Job creation - Acceptance of public
Wu et al. (2018)	Triangular fuzzy number, AHP and cumulative prospect theory	- Capital cost - Cost of operation and maintenance - Electricity cost - Payback period - Market potential	- Land use - Ecological effect	- Maturity - Reliability - Efficiency - Resource availability	- Job creation - Acceptance of public - Benefits for society
Karakaş and Yildiran (2019)	Fuzzy AHP	- Capital cost - Cost of operation and maintenance	- Particle emission - Land Use	- Efficiency	- Job creation - Acceptance of public - Accident-related fatality - Primary energy and technology dependence

Studies	Method	Economic Criteria	Environmental Criteria	Technical Criteria	Social/Political Criteria
Solangi et al. (2019)	Delphi-AHP and fuzzy TOPSIS	- Capital cost - Cost of operation and maintenance - Energy potential - Electricity production cost - Plant service life	- CO_2 - Noise - Land use - Ecological impact	- Efficiency - Capacity factor - Maturity - Expertise of labor - Condition of climate - Reliability/Feasibility	- Acceptance of public - Job creation - Security of energy supply - Institutional arrangement - Mechanism of regulation
Zhang et al. (2019)	Extended TODIM	- Capital cost - Operation life	- Land use - CO2 emission reduction	- Maturity - Technology efficiency	- Job creation
Abdel-Basset et al. (2020)	Triangular neutrosophic numbers, AHP, VIKOR and TOPSIS	- Principal costs - Investment and maintenance cost - Accessibility of funds - Payback duration - Prospective market	- Land use - Contaminant release - Waste - Ecosystem impact - Fuel use - Insulation strength Material intensity	- Maturity - Efficiency and continuity - Operational life	- Compliance with national energy policy - Acceptance of public - Impact on labour
Alkan and Albayrak (2020)	Fuzzy Entropy, fuzzy COPRAS and fuzzy MULTIMOORA	- Operation life - LCOE - Incentives - Payback period	- Land use - GHG emissions	- Regional energy efficiency - Regional electricity production potential - Regional primary energy saving rate - Regional technology maturity)	- Acceptance of public - Green job creation
Derse and Yontar (2020)	SWARA and TOPSIS	- Cost - Energy output - Operation life - Delivery time - Coordination with the external environment - Incentives	- Land use - Noise - GHG emissions - Ecological impact - Water use - Waste - Sustainability and predictability of resources	- Maturity - Reliability - Efficiency - Modularity in production and installation	- Job creation - Low risk of failure/accident - Social acceptability
Yilan et al. (2020)	MAUT and WSM	- LCOE	- CO_2 emission - Ozone depletion - Natural land transformation	- Efficiency - Flexibility - Electricity mix share - Capacity factor	- Job creation - Social acceptance - Fatalities caused by accidents - Reliance on primary energy and technology

Neutrosophic Set

$\tilde{v} = \langle (v_1, v_2, v_3); \alpha, \theta, \beta \rangle$ denotes a single valued triangular neutrosophic set including falsity membership function $FM_v(x)$, indeterminate membership $ID_v(x)$, and truth membership $TM_v(x)$ (Abdel-Basset et al., 2020):

$$TM_v(x) = \begin{cases} \alpha_v \left(\frac{x - v_1}{v_2 - v_1} \right) & \text{if } v_1 \le x \le v_2 \\ \alpha_v & \text{if } x = v_2 \\ \alpha_v \left(\frac{v_3 - x}{v_3 - v_2} \right) & \text{if } v_2 \le x \le v_3 \\ 0 & \text{otherwise} \end{cases} \quad (1)$$

$$ID_v(x) = \begin{cases} \frac{(v_2 - x)}{(v_2 - v_1)} \theta_v & \text{if } v_1 \le x \le v_2 \\ \theta_v & \text{if } x = v_2 \\ \frac{(x - v_3)}{(v_3 - v_2)} \theta_v & \text{if } v_2 < x \le v_3 \\ 1 & \text{otherwise} \end{cases} \quad (2)$$

$$FM_v(x) = \begin{cases} \frac{(v_2 - x)}{(v_2 - v_1)} \beta_v & \text{if } v_1 \le x \le v_2 \\ \beta_v & \text{if } x = v_2 \\ \frac{(x - v_3)}{(v_3 - v_2)} \beta_v & \text{if } v_2 < x \le v_3 \\ 1 & \text{otherwise} \end{cases} \quad (3)$$

Plithogenic PIPRECIA

Steps for Plithogenic PIPRECIA method are presented as (Ulutaş et al., 2021):

Step 1: RES selection criteria are identified, and experts rank them from the most significant to the least significant.

Step 2: Beginning with the second criterion, the *j*th criterion and the *j*–1th criteria are compared and plithogenic relative importance ($\tilde{d}_j$) values being used in this comparison. Table 2 indicates these plithogenic values.

Table 2. Linguistic Scale (Adapted from Abdel-Basset et al., 2020)

Linguistic Variable	Triangular Neutrosophic Scale (TNS)
Very Weakly Significant (VWKS)	((0.10,0.30,0.35), 0.10, 0.20, 0.15)
Weakly Significant (WKS)	((0.15,0.25,0.10), 0.60, 0.20, 0.30)
Fairly Weakly Significant (FWKS)	((0.40,0.35,0.50), 0.60, 0.10, 0.20)
Equal Significant (EQS)	((0.65,0.60,0.70), 0.80, 0.10, 0.10)
Strong Significant (STSG)	((0.70,0.65,0.80), 0.90, 0.20, 0.10)
Very Strongly Significant (VSTS)	((0.90,0.85,0.90), 0.70, 0.20, 0.20)
Absolutely Significant (ABS)	((0.95,0.90,0.95), 0.90, 0.10, 0.10)

Step 3: A contradiction degree acquires better precision for plithogenic aggregation operations (Smarandache, 2017), thus, it is determined between dominant criterion value and each criterion (Smarandache, 2018). Thus, this value (*c*: *V*×*V*®[0,1]) is identified.

Step 4: Opinions of all decision-makers are integrated with Equation 4.

$$\left(\left(v_{i1}, v_{i2}, v_{i3}\right), 1 \le i \le n\right) \wedge_p \left(\left(t_{i1}, t_{i2}, t_{i3}\right), 1 \le i \le n\right) = \left(v_{i1} \wedge_F t_{i1}, \frac{1}{2}\left(v_{i2} \wedge_F t_{i2}\right) + \frac{1}{2}\left(v_{i2} \vee_F t_{i2}\right), v_{i3} \vee_F t_{i3}\right), \; 1 \le i \le n \tag{4}$$

where Ù$_F$ and Ú$_F$ present the fuzzy t-norm and t-conorm, respectively.

Step 5: As follows, the neutrosophic numbers ($\tilde{d}_j$) are converted to crisp numbers (d_j):

$$U(v) = \frac{1}{9}\left(a_1 + b_1 + c_1\right) \times \left(2 + \alpha - \theta - \beta\right) \tag{5}$$

Step 6: The decision-makers' criteria rankings are integrated with the geometric mean to provide the final ranking of the criteria.

Step 7: g_j coefficient is obtained as:

$$g_j = \begin{cases} 1 & j = 1 \\ 2 - d_j & j > 1 \end{cases} \quad (6)$$

Step 8: z_j value is calculated as:

$$z_j = \begin{cases} 1 & j = 1 \\ \frac{z_{j-1}}{g_j} & j > 1 \end{cases} \quad (7)$$

Step 9: The criteria weights (w_j) are computed as:

$$w_j = \frac{z_j}{\sum_{k=1}^{n} z_k} \quad (8)$$

APPLICATION

The criteria utilized in RES selection are compared and ranked in this study. For the examination of the criteria, three experts' opinions were collected. Six criteria were determined by experts. These six criteria are as follows: Energy Efficiency (EE), Capacity Factor (CF), Investment Cost (IC), Electricity Generation Cost (EGC), CO_2 Emission (CE), and Social Criteria (SC). Experts have ranked these criteria with respect to their significance. The experts' criteria rankings are presented in Table 3.

Table 3. The criteria rankings of the experts

Experts Criteria	Exp-1	Exp-2	Exp-3
EE	1	1	1
CF	6	5	5
IC	2	3	2
EGC	3	2	3
CE	4	6	4
SC	5	4	6

Plithogenic values are assigned to each criterion by each expert, commencing with the second criterion to compare the criteria. The criteria comparisons of Expert 1 are presented in Table 4.

Table 4. The criteria comparisons of expert 1

Criteria	Rankings	Criteria	Linguistic Values	TNS
EE	1	EE	-	-
CF	6	IC	ABS	((0.95,0.90,0.95), 0.90, 0.10, 0.10)
IC	2	EGC	STSG	((0.70,0.65,0.80), 0.90, 0.20, 0.10)
EGC	3	CE	WKS	((0.15,0.25,0.10), 0.60, 0.20, 0.30)
CE	4	SC	FWKS	((0.40,0.35,0.50), 0.60, 0.10, 0.20)
SC	5	CF	VWKS	((0.10,0.30,0.35), 0.10, 0.20, 0.15)

Each criterion's contradiction degree is taken as 1/6. Then, equation 4 is utilized to combine the judgments of all decision-makers. Equation 5 is utilized to transform aggregated plithogenic values of criteria into crisp numbers. Table 5 shows crisp numbers (d_j) and aggregated plithogenic values ($\tilde{d}_j$) of criteria.

Table 5. Crisp numbers and aggregated plithogenic values of criteria

Criteria	$\tilde{d}_j$	d_j
EE	-	-
IC	((0.360,0.550,0.883), 0.532, 0.100, 0.279)	0.429
EGC	((0.295,0.488,0.855), 0.532, 0.125, 0.279)	0.387
CE	((0.285,0.600,0.899), 0.531, 0.200, 0.372)	0.388
SC	((0.098,0.388,0.752), 0.146, 0.150, 0.299)	0.233
CF	((0.073,0.363,0.675), 0.229, 0.200, 0.375)	0.204

The geometric mean is used to integrate the expert rankings of the criteria. The criteria weights are then found with Equations 6-8. Table 6 shows the findings of plithogenic PIPRECIA.

Table 6. Plithogenic PIPRECIA's findings

Criteria	Rankings by Geometric Mean	d_j	g_j	z_j	w_j
EE	1		1	1	0.401
IC	2	0.429	1.571	0.637	0.256
EGC	3	0.387	1.613	0.395	0.158
CE	4	0.388	1.612	0.245	0.098
SC	5	0.233	1.767	0.139	0.056
CF	6	0.204	1.796	0.077	0.031

According to Table 6, the order of the criteria is as follows: EE, IC, EGC, CE, SC and CF. Accordingly, the most significant criterion was found as EE, that is, Energy Efficiency. The least significant criterion was determined as CF (Capacity Factor).

FUTURE RESEARCH DIRECTIONS

Few criteria have been taken into account regarding RES in this study. The criteria used in this study is limited to six criteria. In future studies, more criteria can be included to the analysis. In addition, data were obtained from a small number of decision-makers in this study. Future studies can carry out a more detailed study by getting opinions from more experts.

CONCLUSION

Energy has been a vital input for industry for ages, and it has been used extensively by economies to maintain progress. Initially, fossil fuels were the ideal option for energy consumption since they are easily available and less expensive than alternative resources. As a result, the majority of countries throughout the world rely on fossil fuels to supply their energy demands. However, relying on fossil fuels as a major source of energy has had a variety of negative environmental repercussions, such as air pollution and climate change.

When compared to fossil fuels, RES are less hazardous to the environment. Turkey's geographic position offers significant potential for RES, particularly solar energy. While RES are superior to fossil fuels, renewable energy potential is different for every country, making it necessary to determine which RES is ideal for the country. MCDM techniques are regarded to be particularly useful in researching a variety of energy concerns and identifying routes to a more sustainable future.

Plithogenic PIPRECIA method was used in this study to evaluate and rank the criteria in the RES selection. The reason for using the Plithogenic PIPRECIA method is that this method is an effective method in unifying the judgments of decision-makers. The study results suggested that the most significant criterion is "Energy Efficiency". According to the same results, the least significant criterion was determined as "Capacity Factor".

REFERENCES

Abdel-Basset, M., Gamal, A., Chakrabortty, R. K., & Ryan, M. J. (2021). Evaluation approach for sustainable renewable energy systems under uncertain environment: A case study. *Renewable Energy*, *168*, 1073–1095. doi:10.1016/j.renene.2020.12.124

Abdel-Basset, M., Mohamed, R., Zaied, A. E. N. H., Gamal, A., & Smarandache, F. (2020). Solving the supply chain problem using the best-worst method based on a novel Plithogenic model. In *Optimization Theory Based on Neutrosophic and Plithogenic Sets* (pp. 1–19). Academic Press. doi:10.1016/B978-0-12-819670-0.00001-9

Al Garni, H., Kassem, A., Awasthi, A., Komljenovic, D., & Al-Haddad, K. (2016). A multicriteria decision making approach for evaluating renewable power generation sources in Saudi Arabia. *Sustainable Energy Technologies and Assessments*, *16*, 137–150. doi:10.1016/j.seta.2016.05.006

Algarin, R. A., Llanos, A. P., & Castro, A. O. (2017). An analytic hierarchy process based approach for evaluating renewable energy sources. *International Journal of Energy Economics and Policy*, *7*(4), 38–47.

Alkan, Ö., & Albayrak, Ö. K. (2020). Ranking of renewable energy sources for regions in Turkey by fuzzy entropy based fuzzy COPRAS and fuzzy MULTIMOORA. *Renewable Energy*, *162*, 712–726. doi:10.1016/j.renene.2020.08.062

Büyüközkan, G., & Güleryüz, S. (2016). An integrated DEMATEL-ANP approach for renewable energy resources selection in Turkey. *International Journal of Production Economics*, *182*, 435–448. doi:10.1016/j.ijpe.2016.09.015

Büyüközkan, G., & Güleryüz, S. (2017). Evaluation of Renewable Energy Resources in Turkey using an integrated MCDM approach with linguistic interval fuzzy preference relations. *Energy*, *123*, 149–163. doi:10.1016/j.energy.2017.01.137

Çelikbilek, Y., & Tüysüz, F. (2016). An integrated grey based multi-criteria decision making approach for the evaluation of renewable energy sources. *Energy*, *115*, 1246–1258. doi:10.1016/j.energy.2016.09.091

Derse, O., & Yontar, E. (2020). Determination of the Most Appropriate Renewable Energy Source by SWARA-TOPSIS Method. *Journal of Industrial Engineering*, *31*(3), 389–419.

Erdogan, M., & Kaya, I. (2015). An integrated multi-criteria decision-making methodology based on Type-2 fuzzy sets for selection among energy alternatives in Turkey. *Iranian Journal of Fuzzy Systems*, *12*(1), 1–25.

IEA. (2016). *Energy and air pollution: World energy outlook special report 2016*. Available online: http://pure.iiasa.ac.at/id/eprint/13467/1/WorldEnergyOutlookSpecialReport2016EnergyandAirPollution.pdf

Ijadi Maghsoodi, A., Ijadi Maghsoodi, A., Mosavi, A., Rabczuk, T., & Zavadskas, E. K. (2018). Renewable energy technology selection problem using integrated h-swara-multimoora approach. *Sustainability*, *10*(12), 4481. doi:10.3390u10124481

Karakaş, E., & Yildiran, O. V. (2019). Evaluation of renewable energy alternatives for Turkey via modified fuzzy AHP. *International Journal of Energy Economics and Policy*, *9*(2), 31–39. doi:10.32479/ijeep.7349

Kuleli, P. B., Albayrak, Y. E., & Erensal, Y. C. (2015). Renewable energy perspective for Turkey using sustainability indicators. *International Journal of Computational and Intelligent Systems*, *8*(1), 187–197.

Kumar, A., Sah, B., Singh, A. R., Deng, Y., He, X., Kumar, P., & Bansal, R. C. (2017). A review of multi criteria decision making (MCDM) towards sustainable renewable energy development. *Renewable & Sustainable Energy Reviews*, *69*, 596–609. doi:10.1016/j.rser.2016.11.191

Lee, H. C., & Chang, C. T. (2018). Comparative analysis of MCDM methods for ranking renewable energy sources in Taiwan. *Renewable & Sustainable Energy Reviews*, *92*, 883–896. doi:10.1016/j.rser.2018.05.007

Martín-Gamboa, M., Iribarren, D., García-Gusano, D., & Dufour, J. (2017). A review of life-cycle approaches coupled with data envelopment analysis within multi-criteria decision analysis for sustainability assessment of energy systems. *Journal of Cleaner Production*, *150*, 164–174. doi:10.1016/j.jclepro.2017.03.017

McCollum, D., Gomez Echeverri, L., Riahi, K., & Parkinson, S. (2017). *Sdg7: Ensure access to affordable, reliable, sustainable and modern energy for all.* Available online: http://pure.iiasa.ac.at/id/eprint/14621/1/SDGs-interactions-7-clean-energy.pdf

Sinha, A., Shahbaz, M., & Balsalobre, D. (2017). Exploring the relationship between energy usage segregation and environmental degradation in N-11 countries. *Journal of Cleaner Production, 168*, 1217–1229. doi:10.1016/j.jclepro.2017.09.071

Smarandache, F. (2017). *Plithogeny, plithogenic set, logic, probability, and statistics*. Pons Publishing House.

Smarandache, F. (2018). Plithogenic Set, an extension of crisp, fuzzy, intuitionistic fuzzy, and neutrosophic sets-revisited. *Neutrosophic Sets Syst., 21*, 153–166.

Solangi, Y. A., Tan, Q., Mirjat, N. H., Valasai, G. D., Khan, M. W. A., & Ikram, M. (2019). An integrated Delphi-AHP and fuzzy TOPSIS approach toward ranking and selection of renewable energy resources in Pakistan. *Processes (Basel, Switzerland)*, 7(2), 118. doi:10.3390/pr7020118

Sugiawan, Y., & Managi, S. (2019). New evidence of energy-growth nexus from inclusive wealth. *Renewable & Sustainable Energy Reviews, 103*, 40–48. doi:10.1016/j.rser.2018.12.044

Ulutaş, A., Topal, A., Karabasevic, D., Stanujkic, D., Popovic, G., & Smarandache, F. (2021, August). Prioritization of logistics risks with plithogenic PIPRECIA method. In *International Conference on Intelligent and Fuzzy Systems* (pp. 663-670). Springer.

Uysal, F. (2011). Graph Theory and Matrix Approach for the Selection of Renewable Energy Alternatives in Turkey. *Istanbul University Econometrics and Statistics e-Journal, 13*, 23-40.

Wu, Y., Xu, C., & Zhang, T. (2018). Evaluation of renewable power sources using a fuzzy MCDM based on cumulative prospect theory: A case in China. *Energy, 147*, 1227–1239. doi:10.1016/j.energy.2018.01.115

Yilan, G., Kadirgan, M. N., & Çiftçioğlu, G. A. (2020). Analysis of electricity generation options for sustainable energy decision making: The case of Turkey. *Renewable Energy, 146*, 519–529. doi:10.1016/j.renene.2019.06.164

Zhang, L., Xin, H., Yong, H., & Kan, Z. (2019). Renewable energy project performance evaluation using a hybrid multi-criteria decision-making approach: Case study in Fujian, China. *Journal of Cleaner Production, 206*, 1123–1137. doi:10.1016/j.jclepro.2018.09.059

ADDITIONAL READING

Burke, M. J., & Stephens, J. C. (2018). Political power and renewable energy futures: A critical review. *Energy Research & Social Science*, *35*, 78–93. doi:10.1016/j.erss.2017.10.018

Gomathy, S., Nagarajan, D., Broumi, S., & Lathamaheswari, M. (2020). Plithogenic sets and their application in decision making. *Neutrosophic Sets and Systems*, *38*, 453–469.

Lee, H. C., & Chang, C. T. (2018). Comparative analysis of MCDM methods for ranking renewable energy sources in Taiwan. *Renewable & Sustainable Energy Reviews*, *92*, 883–896. doi:10.1016/j.rser.2018.05.007

Letcher, T. M. (2019). Why do we have global warming? In *Managing global warming* (pp. 3–15). Academic Press. doi:10.1016/B978-0-12-814104-5.00001-6

Lu, Y., Shao, M., Zheng, C., Ji, H., Gao, X., & Wang, Q. G. (2020). Air pollutant emissions from fossil fuel consumption in China: Current status and future predictions. *Atmospheric Environment*, *231*, 117536. doi:10.1016/j.atmosenv.2020.117536

Martin, N., & Smarandache, F. (2020). Introduction to Combined Plithogenic Hypersoft Sets. *Neutrosophic Sets and Systems*, *35*, 503–510.

Smarandache, F., & Abdel-Basset, M. (Eds.). (2020). *Optimization Theory Based on Neutrosophic and Plithogenic Sets*. Academic Press.

Stojčić, M., Zavadskas, E. K., Pamučar, D., Stević, Ž., & Mardani, A. (2019). Application of MCDM methods in sustainability engineering: A literature review 2008–2018. *Symmetry*, *11*(3), 350. doi:10.3390ym11030350

KEY TERMS AND DEFINITIONS

Air Pollution: Air pollution is caused by any material that contaminates environment and modifies the natural qualities of the atmosphere negatively.

Energy Resource: A source of energy is anything that can generate heat, power life, move items, or generate electricity.

Fossil Fuel: It is a type of fuel that is created in the soil from the decomposition of plants or animals. Fossil fuels include coal, oil, and natural gas.

PIPRECIA: Compared to the SWARA method, it is an MCDM method that can be used effectively in group decision-making problems.

Plithogenic Aggregators: They are used to aggregate decision makers' judgments (neutrosophic numbers).

Plithogenic PIPRECIA: A method formed by the combination of PIPRECIA, which is one of the MCDM methods, and Plithogenic aggregators.

Pollution: Pollution occurs when toxins are introduced into the natural environment and cause harm.

Sustainable Development Goals: The Sustainable Development Goals (SDGs) are a collection of 17 interrelated global targets aimed at ensuring that everyone has a better life in the future. The Sustainable Development Goals (SDGs) were created in 2015, with the intention of fulfilling them by 2030.

Chapter 6
Investigating the Viability of Implementing Electric Freight Vehicles in Morocco:
Using an Integrated SWOT PESTEL Analysis in Combination With Analytic Hierarchy Process

Rim Bakhat
University of Abdelmalek Essaadi, Morocco

Said Marroun
University of Abdelmalek Essaadi, Morocco

ABSTRACT

The electric vehicle segment is gaining momentum around the globe, and Morocco will not be the exception in this regard. The present study serves to look into the question of the current and future electricity needs of this segment of the means of transport. The main contribution is preparing the necessary adaptations in the frame of electricity production capacity at the national level. This chapter aims to highlight the enablers to be seized and the main barriers to be overcome by the use of an integrated SWOT-PESTEL analysis in combination with the analytical hierarchy process. First, the SWOT-PESTEL framework is dedicated to identifying the main criteria that enable and hinder the viability of implementing electric freight vehicles (EFV) in Morocco from a sustainability perspective. Afterwards, the quantification process of the output is realized through the application of the AHP method.

DOI: 10.4018/978-1-6684-2472-8.ch006

INTRODUCTION

The increase of the size of cities to being denser in terms of population, the vast number of companies or projects appearing over last decades only poses the question about the number of resources consumed for this population's blast and the increase of the traffic congestion which could lead to multiple social and environmental problems. Urban logistics has become essential in the last few decades (**İmre et al., 2021**). It inhabits a gradually significant position across the sustainable development theme, and it is inevitable to say that urban logistics presents the pillar of sustainable development. Yet, the continuous progress of the problems would threaten the advancement of the communities, especially in terms of regional scale (**İmre et al., 2021**). It is commonly acknowledged that researchers have been studying sustainability in transportation for a considerable amount of time. More precisely, the research on one of the ways to achieve sustainability in transportation, which is the use of electric vehicles in urban freight transportation, has been growing in popularity in the last decade (**Wang et al., 2018**). By definition, Urban Freight Transport (UFT) is a part of freight transport that concerns the movement of goods using commercial vehicles in urban areas (**Dablanc, 2009**). This emanates from the fact that sustainability in UFT is related to three major performance categories. The first category is economic performance, which includes economic indicators such as transport cost, punctuality and inventory levels. The second category is environmental performance, in which we find pollution, energy consumption and noise indicators. The third pillar of this sustainability is a social performance, which includes other indicators of job creation as an example (**Moufad & Jawab, 2018**). However, to achieve this sustainability in UFT, experts suggest that one of the factors of this objective is the exploitation of electricity instead of fossil fuels as an energy source. Electric vehicles for passenger transport have gained much traction in the last ten to twenty years, the technology in battery development has been increasing every year, and the popularity among the usual consumers has reached record numbers (**İmre et al., 2021**). Electric Commercial Vehicles (ECVs) are battery-powered vehicles used to transport goods, haven't gotten the same amount of attention as electric vehicles for passengers (**Schulte & Ny, 2018**). In this sense, ECVs can be powered partially or fully by electricity and are expected to influence the sustainability in UFT by their limited greenhouse gases emissions, decreasing noises, and their low-level energy consumption (**Wang et al., 2018**). Thus, ECVs have made their entrance to the public market in the shape of battery-powered trucks that can reach 800 km range on one charge; some examples are the Tesla Semi and the Nikola One (**Schulte & Ny, 2018**). In Morocco, urban logistics are still characterized by a traditional network of points of sale, with nearly 200,000 points of sale, 40% of them are small grocery stores, according to a report by the Ministry of Transportation and Equipment and Logistics. The majority of distribution in Morocco is carried out through traditional trade involving a multiplicity of uncoordinated actors. Transport is mainly organized by the shipper (production units, assembly plants, packaging or consolidation / unbundling warehouses, wholesalers) on their account, with a filling rate resulting in an overstaffing of vehicles in circulation in the cities (**Kammas & Zendal, 2017**). Being aware of the significance of urban logistics in modern societies, the traditional urban network leads to weak management of freight transport in urban areas. According to the High Commission for Planning, the flow of goods that pass through the region of Casablanca is about 50.8%. At this occupancy rate, most problems can arise, such as road congestion, high pollution levels and low UFT performance (**Elhasbi et al., 2015**). Here, the fleet of vehicles in the Moroccan UFT is constituted of Internal Combustion Engine (ICE) vehicles (light and heavy-duty trucks). Very importantly, the penetration rate of ECVs is weak, and the city policy must, therefore, well integrate transport and logistics practices. However, the question of sustainability in transport has been embedded in the national strategies. According to the Strategy for the Logistics Competitiveness Development, Morocco intends to lower its emissions of CO2 by 35% and create 96,000 jobs within the sector of transport by the year 2030 (**Elhasbi et al., 2015**). Elhasbi et al., (**2015**) made a study about sustainable transport in Morocco, and confirmed that the question of sustainability in freight transport is far from being a priority for freight transport providers since the market and competition pressure them to prioritize costs and the level of service.

Put concisely, transportation is a theme that has attracted much attention. Having roots in sustainable development, it actively revolves around contributing to life quality enhancement, pollution reduction, and noise underproduction. Although the discussion among scholars and professionals on the implementation of electric vehicles in UFT in national debates and scientific seminars, there is a significant lack of frameworks to study the feasibility of this adoption and its effects on promoting sustainability in the transportation sector in Morocco. For this reason, the present chapter develops a constructive framework that addresses all aspects of electric mobility to fill in the gaps in the literature. The proposed approach evaluates the leading and the lagging factors that hinder or facilitate the implementation of ECVs. In this context, the present work investigates sustainable mobility in Morocco to assess the potential of the Moroccan market and identify the most suitable alternative solutions. Among other things, the study tackles the regulatory aspect, pricing, standards and marketing. Therefore, the research questions are addressed as follows:

Question One: What are the crucial stakeholders' attitudes regarding the EFV adoption in Morocco? Are stakeholders ready to make the transition?

Question Two: What potential policy measures are in place to encourage the adoption of ECVs in Morocco?

The main aim of the present chapter is to evaluate the process by the use of the contemporary strategic management methods and Multi-Criteria Decision Making (MCDM) approach, mainly the SWOT-PESTEL analysis and Analytical Hierarchy Process (AHP). The former is a combination of PESTEL, which is a study of the environment and an analysis of the factors that will stem from the six perspectives (politics, economy, society, technology, environment and legal), and SWOT analysis which divides the environment into two parts, an internal part (which contains strengths and weaknesses) and an external one (which includes opportunities and threats). The integration of the two analysis approaches enables covering the research matter from the main points of view (**Sansa et al., 2021**). The latter method is AHP, and it serves as a weighting method that results in a hierarchy in terms of the importance of the proposed factors according to the experts that participated in the study (**Tsangas et al., 2019**). At the stake here is the fact that several countries get caught up in the hype of the ECVs implementation craze and neglect that it is a ramified policy decision that encompasses several aspects (**İmre et al., 2021**). Although these dimensions accrue to well-planned and feasible policies initiatives, most governments fail to acknowledge and admit that their experience of urban freight transport is minimal, and most decisions apply haphazardly. Thus, this chapter will help the decision-makers in Morocco to consider the essential criteria for implementing the ECVs in UFT and study the hinder effects in the long term. Since sustainability appears to be a vital element of the strategic plan of Morocco, the proposed framework will highlight the contributions of ECVs to more sustainable transportation and adapt the new regulations. Increasingly enough, in the present chapter, we will study in the first place the feasibility of implementing the ECVs in the Moroccan market by unveiling the opportunities and the limitations of this mode of transport and then projecting the potential effect of this implementation on increasing the sustainability of UFT in Morocco.

The rest of the chapter is organized as follows; Section 2 highlights the research background of the subject. Section 3 presents the methodology. Section 4 enumerates the suggested solutions to overcome the issues found in the study. Section 5 lists the obtained results and discussions for future studies. Then we conclude the chapter in Section 6.

Background

As discussed in the introduction, the main reason to keep the advancement of sustainable development is the motivation aspects that push the researchers and professionals alike to conduct various scientific researches in the urban logistics sector by conducting solid work that would-as result- help in the decrease of the negative impact of urbanization economically, socially, and environmentally. What is at issue here is that employing urbanization as a policy might offer communities a quite enticing advantage. In our case, the implementation of the Urban Freight Transport (ECVs) is seen as the silver bullet to solve intricacies to keep the advancement of sustainable development regionally and nationally. To ground the rest of the sections in this chapter, we vision it is obvious to state the common traps processed during the urbanization of the transport sector. The authors go into a detailed explanation of the impact of the UFT/ ECVs in modern communities and their crippling chances for progress. The different veteran stakeholders affecting the decision-making process in this sector are also enumerated. Afterwards, analyse the feasibility of implementing the ECVs in UFT with the support of SWOT-PESTEL analysis and the AHP method. The findings communicate the potential of ECVs in promoting sustainability in the transport sector in Morocco, as discussed in the next section.

URBAN FREIGHT TRANSPORT

The prime role of transportation is to ship goods and convey passengers from one point to another (Gurtu et al., 2019). Nevertheless, transportation activity remains one of the main contributions to climate change since it is estimated that it is the root cause for 25% of CO2 emission worldwide and the biggest polluter after the energy sector (IEA, 2009; European Commission, 2016). In addition, transportation is responsible for the high-temperature variability, rain disturbances, and wind patterns. These crucial facts raised the communities' awareness about the negative impacts on the total quality of life. Such emphasizes the need for the transportation systems to be sustainable; if addressed correctly. Traffic congestion remains one of the problems degrading the quality of life on economic, social, and environmental levels (Dimitrakopaulos et al., 2020).

Freight transportation consists of using commercial vehicles to transport goods in cities; these vehicles are categorized into light-size vehicles, medium-size commercial vehicles, and; heavy-size commercial vehicles (**European Parliament, 2007**). The UFT is divided into five markets; retail, mail delivery, hotel and restoration delivery, construction, waste disposal. These markets differ by features as shown in **Table 1.**

Table 1. Review of UFT markets source

UFT Markets	Features
Retail: Freight transportation includes moving goods to retail spaces.	Reliance on medium to large size freight vehicles. Numerous deliveries and multiple suppliers.
Mail delivery: Delivering mail packages and letters, While also offering special delivery services.	Governmental and non-governmental postal service providers. Multiplicity of beneficiaries.
Hotel and restoration delivery: Moving food and products to hotels and restaurants	Just-in-time delivery of food. Consistent ordering.
Construction: Transportation of construction raw materials and equipment to workshops.	Traditional management of transportation. Independent and atomic transport providers.
Waste Disposal: Consists of collecting and disposing of city waste.	Mixed cargo (no waste separation). Multiple stops but predetermined paths.

Source: (Wang et al., 2018)

In urban agglomerations, most citizens use their vehicles to fulfil their daily needs, which lead to traffic congestion, noise, and pollution increase in the local atmosphere. This phenomenon is exacerbated due to the unavailability of sustainable transport (electric vehicles). According to WBSCD (2004) report, it is foreseen that the average growth of personal vehicles will be up to 1.7% each year till 2050. Moreover, several studies have demonstrated that a vehicle user produces at least an average of 150g of CO2 emissions per kilometre while the user is responsible for only 80g if he takes a mode of public transport. What can be deduced here is that along with the efficient use of transportation, the level of pollution will be lessened down. Another option to reduce the emission of GHG in the long term is the use of electric vehicles. Brand et al., (2012) explained through their model of the life cycle of carbon emission in the UK transport system, that the use of electric vehicles offers an effective strategy to mitigate the climate change externalities. The main actors in the urban freight transport sector are consumers (citizens and visitors), shipper, retailers, and wholesalers, in addition to freight carriers who usually use third-party providers (3PLs) (Wang et al., 2018). However, the intricacy of the UFT system intensifies with the involution of supplementary stakeholders carried by the system of the electric vehicle. Figure 1 exemplifies the main stakeholders in the urban freight system in Morocco and the mutuality among these stakeholders.

Basically, Governments are confronted today with an increasing number of projects of permanent innovation and continuous adaptation in the face of a changing and plural environment. This multitude of changes contributes to a risk of dispersal of the forces present (energy, skills, resources) and of failure due to the progressive lack of respondent among actors unable to offer new services as frequently and quickly. In the case of Morocco, a need remains for developing both technical and service training programs and invest in scientific research and innovation, improving skills by adapting the training of the public-private institutes and universities for technological development, and finding synergies between the manufacturing sector and other sectors, especially logistics and infrastructure. Similarly, the government ought to speed up the establishment of the institutional, regulatory, budgetary, training and support provisions necessary to promote the development of the transport and logistics sector and the creation of jobs for young people by giving a suitable place for innovative ideas and entrepreneurship. Nevertheless, among these adjustments, the integration of the local communities and villages in the development of electric vehicles within the urban freight transport remains necessary by capitalizing on studies and evaluations of projects to further increase the visibility of the sector to attract investment.

Figure 1. Interdependencies among stakeholders in urban freight system in Morocco
Source: Authors' own illustration

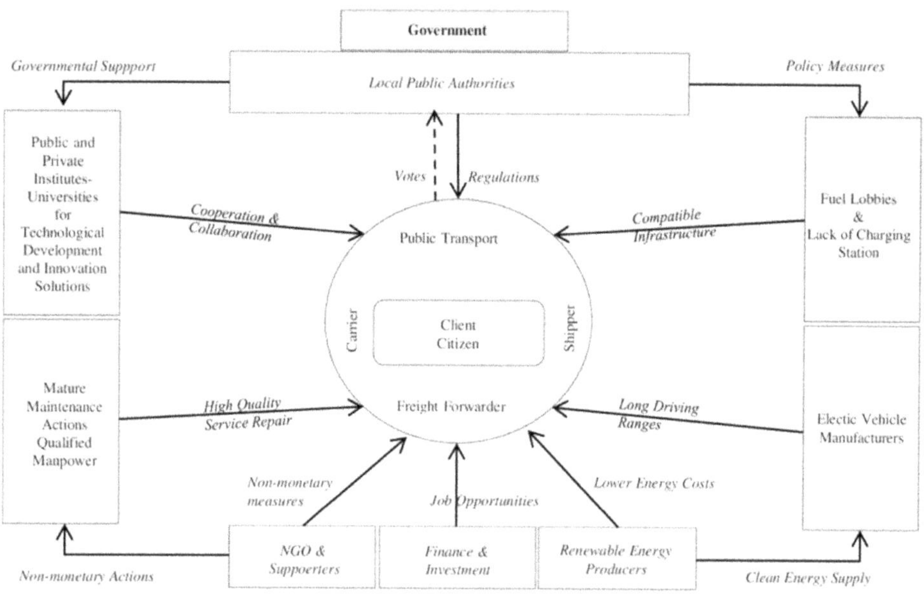

Furthermore, these multiple changes increase the risk of dilution of central skills (controls, coordination, and logistical support) and reinforce the complexity of projects by mobilizing more and more actors. In this case, there is a high risk of loss of meaning, the policy-maker no longer having a clear and precise reading of the situation, and a lack of consistency in the actions to be taken to achieve the fixed objectives. This delicate situation of uncertainty often leads those in charge to go to the most urgent, with risks of precipitation and the development of imperfect solutions. Efficient adoption of EFVs entails significant adjustments in actions. The unscheduled change leaves plenty of room for emerging phenomena and intuition. Identify the political issues, the risk of ambiguities, and negotiate the acceptable agreements to go in the direction of the general interest.

It is worth noting that there is no definition of sustainable logistics, as there may be for sustainable transport. However, logistics sustainability is characterized by its final results. In other words, we can say that logistics is sustainable if the transport is, even though we institute logistics as the key to sustainable mobility. The place of transport will be different, depending on the context, logistics and production. Concerns about sustainability are elements that will affect the operating environment of the communities. They are, therefore, required to adjust the logistical processes and modify the role of the transport. The place of transport will be different, insofar as the margins of action of governments in favour of sustainability will not be identical for all, some governments being more constrained than others. The policy-makers have to consider the interactions among logistics, transport, public administrations, and organisations at the micro and macro-level of the country.

Electric Commercial Vehicles

Electric Commercial Vehicles differ mainly by the source of energy used to power them. For instance, Battery Electric Vehicles (BEVs) use batteries as their source of energy, and Hybrid Electric Vehicles rely on two sources of propulsion, the Internal Combustion Engine and electric motors[1]. In this study, the authors will focus mainly on BEVs and their potential use in UFT. Energy use efficiency is another challenge. Battery Electric Vehicles, particularly lithium-ion batteries, are characterized by a high level of reliability. This feature allowed them to be the cleanest clean mode of transportation and is available on the market for sale. Meanwhile, the limitations to this type of vehicle are their short-range, high selling price and insufficient charging stations (Chan et al., 2009). In terms of the competitiveness of BEVs against diesel vehicles, multiple studies have been carried on that subject. Another way of expressing this is that Feng & Figliozzi (2013) found that electric trucks are an economically competitive advantage in cases of severe usage. Hence, Lee et al., (2013) confirm that the total cost of ownership (TCO) of traditional diesel trucks is 22% higher than electric delivery trucks. However, Zhou et al., (2017) realized contradictory results and; that even though electric trucks have low greenhouse gas emissions; their TCO is higher than traditional diesel trucks. The future, on the other hand, seems promising for battery electric vehicles. In a more recent study, Çabukoglu (2018) conducted a case study in Switzerland on the potential of heavy-duty vehicles electrification like freight delivery trucks. *Economically,* Battery based electric vehicles have reportedly four times lowers energy costs in comparison with diesel trucks (Feng & Figliozzi, 2013). The maintenance cost of BEVs is also 20% to 30% lower than those of traditional internal combustion engine vehicles (Taefi et al., 2015). In addition, maintenance costs are usually higher with time in the case of diesel vehicles. Besides the energy efficiency use, the BEVs have an additional advantage in terms of strategic planning. However, if the goal is to evaluate BEVs' costs, thus what should be taken into consideration is the replacement of their batteries which is required regularly with intensive use (Pelletier et al., 2016). From a long-term perspective, researchers expect that the competitiveness of BEVs will increase because of the incremental improvement of their operational performance with time. The purchase price is significantly high compared with diesel trucks and is expected to decrease with mass production and the growing consciousness of the environmental effects of fuel-powered vehicles (Quak & Nesterova, 2014). This study also examines the impact of the development and increase of electric mobility on the environment. Indeed, who says electric vehicles says the increase in electricity needs. Many projects reported the *environmental* efficiency of ECVs. For example, Posten Norge (NO)[2] claimed that it is possible to reduce up to 2.1 tons of CO2 emission per truck per year by using BEVs rather than diesel traditional trucks. Tesco (UK) has 15 Modec BEVs on their fleet, and they reported that each electric vehicle reduces a quantity of up to 15 tons of CO2 per year. UPS announced a 20% reduction of CO2 emissions according to their use of electric-powered vehicles. Moreover, BEVs proved that there are almost zero local emissions of NOx emitted by battery-powered vehicles. Therefore, the air quality might improve due to the adoption of electric vehicles (Nesterova et al., 2013). Göhlich et al (2021) realized a comparative study between the performance of diesel and electric vehicles in waste collection, and they found that greenhouse gases emissions are lowered by approximately 27%. The electrification of the vehicles fleet will lead to an increase of 18% and 30% in the total cost of ownership. Another essential aspect of the impact of electric vehicles on the environment is the reduced noise nuisance since electric vehicles are mostly silent (the manufacturers of electric vehicles implement safety sounds into the trucks for safety reasons) (Quak & Nesterova, 2014). One of the barriers to adopting ECVs in UFT is their limited range. Some of the UFT categories (i.e. retail) require frequent deliveries and demand long-range vehicles. However, with the development of battery technology, the range of electric vehicles is getting larger. Another limitation of the use of ECVs is their payload. The UFT requires several cases of heavy-duty trucks that are often difficult to implement because of the exceeding weight of the batteries (Taefi et al., 2016).

METHODOLOGY

To be resilient in the market place, organizations must determine, conduct and apply proven methodologies and management best initiatives. When an organization tries to lead management practices without perceiving future ramifications, it will be impossible to articulate an execution strategy due to the ambiguity in collected data and this result in a high probability of failure. SWOT and PESTEL analysis are proven methods to deal with numerous issues and lead to unsophisticated reliable results. The first is a strategic tool capable of conducting the sustainability issues, economic, social, environmental, political, legal, and technical, that have a potential impact on the organization's operations. Besides, it has been recently applied in several domains such as in the energy waste (Song et al., 2017), productivity management (Pan et al., 2019), hydrocarbons sector (Tsangas et al., 2019), and healthcare waste management (Thakur. 2021). The latter is a strategic technique used to combine the strengths and weaknesses of a study concerning the opportunities and threats found in its background (Büyüközkan et al., 2021). It has been also combined with the MCDM methods like Analytic Network Process (ANP) (Zhang & Paudel. 2021) and Analytic Hierarchy Process (Ali et al., 2021). Furthermore, AHP is applied in the present research study as a weighting technique to quantify and assess the SWOT-PESTEL analysis output since the structure of the assessment problem (indicators) is hierarchal. The quantification of the criteria is processed by exploiting the academic online AHP Calculator based system (Geopel. 2018). For the very first time, the integrated SWOT-PESTEL analysis in combination with AHP is used in sustainable transport area to investigate the viability of implementing the ECVs.

SOLUTIONS AND RECOMMENDATIONS

Potential of the ECVs Adoption in Morocco

By its commitments expressed to the international community at the Earth Summits in Rio de Janeiro in 1992 and Johannesburg in 2002, Morocco has embarked on a new mission of sustainable development by putting in place foundations aimed at ensuring economic development that respects the environment and social equity. Committed to this dynamic, Morocco is positioning itself as an actor invested in global and regional agendas that focus on a sustainable and inclusive green economy. This commitment was enshrined in 2014 by the adoption of framework law N ° 99.12 on the National Charter for the Environment and Sustainable Development and in 2017 by the adoption by the Council of Ministers of the National Strategy for Sustainable Development which infused several sector strategies that must further integrate sustainability parameters into their objectives (ESEC, 2018). Figure 2 illustrates the potential of the sustainable transport market in Morocco, where three scenarios were considered. These scenarios lead to very different potentials, depending mainly on the involvement of the *State*. For the optimistic scenario, 425,704 vehicles will be adopted in 2030 if the State is fully involved in the implementation of strategic recommendations. Concerning the optimistic scenario, 257,777 vehicles will be exploited in 2030 if state involvement is medium. In the pessimist scenario, only 11,391 vehicles will be adopted if the Moroccan market follows the global growth of the stock of sustainable vehicles without any *government* intervention.

Figure 2. Scenarios of the sustainable transport mobility in Morocco Source: (Sunergia Studies - Nevolys Consulting. 2019)

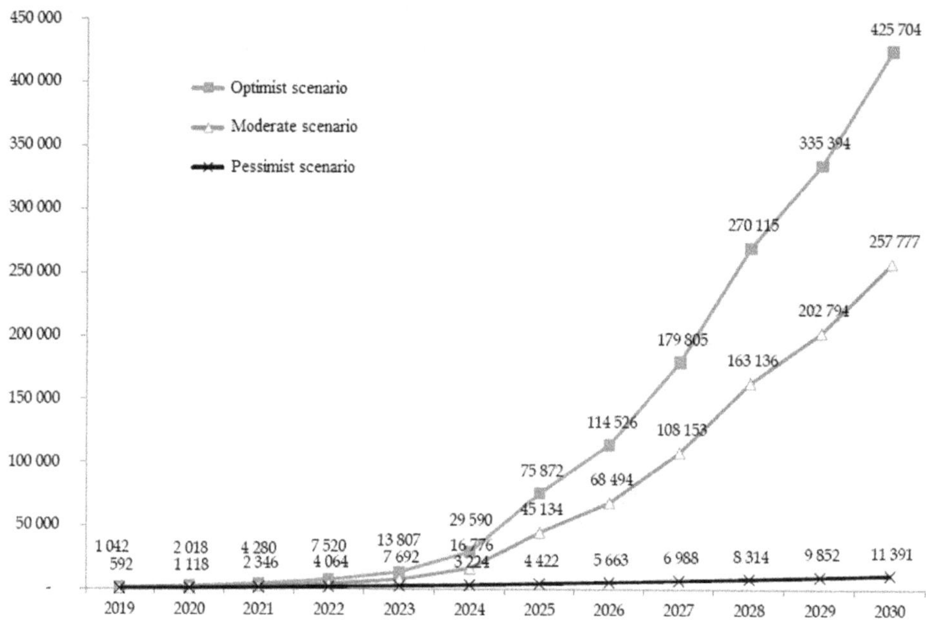

Therefore, the economic and social opportunities of the Moroccan transport sector, but also the threats weighing on them, require a new strategic approach to urban freight. In this context, the present chapter investigates sustainable mobility in Morocco to assess the potential of the Moroccan market and identify the most suitable alternative solutions. The proposed framework regulates all the essential issues of adopting the ECVs in Morocco. The SWOT-PESTEL analysis, as given in Table 2, illustrates the leading and lagging factors of implementing electric vehicles, which are established as sustainability indicators.

Table 2. SWOT-PESTEL matrix of the electric vehicles implementation Source: Authors' own elaboration

SWOT PESTEL	Strengths	Weaknesses	Opportunities	Threats
Political	**P1.** Draft Strategy for Energy Efficiency 2030	**P2.** Insufficient fast charging stations	**P3.** National Strategy for Sustainable Development by 2030	**P4.** Resistance from fossil fuel providers
Economic	**E1.** Global market orientation towards EVs including China	**E2.** Sale price of the EVs	**E3.** Modular production platform	**E4.** Traditional network of points of sale
Social	**S1.** General Acceptance	**S2.** Absence of the engine noise	**S3.** Emergence of new professions	**S4.** Short distance of the green miles project
Technological	**T1.** Success of electric passenger cars worldwide	**T2.** Lack of skilled labour for maintaining the EVs	**T3.** Development of battery technology	-
Environmental	**V1.** Establishment of an automotive ecosystem in Morocco	**V2.** The source of the electricity is still fossil fuel stations	**V3.** Strong compatibility and in line with the development of renewable energies	-
Legal	-	**L1.** Absence of a regulatory framework on sustainable mobility	**L2.** State exemplarity (10% of new acquisitions by 2019)	-

The study aimed to highlight the opportunities to be seized and the main challenges to be met. The SWOT-PESTEL analysis of the sustainable mobility market in Morocco synthesized the vital factors that impact the ECVs adoption decision and at the same time respected the sustainability dimensions. The forces encompass the proactive policy of the State through mainly the National Strategy for Sustainable Development by 2030, the establishment of an automotive ecosystem in Morocco, memorandum of understanding relating to the development of an electric transport ecosystem in Morocco by the Chinese group "BYD Auto Industry, the fact of the electric vehicle mobility all over the world, and the general acceptance of the electric vehicle by the customer. For weaknesses, some factors have been gathered such as the insufficient fast-charging stations on the national road network; the limited incentives that do not significantly reduce the selling prices of EVs; the lack of qualified labour for the servicing and maintenance of EVs; the treatment of sustainable mobility apart from the energy transition; the absence of a regulatory framework on sustainable mobility and sale of energy by the operators of recharging stations; and the absence of engine noise which affects the general acceptance of the EVs. The opportunities consist of the Energy Efficiency Strategy Project 2030, the modular production platform within the *Renault* and *Peugeot* factories in Morocco, the emergence of new professions, the guidance from manufacturers installed in Morocco (*Renault*, *Peugeot*) towards the development of battery technology of the electric vehicle, the strong compatibility and linked to the development of renewable energies (solar energy in particular) in the territory, and the exemplary nature of the State (10% of new acquisitions by 2019). The threats are about the resistance from fossil fuel providers and lobbyists, the traditional network of points of sale and possible impact on the spare parts and after-sales service market, and insufficient fast-charging stations on the national road network (Tangier-Casablanca).

Pairwise Comparison of the PESTEL Analysis

The research is about anticipating and planning the necessary adaptations regarding the ECVs adoption at the national level. The main aim of the empirical study is to support and drive this acceleration of the development of electric mobility in Morocco. To reap the above-mentioned advantages, stakeholders with variable professional and scientific backgrounds were called to pairwise compare the priority of the six components of the PESTEL analysis via a well-thought-out questionnaire, due to their deep impact in decision-making, from the policy scope. These were an academician who is in charge of the training programs and the scientific research to improve the future graduates' skills, a logistician who is an expert in supply chain management and logistics management working for an automotive company, and an economy expert who is working as a financial adviser and interested in the electric mobility in Morocco, an environmentalist is an active member in a non-profit organization that make significant contributions to the sustainability issues, a fleet manager who is a transport expert working for a company implemented in MedPort Tangier terminal, and an urban freight expert, who works in public regional administration and he is in charge of sustainable urban transport planning. It is particularly important to point out here that the group of the designated experts intended to embrace the main pillars of sustainability. The main reason behind selecting specialists with different backgrounds is to obtain the maximum points of view to construct diverse pairwise comparison matrices and generate valid results. The collected judgments are given in Table 3-4 and Figure 3.

The obtained results illustrate that there is a significant variety of importance regarding the sustainability pillars. It is very crucial to note that the political issues vs. legal issues or political issues vs. economic issues or social issues vs. environmental issues or social issues vs. technical issues or economic issues vs. environmental issues or political issues vs. technological issues encompasses at least one judgement of equal importance. For the rest of the issues, there was a divided point of view from the decision-makers. For example, five of them selected the first side, and the other one prioritized the second side regarding the economic issues vs. technical issues. It is worth noting that environmental issues are of higher significance compared to technical and legal issues. This means that there is a prominent agreement among the part involved in the assessment of investigating the viability of implementing the ECVs in Morocco.

Table 3. Stakeholders' perspectives on the sustainability issues of ECVs adoption in Morocco Source: Authors' own elaboration

Sustainability Indicators		Academician		Logistician		Environmentalist		Economy expert		Fleet manager		Urban freight expert	
1	Political	4.6%	6	7.5%	5	7.4%	5	26.2%	2	13.4%	3	5.2%	6
2	Economic	26.2%	2	27.6%	2	7.8%	3	31.5%	1	49.1%	1	5.5%	4
3	Social Issues	22.6%	3	12.7%	3	22.2%	2	8.9%	5	4.4%	6	15.8%	2
4	Technical	10.5%	4	9.2%	4	4.5%	6	13.3%	4	16.1%	2	15.7%	3
5	Environmental	29.5%	1	37.3%	1	50.3%	1	5.9%	6	5.2%	5	52.2%	1
6	Legal	6.7%	5	5.8%	6	7.8%	4	14.2%	3	11.8%	4	5.5%	4

Figure 3. Pairwise comparison results of the sustainability indicators Source: Authors' own elaboration

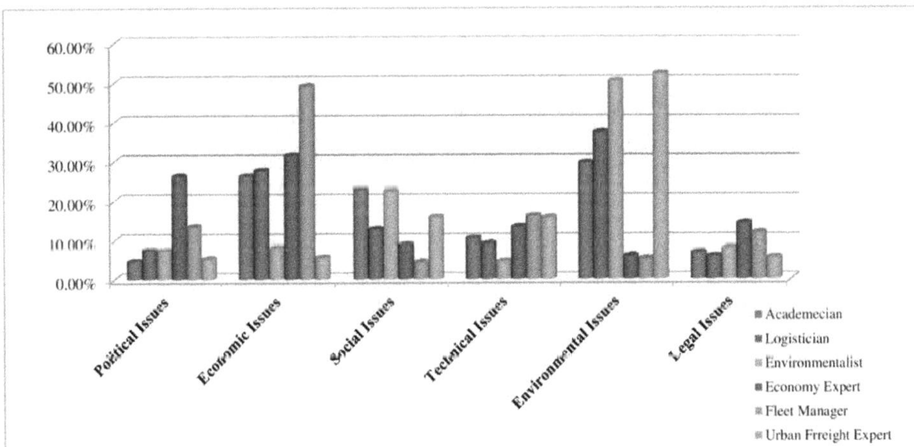

Leading and Lagging Factors' Quantification

The quantification of the leading and lagging factors (enabler and barriers) of the electric vehicle adoption in Morocco were processed by exploiting the academic online AHP Calculator based system. The latter was initially created by the Business Performance Management Singapore Group to facilitate the calculation of the AHP method, especially when it comes to multiple criteria with hierarchical structure. The twenty (20) factors were inputs as criteria, and one hundred-ninety pairwise comparisons were assessed for each evaluator case. The collected judgements from the decision-makers concerning the six pillars of sustainability were utilized. The sub-criteria regarding similar issues were given equal importance. However, the calculation of the Consistency Rate (CR) for each evaluation case was mandatory to study the reliability and the feasibility of the collected judgments. Accordingly, all the rates in this research study are under 10%, which prove the consistency of the stakeholders' responses. Table 5 presents the factors quantification results for each expert with the support of the AHP Calculator (Geopel. 2018). Essentially, the assessment of the academician and the fleet manager judgments ranked the source of the electricity issue in the first place. For the logistician, the emergence of new professions is the priority. Concerning the environmentalist, the absence of a regulatory framework on sustainable mobility is an essential element. The economist supposes that the sale price of EVs is crucial. Meanwhile, urban freight transport emphasizes the necessity of considering the development of battery technology.

Table 4. Judgements summary review Source: Authors' own elaboration

Pairwise Comparisons		Category A of Extreme Importance	Category A of Very Strong Importance	Category A of Strong Importance	Category A of Moderate Importance	Both Categories are of Equal Importance	Category B of Moderate Importance	Category B of Strong Importance	Category B of Very Strong Importance	Category B of Extreme Importance
Category A	**Category B**									
Political Issues	Economic Issues	0	0	0	0	**3**	0	2	1	0
Political Issues	Social Issues	0	0	1	1	**1**	1	2	0	0
Political Issues	Technical Issues	0	0	0	2	**2**	2	0	0	0
Political Issues	Environmental Issues	0	0	1	1	**0**	0	2	2	
Political Issues	Legal Issues	0	0	0	0	**5**	1	0	0	0
Economic Issues	Social Issues	0	1	1	0	**2**	2	0	0	0
Economic Issues	Technical Issues	0	0	**2**	**3**	**0**	**1**	0	0	0
Economic Issues	Environmental Issues	0	0	1	1	**2**	0	0	2	0
Economic Issues	Legal Issues	0	0	1	3	**2**	0	0	0	0
Social Issues	Technical Issues	0	0	0	2	**3**	1	0	0	0
Social Issues	Environmental Issues	0	0	0	0	**3**	0	2	1	0
Social Issues	Legal Issues	0	0	1	3	**1**	1		0	0
Technical Issues	Environmental Issues	0	0	1	1	**0**	1	3	0	0
Technical Issues	Legal Issues	0	0	0	4	**1**	1	0	0	0
Environmental Issues	Legal Issues	0	1	3	0	**0**	2	0	0	0

*Table 5. Enablers and barriers of ECVs implementation quantification for each stakeholder with the support of AHP Calculator (***Geopel. 2018***) Source: Authors' own elaboration*

Decision-maker	Academician		Logistician		Environmentalist		Economy Expert		Fleet manager		Urban Freight Expert	
Number of Comparisons	190		190		190		190		190		190	
Consistency Rate (CR)	9.80		9.50		9.70		9.40		9.90		9.50	
Sustainability Indicators	Weights%	Rank	Weights%	Rank	Weights%	Rank	Weights%	Rank	Weights%	Rank	Weights%	Rank
P1. Draft Strategy for Energy Efficiency 2030	1.6%	17	1.3%	19	7.4%	7	7.6%	4	2.8%	12	1.0%	20
P2. Insufficient fast charging stations	5.7%	9	4.2%	8	3.7%	11	2.2%	19	1.4%	20	2.6%	15
P3. National Strategy for Sustainable Development by 2030	2.7%	13	1.5%	17	6.7%	8	4.9%	10	1.9%	18	3.4%	12
P4. Resistance from fossil fuel providers	8.4%	5	2.3%	15	3.2%	13	2.6%	17	1.7%	19	1.8%	16
E1. Global market orientation towards EVs including China	2.5%	14	2.0%	16	4.3%	10	2.4%	18	7.1%	6	3.1%	14
E2. Sale price of the EVs	1.9%	16	1.2%	20	2.3%	15	10.1%	**1**	2.4%	16	1.5%	18
E3. Modular production platform	1.2%	20	1.5%	18	1.6%	19	5.6%	6	2.5%	15	1.7%	17
E4. Traditional network of points of sale	5.6%	10	3.7%	10	1.4%	20	5.4%	8	2.7%	13	3.1%	13
S1. General Acceptance	1.3%	19	2.9%	13	1.6%	17	3.7%	15	1.9%	17	5.1%	10
S2. Absence of the engine noise	1.3%	18	2.9%	12	1.9%	16	1.9%	20	4.4%	11	1.4%	19

continues on following page

Table 5. Continued

Decision-maker	Academician		Logistician		Environmentalist		Economy Expert		Fleet manager		Urban Freight Expert	
S3. Emergence of new professions	6.1%	8	12.5%	**1**	1.6%	18	3.9%	14	2.6%	14	8.8%	3
S4. Short distance of the green miles project	5.1%	11	9.9%	4	7.9%	4	4.6%	11	7.2%	5	5.0%	11
T1. Success of electric passenger cars worldwide	1.9%	15	2.7%	14	3.4%	12	2.7%	16	6.0%	8	5.1%	9
T2. Lack of skilled labour for maintaining the ECVs	7.0%	6	11.9%	2	2.4%	14	4.0%	13	5.8%	9	9.4%	2
T3. Development of battery technology	9.0%	3	11.5%	3	7.8%	5	5.8%	5	10.0%	3	12.4%	**1**
V1. Establishment of an automotive ecosystem in Morocco	8.4%	4	6.2%	7	10.0%	2	5.4%	9	6.2%	7	7.4%	5
V2. The source of the electricity is still fossil fuel stations	10.0%	**1**	7.0%	6	8.3%	3	8.7%	3	10.1%	**1**	7.7%	4
V3. Strong compatibility and in line with the development of renewable energies	9.7%	2	7.4%	5	7.4%	6	9.0%	2	10.0%	2	6.2%	7
L1. Absence of a regulatory framework on sustainable mobility	6.6%	7	3.5%	11	10.8%	**1**	5.5%	7	7.9%	4	7.2%	6
L2. State exemplarity	4.0%	12	3.8%	9	6.2%	9	4.2%	12	5.3%	10	6.1%	8

Table 6 presents the obtained results of the SWOT-PESTEL analysis via the calculation of the geometric means of the inputs. The factors of adopting electric vehicles in Morocco are segmented into two main categories. The first category encompasses the strengths and opportunities, which hold positive values. Meanwhile, the second category embraces weakness and threats, which has negative values. **Figure 4** illustrates the value size indicators of leading factors that enable the implementation of the EVs in Morocco and the lagging factors that hinder the latter, respectively.

Table 6. SWOT-PESTEL analysis results with the combination of AHP Source: Authors' own elaboration

Leading and Lagging Factors	Geometric Mean (%)	SWOT	Value (%)
P1. Draft Strategy for Energy Efficiency 2030	2.63%	S	2.63%
P2. Insufficient fast charging stations	2.99%	W	-2.99%
P3. National Strategy for Sustainable Development by 2030	3.08%	O	3.08%
P4. Resistance from fossil fuel providers	2.81%	T	-2.81%
E1. Global market orientation towards EVs including China	3.23%	S	3.23%
E2. Sale price of the EVs	2.40%	W	-2.40%
E3. Modular production platform	2.02%	O	2.02%
E4. Traditional network of points of sale	3.31%	T	-3.31%
S1. General Acceptance	2.45%	S	2.45%
S2. Absence of the engine noise	2.09%	W	-2.09%
S3. Emergence of new professions	4.71%	O	4.71%
S4. Short distance of the green miles project	6.36%	T	-6.36%
T1. Success of electric passenger cars worldwide	3.36%	S	3.36%
T2. Lack of skilled labour for maintaining the ECVs	5.93%	W	-5.93%
T3. Development of battery technology	9.13%	O	9.13%
V1. Establishment of an automotive ecosystem in Morocco	7.11%	S	7.11%
V2. The source of the electricity is still fossil fuel stations	8.56%	W	-8.56%
V3. Strong compatibility and in line with the development of renewable energies	8.17%	O	8.17%
L1. Absence of a regulatory framework on sustainable mobility	6.54%	W	-6.54%
L2. State exemplarity	4.84%	O	4.84%

Figure 4. Sustainability Indicators values size Source: Authors' own elaboration

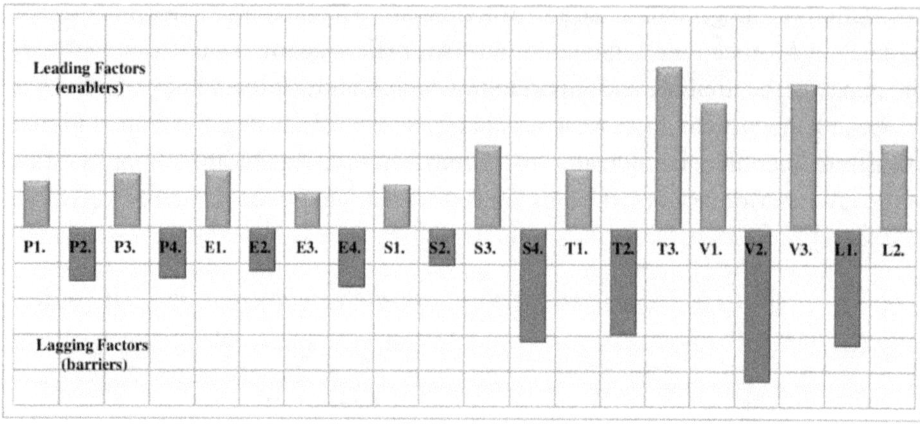

DISCUSSION AND FUTURE RESEARCH DIRECTIONS

The transportation industry has known after the explosion of global competition, several opportunities for development. It takes an important place and a rapid growth due to its role in improving performance communities. The findings depict that besides the lack of information on policy measures and support strategies, a willingness of Morocco to move towards a cleaner and environmentally friendly mobility has increased. The analysis of the obtained results shows that the leading factors outnumber the lagging ones in terms of the size of the values, which indicates that the implementation of electric vehicles in the freight transportation market in Morocco is viable. The quantification of the factors leads to the same results, where the strengths and opportunities of the EVs adoption have gained traction from the stakeholders, more than the weaknesses and the threats. The interviewed experts from different backgrounds prioritized the environmental issues when asked about the sustainability issues related to the ECVs adoption in Morocco. What can be deduced here is that the Sustainability matter would be highly influenced by environmental factors (enablers and barriers). For instance, the factor V3 (*strong compatibility and in line with the development of renewable energies*) is- according to the experts- a promising factor that may contribute to the facilitation of the implementation in question. On the other hand, the factor V2 (*source of the electricity is still fossil fuel stations*) has the highest value on the side of the barrier. The analysis showed that the local and regional public authorities should pay more attention to the proportions of energy sources that are clean. The second set of indicators that got high-value sizes were the technical factors. For example, an opportunity that ought to be exploited is the T3 factor (*increased development of battery technology in recent years*). The production of ECVs' batteries has grown significantly around the world. Morocco has the potential of hosting companies in this sector to add them to the already existing ecosystem of its automobile industry. Another technical factor but on the side of the barrier is the T2 (*lack of skilled labour for maintaining ECVs*). With the growing penetration of electric vehicles to the Moroccan market, institutions are required to train their students on how to deal with automotive electrical systems. Manufacturers could also offer similar training to their employees to overcome the challenge of unskilled labour and softly solve technical problems without generating additional costs and unwanted waste. The legal indicators are also among the important factors that should be considered. Firstly, the L1 factor (*absence of a regulatory framework on sustainable mobility and energy sales by charging stations operators*) is the third-highest lagging factor. Regulators are invited to create a legal framework to facilitate electric mobility in the freight transportation sector. On the upside, the state exemption of some taxes on electric vehicles is seen as an enabler that should encourage companies to implement more ECVs in their transportation fleet. The economic and political factors are in the middle in terms of value size. The experts prioritized approximately the same types of criteria. In other words, the evaluators are dotted with the same importance but not as crucial as the previous environmental and technical factors. The obtained results are rational given that the aforementioned factors present the current and overall political and economic directions of Morocco (*the State's orientation for adopting sustainable production methods and the National strategy for sustainable development*). In addition to the fact that the political view and the economic strategy are usually interconnected, it is only fair to consider them as equals in terms of their effect on the implementation of ECVs in freight mobility. The social factors are also figured among the essential elements for the evaluators. The highest indicator is S3 (The emergence of new professions), which is considered an opportunity because the implementation of ECVs in freight transportation will create new jobs. That would start from manufacturing lines to maintenance workshops. On the flipside, S4 (Short distance of the green miles project) is considered a barrier that would be a challenge that companies and authorities are required to deal with. After discussing the results of the SWOT-PESTEL analysis with the combination of AHP, the essential factors that enable and hinder the implementation of EVCs in freight transportation are summarized in Table 7.

The main objective of the transport activity is to minimize its energy consumption and avoid congestion on the main roads. Monitoring of various key points such as the vehicle fill rate, the number of group deliveries, the share of alternative modes of transport to the road, the number of tons of carbon dioxide emitted in relation to the mode of transport used, kilometres travelled, weight transported is essential. Sustainable logistics therefore aims to disseminate sustainable development, while minimizing its environmental and social impacts. These goals can span the entire supply chain and results in various performance effects that need to be assessed. Sustainable transportation strategy aims to improve the social, economic, and behavioural status of sustainable transportation while respecting the environment for green and healthy future. The impact of transportation on the environment could be reduced by achieving a set of objectives that directly or indirectly relate to the environment and thus the sustainable transportation strategy could be achieved. Reducing the carbon footprint by using environmentally friendly transportation such as bicycles or walking which have massive benefits, such as reducing traffic congestion, reducing greenhouse gas emissions, reducing transport costs, and many other benefits. Not all these solutions are one hundred per cent efficient, but they contribute in one way or another to the improvement of the sustainable transportation strategy.

Table 7. Major leading and lagging factor Source: Authors' own elaboration

Major Leading factors	Major Lagging factors
T3. Increased development of battery technology in recent years	**V2.** The source of the electricity is still fossil fuel stations
T3. Increased development of battery technology in recent years	**L1.** Absence of a regulatory framework on sustainable mobility and energy sales by charging stations operators
V1. Establishment of an automotive ecosystem in Morocco	**T2.** Lack of skilled labour for maintaining ECVs
L2. State exemplarity	**S4.** Short distance of the green miles project

CONCLUSION

The analysis of the study has shown that there is a wide variety of leading factors, which could facilitate the implementation of electric freight vehicles in Morocco. There is also a consensus between the different stakeholders on the importance of sustainability in the transportation sector. However, some constraints will stand in the way of this implementation. The interviewed experts raised their concerns about the environment and considered that the development of renewable energy stations is an enabler of clean transport adoption. The PESTEL-SWOT analysis represents a powerful tool that helped the authors in pinning down the main factors linked to the research question internally and externally. Moreover, it has yielded a set of leading indicators to be exploited and other lagging indicators to be alleviated. The AHP method was applied to quantify the factors revealed in the SWOT-PESTEL analysis.

The limits of this study emerge from two major aspects. On one hand, it does not include the perspectives of other stakeholders like policymakers that are strongly involved in decision making, especially, when it comes to the implementation of new technologies. The inclusion of other stakeholders will be for future research to implement. On the other hand, in terms of the methods used, pairwise comparisons of indicators have resulted in important insights for the study. However, it would be richer if the comparisons were between each individual indicator and the others and that might cost more time and energy to realise.

REFERENCES

Ali, E. B., Agyekum, E. B., & Adadi, P. (2021). Agriculture for Sustainable Development: A SWOT-AHP Assessment of Ghana's Planting for Food and Jobs Initiative. *Sustainability*, *13*(2), 628. doi:10.3390u13020628

Brand, C., Tran, M., & Anable, J. (2012). The UK transport carbon model: An integrated life cycle approach to explore low carbon futures. *Energy Policy*, *41*, 107–124. doi:10.1016/j.enpol.2010.08.019

Büyüközkan, G., Mukul, E., & Kongar, E. (2021). Health tourism strategy selection via SWOT analysis and integrated hesitant fuzzy linguistic AHP-MABAC approach. *Socio-Economic Planning Sciences*, *74*, 100929. doi:10.1016/j.seps.2020.100929

Çabukoglu, E., Georges, G., Küng, L., Pareschi, G., & Boulouchos, K. (2018). Battery electric propulsion: An option for heavy-duty vehicles? Results from a Swiss case-study. *Transportation Research Part C, Emerging Technologies*, *88*, 107–123. doi:10.1016/j.trc.2018.01.013

Chan, C. C., Bouscayrol, A., & Chen, K. (2009). Electric, hybrid, and fuel-cell vehicles: Architectures and modeling. *IEEE Transactions on Vehicular Technology*, *59*(2), 589–598. doi:10.1109/TVT.2009.2033605

Dablanc, L. (2009). Freight transport for development toolkit: Urban freight (No. 57971). The World Bank.

Dimitrakopoulos, G. J., Uden, L., & Varlamis, I. (2020). *The future of intelligent transport systems*. Elsevier.

Economic, Social and Environmental Council Rapport. (2018). *L'économie bleue: pilier d'un nouveau modèle de développement du Maroc* [The blue economy: pillar of a new development model for Morocco]. Available online: https://www.cese.ma/media/2020/11/Rapport-AS38-VF-2.pdf

Electric and hybrid car in Morocco. (2019). *Study on sustainable mobility in Morocco.* Group Sunergia. Available online: https://groupe-sunergia.com/wp-content/uploads/2019/11/Potentiel-3.png

Elhasbi, A., Jami, J., & Kammas, S. (2015). Sustainable transport in morocco: What contingency factors for which maturity level? Case of road haulage" RH" service providers in the region of Casablanca metropolis. *European Scientific Journal, 11*(29).

European Parliament and of the Council. (2007). *Establishing a Framework for the Approval of Motor Vehicles and their Trailers, and of Systems, Components and Separate Technical Units Intended for such Vehicles.* Available online: https://www.legislation.gov.uk/eudr/2007/46/annex/I/division/2/2020-01-31

European strategy for low emission mobility, (2016). *Annexe 1 historical development activity inn transport activity and energy use and emissions*. Available online: https://ec.europa.eu/transparency/documents-register/detail?ref=COM(2016)501&lang=en

Feng, W., & Figliozzi, M. (2013). An economic and technological analysis of the key factors affecting the competitiveness of electric commercial vehicles: A case study from the USA market. *Transportation Research Part C, Emerging Technologies, 26*, 135–145. doi:10.1016/j.trc.2012.06.007

Goepel, K. D. (2018). Implementation of an online software tool for the analytic hierarchy process (AHP-OS). *International Journal of the Analytic Hierarchy Process, 10*(3). Advance online publication. doi:10.13033/ijahp.v10i3.590

Göhlich, D., Nagel, K., Syré, A. M., Grahle, A., Martins-Turner, K., Ewert, R., Miranda Jahn, R., & Jefferies, D. (2021). Integrated approach for the assessment of strategies for the decarbonization of urban traffic. *Sustainability, 13*(2), 839. doi:10.3390u13020839

Gurtu, A. (2019). A pioneering approach to reducing fuel cost and carbon emissions from transportation. *Transportation Journal, 58*(4), 309–322. doi:10.5325/transportationj.58.4.0309

İmre, Ş., Çelebi, D., & Koca, F. (2021). Understanding barriers and enablers of electric vehicles in urban freight transport: Addressing stakeholder needs in Turkey. *Sustainable Cities and Society, 68*, 102794. doi:10.1016/j.scs.2021.102794

International Energy Agency. (2009). *Transport and CO2 moving toward sustainability. Paris IEA declaration.* General United Nations Publications. Available online: https://www.iea.org/news/transport-energy-and-co2-moving-toward-sustainability

Kammas, S., & Zendal, S. (2017). La logistique urbaine durable «LUD»: Concepts, état des lieux à Tanger (nord du Maroc), vers un modèle conceptuel de mise en œuvre dans les pays en développement [Sustainable urban logistics "LUD": Concepts, inventory in Tangier (northern Morocco), towards a conceptual model for implementation in developing countries]. *Revue des Etudes et Recherche en Logistique et Développement, 2*, 1–25.

Lee, D.-Y., Thomas, V. M., & Brown, M. A. (2013). Electric Urban Delivery Trucks: Energy Use, Greenhouse Gas Emissions, and Cost-Effectiveness. *Environmental Science & Technology, 47*(14), 8022–8030. doi:10.1021/es400179w PMID:23786706

Moufad, I., & Jawab, F. (2018). The Determinants of the performance of the urban freight transport-An empirical Analysis. In *International Colloquium on Logistics and Supply Chain Management (LOGISTIQUA)* (pp. 99-104). IEEE.

Nesterova, N., Quak, H., Balm, S., Roche-Ceraso, I., & Tretvik, T. (2013). State of the art of the electric freight vehicles implementation in city logistics. *FREVUE Project Deliverable D, 1*.

Pan, W., Chen, L., & Zhan, W. (2019). PESTEL analysis of construction productivity enhancement strategies: A case study of three economies. *Journal of Management Engineering, 35*(1), 05018013. doi:10.1061/(ASCE)ME.1943-5479.0000662

Pelletier, S., Jabali, O., & Laporte, G. (2016). 50th Anniversary Invited Article—Goods Distribution with Electric Vehicles: Review and Research Perspectives. *Transportation Science, 50*(1), 3–22. doi:10.1287/trsc.2015.0646

Quak, H., & Nesterova, N. (2014). Towards Zero Emission Urban Logistics: Challenges and Issues for Implementation of Electric Freight Vehicles in City Logistics. In C. Macharis, S. Melo, J. Woxenius, & T. V. Lier (Eds.), *Transport and Sustainability* (Vol. 6, pp. 265–294). Emerald Group Publishing Limited.

Sansa, M., Badreddine, A., & Romdhane, T. B. (2021). Sustainable design based on LCA and operations management methods: SWOT, PESTEL, and 7S. In Methods in Sustainability Science (pp. 345-364). Elsevier.

Schulte, J., & Ny, H. (2018). Electric road systems: Strategic stepping stone on the way towards sustainable freight transport? *Sustainability, 10*(4), 1148. doi:10.3390u10041148

Song, J., Sun, Y., & Jin, L. (2017). PESTEL analysis of the development of the waste-to-energy incineration industry in China. *Renewable & Sustainable Energy Reviews, 80*, 276–289. doi:10.1016/j.rser.2017.05.066

Taefi, T. T., Kreutzfeldt, J., Held, T., Konings, R., Kotter, R., Lilley, S., & Nyquist, C. (2016). Comparative analysis of european examples of freight electric vehicles schemes—a systematic case study approach with examples from Denmark, Germany, the Netherlands, Sweden and the UK. In Dynamics in logistics (pp. 495-504). Springer.

Thakur, V. (2021). Framework for PESTEL dimensions of sustainable healthcare waste management: Learnings from COVID-19 outbreak. *Journal of Cleaner Production, 287*, 125562. doi:10.1016/j.jclepro.2020.125562 PMID:33349739

Tsangas, M., Jeguirim, M., Limousy, L., & Zorpas, A. (2019). The application of analytical hierarchy process in combination with PESTEL-SWOT analysis to assess the hydrocarbons sector in Cyprus. *Energies, 12*(5), 791. doi:10.3390/en12050791

Wang, M., Thoben, K. D., Bernardo, M., & Daudi, M. (2018). Diversity in employment of electric commercial vehicles in urban freight transport: A literature review. *Logistics Research, 11*(10), 1–13.

WBCSD. (2004). *Doing Business with the Poor: A field guide*. WBCSD.

Zhang, Z., & Paudel, K. P. (2021). Small-Scale Forest Cooperative Management of the Grain for Green Program in Xinjiang, China: A SWOT-ANP Analysis. *Small-scale Forestry, 20*(2), 221–233. doi:10.100711842-020-09465-2

Zhou, T., Roorda, M. J., MacLean, H. L., & Luk, J. (2017). Life cycle GHG emissions and lifetime costs of medium-duty diesel and battery electric trucks in Toronto, Canada. *Transportation Research Part D, Transport and Environment, 55*, 91–98. doi:10.1016/j.trd.2017.06.019

ADDITIONAL READING

Büyüközkan, G., & Uztürk, D. (2020). Fleet Vehicle Selection for Sustainable Urban Logistics. In *Proceedings of the 2020 The 9th International Conference on Informatics, Environment, Energy and Applications* (pp. 116-120). 10.1145/3386762.3388955

De Vos, J., & El-Geneidy, A. (2022). What is a good transport review paper? *Transport Reviews, 42*(1), 1–5. doi:10.1080/01441647.2021.2001996

Erdelić, T., & Carić, T. (2022). Goods Delivery with Electric Vehicles: Electric Vehicle Routing Optimization with Time Windows and Partial or Full Recharge. *Energies, 15*(1), 285. doi:10.3390/en15010285

Karaaslan, A., & Gezen, M. (2022). The evaluation of renewable energy resources in Turkey by integer multi-objective selection problem with interval coefficient. *Renewable Energy*, *182*, 842–854. doi:10.1016/j.renene.2021.10.053

Lagorio, A., Pinto, R., & Golini, R. (2017). Urban Logistics Ecosystem: A system of system framework for stakeholders in urban freight transport projects. *IFAC-PapersOnLine*, *50*(1), 7284–7289. doi:10.1016/j.ifacol.2017.08.1402

Ziegler, D., & Abdelkafi, N. (2022). Business models for electric vehicles: Literature review and key insights. *Journal of Cleaner Production*, *330*, 129803. doi:10.1016/j.jclepro.2021.129803

KEY TERMS AND DEFINITIONS

Electric Vehicles: The electric vehicle is a vehicle that draws its energy from electric batteries or a fuel cell. It is opposed to the thermal vehicle, which itself is powered by the energy of a fossil fuel or a biofuel. The electric vehicle comprises one or more electric motors.

Logistics: Logistics is one of the most vital functions of the economy and is a service activity which aims to manage the flow of materials by making available and managing resources corresponding to needs, to economic conditions and for a determined quality of service, under satisfactory safety and security conditions.

Multi-Criteria Decision-Making (MCDM) Approach: One of the most well-known branches of decision theory is MCDM. It is a subdivision of the larger class of Operations Research (OR) models that deal with decision problems in the presence of many decision criteria. MCDM could be divided into Multi-Objective Decision Making (MODM) and Multi-Attribute Decision Making (MADM) and each approach has its own distinct features.

Renewable Energies: Renewable energies all energies produced via a so-called “inexhaustible” energy source, which regenerates easily. There are five major families of renewable energies. Solar photovoltaic and thermal energy refers to electricity produced from photovoltaic panels and heat produced via thermal collectors. Wind power is energy created from the wind. There are two types: onshore and offshore. In both cases, the wind turns blades connected to the wind turbine generator, which transforms the mechanical energy of the wind into electrical energy. Hydraulic energy is obtained by harnessing the motive power of water. Energy is created by passing water through turbines, thereby transforming the force of currents into electrical energy. Biomass energy is energy from combustion or anaerobic digestion. Geothermal energy is the use of energy contained in the basement to transform it into a source of heat and electricity. It is inexhaustible, stable, clean energy that is available all year round without interruption.

Sustainability: Sustainability can be defined as the ultimate goal of achieving a fulfilling life while respecting the boundaries of nature, which suggest, therefore, that sustainable development is the process of achieving sustainability.

Sustainability Transport Strategy: Sustainability transport strategy can be defined as when the community tries to provide commuters the means of transport that efficiently meet environmental, economic, and social aspects by mitigating unnecessary impacts and their associated costs, over pertinent space, and time factors.

Transportation: Transportation is the action of transporting someone or something or the process of being transported. Goods transport includes any movement of goods on board any mode of transport: rail, road, river, sea, air, etc. It is measured in tonne-kilometres or, on a given route, in tonnes.

Urban Logistics: Urban logistics covers all the activities inherent in the transport of goods in the city. Located at the crossroads of the challenges of urban development, economic dynamics, and quality of life, it is taking a growing interest in the overall functioning of the city and its management requires efficient rationalization of its components.

ENDNOTES

1 Alternative Fuels Data Center (2018) All-Electric Vehicles. Available online: http://www.afdc.energy.gov/vehicles/electric_basics_ev.html. (accessed on 23 April 2020).

2 Norwegian Postal service.

Chapter 7
Decision-Making for Biomass Harvesting Routing by using the Simulated Annealing

Uğur Atici
Sivas Cumhuriyet University, Turkey

Yunis Torun
Engineering Faculty, Cumhuriyet University, Turkey

Dursun Darendelioğlu
Engineering Faculty, Cumhuriyet University, Turkey

ABSTRACT

Biomass energy is an essential and sustainable type of energy for today and the future because it is produced from renewable sources and is environmentally friendly compared to fossil-based energy. Biomass energy is the only renewable energy source that creates social and economic impact together. It creates added value, provides employment, and creates new tax opportunities in many fields from agriculture to industry, from the transportation sector to the banking insurance sector. While other renewable energy sources cannot be stored, the energy obtained from biomass can be stored. In this aspect, energy production from biomass stands out from other renewable energy sources. One of the crucial handicaps in evaluating biomass as an energy source is the collection of biomass. Biomass is produced in different amounts in different village centers, while country roads are suitable for trucks of different sizes. In this chapter, the multi-capacity vehicle routing problem is modeled for biomass collection in village centers at different production capacities.

DOI: 10.4018/978-1-6684-2472-8.ch007

INTRODUCTION

Due to the rapid increase in the world population in the last century, the required energy is increasing daily (Trivedi et al., 2008). In addition to rapid population growth, technological developments cause the need for energy to increase exponentially from year to year. Developed countries are constantly trying to increase energy supply to meet increasing energy needs (Binswanger, 2001). With the increase in industrial production and the opening of global markets, the consumption of fossil resources, which is the most basic energy, has reached very high levels worldwide (Yergin, 2006). With the decrease in the total reserves of carbon-based fuels in traditional production systems and the emergence of their wastes as global climate change, renewable energy sources have come to the fore (Alniak, 2011).

Unlike fossil fuels, renewable energy sources can be used continuously. Renewable energy sources have disadvantages compared to fossil sources. This disadvantage is transporting raw materials from the source to the power plant, selecting suitable regions. Suitable regions should be determined, and feasibility studies should be carried out to produce renewable energy sources. Limited installation areas and high costs are barriers to renewable energy sources (Nalan et al., 2009).

The use of renewable energy sources has different limits and constraints. The most important of these constraints is that it is difficult to balance while the grid is fed from edible energy sources, and the energy produced cannot be stored. Unlike other renewable energy sources, the energy obtained from biomass can be stored in gas (such as methane).

The installation site is vital so that the biopower plant can be fed continuously with waste in the collection areas. Different studies have been carried out for the types of raw materials used in power plants where various types of waste are used as inputs. While producing biogas from biomass, industrial and agricultural organic wastes are used in energy production by many processes (Schievano et al., 2009). Fermented manure, produced at the end of this biological process, also supports increased agricultural productivity (Ocal, 2013). When energy needs and constraints are considered together, it is necessary to diversify renewable energy sources to solve this problem (Kumar & Jayanti, 2021). Biomass energy has been the primary renewable energy source from the first age to the present (Liserre et al., 2010). Many countries in the world also use biomass energy among renewable energy sources. While Austria provides 13% of its energy from biomass energy, Sweden provides 16% of its energy needed from biomass (Karayilmazlar, 2011).

Boi mass is an essential type of renewable energy source. However, it has disadvantages compared to other renewable energy sources. The biggest problem encountered in renewable energy production is collecting biomass from its source and transporting it to the power plant. The cost of the collection process cannot be more than the return of the energy to be produced. Installation site selection, collection routes, and vehicle types need to be planned cost-effectively. This study focuses on reducing transportation costs in biomass collection. The multi-capacity vehicle routing problem is modeled to minimize transportation costs.

This chapter consists of six sections. Past studies about the considered problem are presented in the second section. The main focus of this chapter is presented in the third section. The classical vehicle routing problem, the road and capacity-constrained vehicle routing problem for biomass, and the simulated annealing algorithm used for solving the problem are included in the fourth section. The data, results, and findings are presented in the fifth section. The obtained numerical results are comprised in the fifth section. The future work and conclusion are presented in the sixth and seventh sections.

BACKGROUND

One of the most crucial problems of century 21st century is global climate change (Jacob & Winner, 2009; Karl & Trenberth, 2003; Misra, 2014). The gaseous emission effect of fossil resources has accelerated environmentally friendly energy resources (Dincer, 2000). To meet the increasing energy need, the resources to be added to the system are selected from renewable energy sources instead of fossil-based production (Aworanti et al., 2017).

Biomass primarily refers to plants or plant-based materials (McKendry, 2002). Biomass material can be used directly by combustion to generate heat or convert it into various biofuels. Biomass harvesting is usually carried out at the same site or immediately after harvest as a land management tool (Naughton-Treves et al., 2007). Biomass harvesting can also be done on fallow lands where harvesting does not occur. The use of uncultivated lands in biomass harvesting is an alternative source of income in agriculture (Kohlheb & Krausmann, 2009). In areas with dense forest cover, it is threatened by insects, disease, drought, and fire. The thinning of dense forest cover is one of the essential tools in combating these threats (Spinelli & Magagnotti, 2010). Another source of biomass is animal excrement and forage crops, which are produced due to animal husbandry and agriculture (Piscioneri et al., 2000). The wastes generated after the meat processing processes in the slaughterhouse can be counted among the biomass inputs. Using biomass resulting from dilution, livestock, and meat processing as a renewable energy source plays an essential role in reducing energy production costs (Vida & Tedesco, 2017). It is obtained by using ethanol, an alcohol type, and biodiesel or methane gas, a fuel type, by using biomass.

With the growing world, the population was increased biomass potential and agricultural supply (Polat, 2021). This increase also reveals the potential for energy production from biomass. Compensating the energy need from different sources is also essential for energy supply security. Diversification of energy sources gives confidence in the energy supply in times of possible crisis (Ajansı, 2016).

In addition to plant wastes, animal wastes also have a high biomass potential. Energy efficiency varies according to the type of waste and its processes (Mutlu, 2019). Different sources affect the amount of energy produced and the production process (Wzorek et al., 2021). While producing biomass energy, which is carbon neutral, the energy produced from combustion is also more efficient than animal wastes (Pahla et al., 2017).

The gasification process in bioenergy production is a technology in which fertilizer and all kinds of organic waste feed the reactor (Mazaheri et al., 2019). It offers a range of benefits, including energy recovery from solid waste, a significant reduction in total waste, and environmental and water pollution (Haugen et al., 2015).

In addition, enzymatic processes in biofuel production also affect the yield of biofuel (Bhushan et al., 2021; Hajinajaf et al., 2021). Processes powered by bacteria and microalgae to reduce the cost of biomass production (Iasimone et al., 2021). In addition, the efficiency of biomass energy can be increased by chemical processes (Jin et al., 2020).

The livestock sector consists of farms of different sizes. Wastes generated in farms due to grazing animals in natural environments vary. The reasons for the different sizes of farm scales and the grazing of animals in the pasture are another problem in front of the total biomass (Salihoglu et al., 2019).

The amount of raw material supply is vital in determining biomass power plant location and the plant power capacity (Liu et al., 2021) (Çepelioğullar & Pütün, 2014). The technical, environmental, economic, and social effects of demand-oriented and participatory approaches and their relationship with other sectors should be analyzed to establish a sustainable bioenergy system (Röder et al., 2020).

While carrying out the feasibility studies, the population in the region, the number of settlements, the physical locations of the settlements, the number of animals, the amount of agricultural waste, and the energy need should be considered (Vitali et al., 2013). Bovine and ovine waste is usually collected after drying (Jayathilakan et al., 2012; Polprasert, 2007). Wastes collected from the region are transported to the power generation plant by vehicles. Animal and agricultural wastes are processed according to energy production processes, and biofuel is obtained. The obtained biofuel is used to produce different types of energy (Ajansı, 2016).

Transportation costs are one of the most critical handicaps in converting biomass, an input in renewable energy sources, into energy (Azevedo et al., 2019). Collecting biomass from its source and transporting it to the processing plant is a crucial cost resource for businesses. Different methods have been developed in the literature to reduce costs. Cultivating innovative biomass types such as microalgal biomass is recommended to reduce harvesting costs (Zhu et al., 2018). The biomass accumulation at its source depends on the amount and type of biomass and potential fuel energy (Toklu, 2017). Waste collection order, equipment used, collection method, and erosion control are issues to consider when collecting agricultural waste (Johnson et al., 2012). It has been reported that the type of transport vehicle and the packing method of the payload should be evaluated to reduce transportation costs (Singh et al., 2010).

Another parameter to be considered in energy production using biomass is storage costs (Voivontas et al., 2001). The supply chain plays a vital role in collection and storage costs. It has been proposed to use mixed-integer linear programming (MILP) to determine the biofuel supply chain, plant location, plant capacity, and operational biofuel production volume (Zhang & Hu, 2013).

Mathematical programming models were proposed to determine the biomass supply's economic potential (Chen, 2016). The distribution of temporal and spatial costs of agricultural wastes influences establishing a bio plant (Kingwell & Abadi, 2014). The main parameters affecting the are the capacity and the distribution biomass (Voivontas et al., 2001).

Route and yield analysis must also be done correctly (Cao, et al., 2021). Given the predetermined supply of biomass resources, a mixed-integer programming model was established to identify the best spots for biomass collection facilities and associated vehicle routes (Shabani & Sowlati, 2013; van Dyken et al., 2010). A hybrid heuristic is developed to address the computational complexity in a variable neighborhood search framework (Cao, Wang, et al., 2021). Integer linear programming formulations of multi-vendor problems need to be calculated when constructing a mathematical model of a biomass plant supplied from different points (Kara & Bektas, 2006). Another method used to solve the deterministic vehicle routing problem is the branch boundary algorithm (Archetti et al., 2007).

In the vehicle routing problem, the sum of the distance from the source points to the demand point is considered. The vehicle routing problem literature was examined in the biomass supply chain problem. While creating the biomass supply chain, the most significant distance of the supply points in priority order is one of the methods used in sequencing. In addition, new heuristic algorithms are proposed in ordering and prioritizing supply points (Li et al., 2017). The expected level of carbon reduction in biomass logistics has been taken into account. A location guidance model is proposed to determine biomass collection stations and routes by using a genetic algorithm (Li et al., 2019). Reducing transportation costs is very important for biopower plants. The vehicle routing problem in biomass collection is necessary for establishing and operating the bioenergy supply chain. A dual-objective MILP was developed, and a

Genetic Algorithm was used to solve the problem (Rabbani et al., 2020). The critical activities in biomass logistics can be summarized as harvesting and collection, storage, and transportation. Parameters such as demand-driven collection, type of biomass (dry, wet), and transport distance are included in the logistics model (Malladi & Sowlati, 2018). In-vehicle routing problems, deterministic versions where all data are known are examined. The dynamic models created by taking instant data from vehicles and supply points can be solved using artificial intelligence techniques by using up-to-date computing technologies (Gendreau et al., 1999).

Vehicle routing problems created for biomass collection are limited in capacity. The Capacity Vehicle Routing Problem (CVRP) is one of the problem types that has been widely studied in the literature. Many heuristics and meta-heuristics have been proposed to solve CVRP. Comparison of the proposed heuristics and meta-heuristics is a separate research topic (Hosseinabadi et al., 2017). Evolutionary algorithms are another method used in VRP models where stochastic demands are met (Niu et al., 2021).

This study aims to consider the vehicle routing problem, which can be applied in different disciplines in biomass collection processes—creating a vehicle routing model for biomass collection and developing a technique for solving the established model. Thus, it is possible to reduce the costs of the enterprises that produce energy using biomass and increase the operating efficiency.

MAIN FOCUS OF THE CHAPTER

As the population increases, the need for energy increases. It meets its energy needs by using different sources. Energy is commonly obtained from fossil fuels. The adverse effects of fossil fuels have become an undeniable reality today. Renewable energy sources are the best alternative to fossil fuels. Unlike fossil fuels, renewable energy sources are inexhaustible and can be used continuously. Biomass energy, one of the renewable energy sources, can be obtained from plants, organic waste, animal feces, and garbage.

Renewable energy sources also have disadvantages compared to fossil sources. In energy production from biomass, this handicap is transporting raw materials from the source to the power plant. The transport process has a cost. It is a decision problem which route will be used in the transportation process and what the capacities of the transfer vehicles will be. Taking the raw material from its source and transporting it to the power plant should be planned effectively. The most significant factor in reducing costs is to keep the costs at an acceptable level while collecting biomass. In literature, biomass collection has been considered a part of the supply chain.

Nevertheless, decision-makers should use cost functions to determine the collection routes and plan the vehicle type suitable. The cost function used should reflect the real-life model. The road types used along the route on the travel times should be calculated while determining the cost function. If a vehicle type suitable for the route is not planned, it should be reflected in the mathematical model as a penalty cost. In addition, vehicles have a load capacity for transport. The payload which can collect must be within the the capacity of the vehicle.

The biggest problem encountered in renewable energy production is collecting biomass from its source and transporting it to the power plant. However, the cost of the collection process cannot be more than the return of the energy to be produced. There are two main cost items of the collection process: the procurement cost, and the other is the transfer cost. The supply cost depends on the contract between the buyer and the seller, factors such as exchange rate policy. The cost of transport directly depends on the efficiency of transport planning. This study focuses on reducing transportation costs in biomass collection. The multi-capacity vehicle routing problem is modeled to minimize transportation costs. The model is solved by using a randomly generated dataset annealing simulation.

METHODOLOGY

This section discusses the vehicle routing problem for planning vehicle routes and optimizing costs in biomass collection. The capacity-constrained vehicle routing problem is examined in which road condition and vehicle capacity constraints are included in the cost function. The biomass supply quantities are considered deterministic. The number of routes and the types of vehicles used for transportation are also previously determined. Unlike previous studies, the VRP cost function is modeled as a combination of distance, road state, and transport vehicle type between supply points and biopower plants. Different capacity types for vehicle types are included as a constraint to the VRP model. The Simulated Annealing solved the proposed model.

The Vehicle Routing Problem

Vehicle routing can be defined as distributing products from warehouses to specific customers in the logistics system and collecting products from customers. When looking at a multi-stage logistics system, the vehicle routing problem comprise manufacturer, commercial goods, transfer vehiclei customers, and demand.

The distribution phase consists of delivering the products purchased from the manufacturer to the customers. At this stage, each vehicle moves one customer to the other, finishes the route, and returns to the center after distributing products as much as each customer's demand. Therefore, the warehouse can be defined as where the vehicles start their route and return. The road has the same standard along the transport vehicles' route in the classical vehicle routing problem.

In the literature, the vehicle routing problem has been examined under four headings: capacity-limited vehicle routing problem, distance-limited vehicle routing problem, vehicle routing problem with multiple warehouses, first distributed and then collected simultaneous and mixed collect-distribute vehicle routing problem, first distribute and then collect vehicle routing problem. The representation of the classical vehicle routing problem is presented in Figure 1.

Figure 1. Classic vehicle routing problem

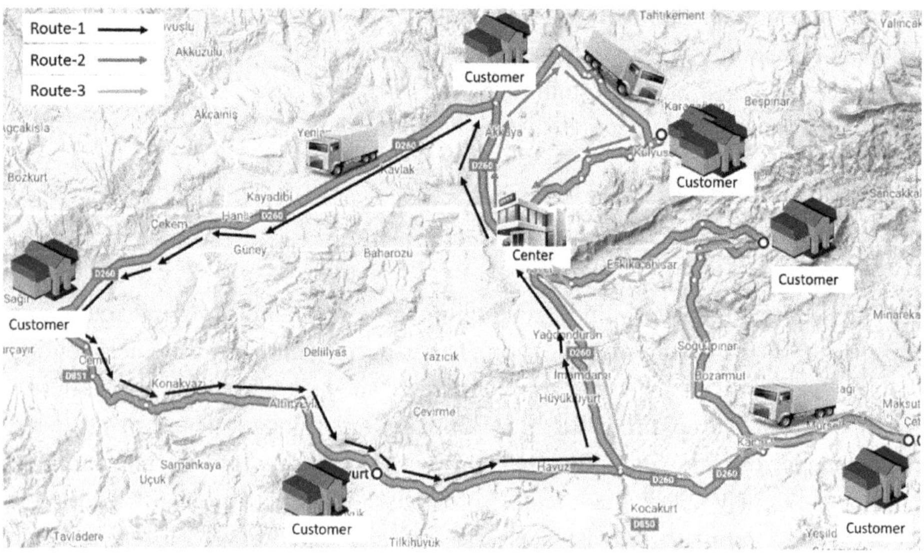

In the capacity-constrained vehicle routing problem, each vehicle has unique capacity. The demand and vehicle capacities of the customers are determined in advance. Vehicles start their movement from the warehouse, return to the warehouse, and end their route. The capacity-constrained vehicle routing problem is modeled using the number of supply points (N), the amount of demand at the node (q), and the number of vehicles used for transportation (K). The demand amount of the customers to be visited by each vehicle should not exceed the vehicle's capacity (V_C). The main point to be considered in this type of problem is that all vehicles start their movement from the warehouse and return the vehicles to the warehouse. Travel time from customer i to customer j is c_{ij}, if vehicle k is moving from customer

i to customer j, x_{ijk} is equal to 1 otherwise x_{ijk} is equal to 0. The objective function of the capacity-constrained vehicle routing problem is presented in Eq.1.

$$\min z \sum_{k=1}^{K} \sum_{ij \in A}^{A} x_{ij} c_{ij} \tag{1}$$

The constraint that determines that each customer is visited by only one vehicle is presented in Eq.2.

$$\sum_{k=1}^{K} \sum_{j \in \Delta^{+} i} X_{ijk} = 1 \tag{2}$$

The constraint limiting sending each vehicle from the starting point to only one customer is shown in Eq.3.

$$\sum_{j \in (0)} X_{0jk} = 1 \tag{3}$$

If a vehicle is visiting a customer, its departure from that customer simultaneously is shown with Eq.4.

$$\sum_{i \in \Delta^{-} j} X_{ijk} - \sum_{i \in \Delta^{+} j} X_{ij} = 0 \; \forall k \in K,\ i \in N \tag{4}$$

Connecting only one node to the warehouse at the end of the routes is modeled with Eq.5.

$$\sum_{i \in \Delta^{-}(n+1)}^{i \in \Delta^{-} j} X_{i,n+1,k} = 1 \; \forall k \in K \tag{5}$$

The constraint created that the amount at the supply points cannot exceed the vehicle capacity is presented in Eq.6.

$$\sum_{i \in \Delta^{-}(n+1)}^{i \in \Delta^{-} j} q_i \sum_{i \in \Delta^{+} i} X_{ijk} \leq C \; \forall k \in K \tag{6}$$

The constraint of variables being positive Eq.7, the binary constraint that takes the value 1 and 0 depending on whether k vehicle goes from customer i to customer j is denoted by Eq.8.

$$X_{ijk} \geq 0 \ \forall k \in K, (i,j) \in A \tag{7}$$

$$X_{ijk} \in (0,1) \ \forall k \in K, (i,j) \in A \tag{8}$$

The maximum distance constraint of each vehicle assigned to the predetermined routes is shown with Eq.9.

$$\sum_{i=0}\sum_{j=0} c_{ijk} X_{ijk} \leq T \tag{9}$$

The Multi-Capacity Vehicle Routing for Biomass

Vehicle routing problems in biomass collection and classical vehicle routing problems have differences in terms of terminology. The route of collecting biomass does not always have the same standards. While some of the roads have asphalt ground, some have dirt ground. While a particular part of the road has the desired width, some are relatively narrow. The roads between the biopower plants established in the countryside and the biowaste supply point are divided into three main categories. These categories are double-lane asphalt road (DLAR), single-lane asphalt road (SLAR), and single-lane dirt road (SLDR). In addition, the route may consist of different road types. While some of the routes are double-lane asphalt roads, some can be single-lane dirt roads. A safety factor is defined for each different road type.

The vehicle capable of traveling safely and economically in traffic on each road type should be planned on the route. If the route consists of more than one road type, the suitable vehicle for the road type with the lowest safety factor should be planned. Different speed and tonnages limits are regulated for different vehicles types on routes. The vehicles' fuel consumption and travel times also change depending on this change. Under these considerations, including the road safety coefficient and the planned vehicle type in the cost function and the length of the road that affects the travel time will make the vehicle routing problem for the biomass more realistic. The illustration of the vehicle routing problem for biomass collection is presented in Figure 2. At the biomass supply points, the amount of waste is modeled similarly to the customer demand in the vehicle routing problem. The vehicle moves from the biopower plant or the regional storage center in biomass collection. The vehicle returns to the starting point by loading from farms.

Figure 2. Vehicle routing problem for biomass collection

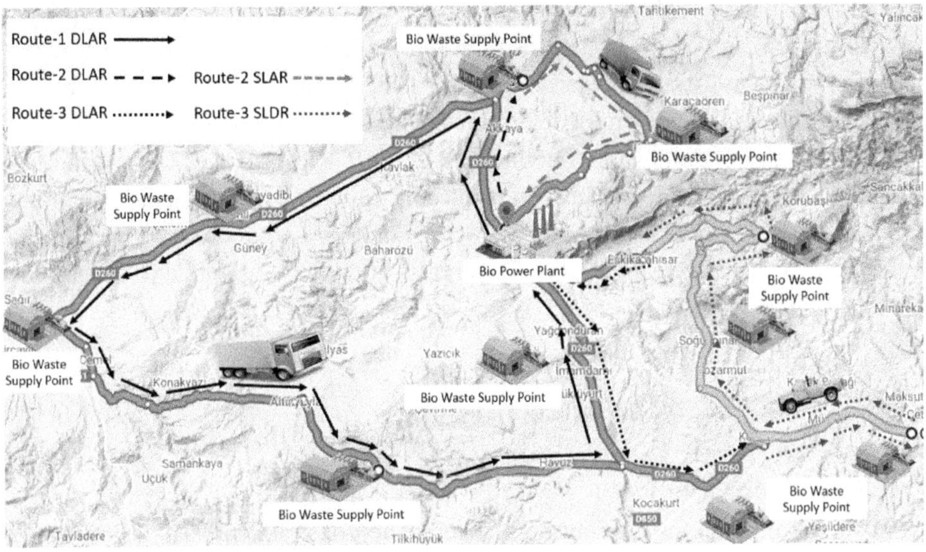

Vehicles with different capacities are used in the vehicle routing problem for biomass collection. Vehicles start action from the biopower plant, visit the biowaste supply points on the route and return to the biopower plant and end route. The amount of biomass to be collected at the supply points and the vehicle capacities are predetermined. Capacity constrained vehicle routing problem, *N* is the number of biowaste supply points, *q* is the amount of waste at the supply point, *K* is the number of vehicles used for collection.

The effect of the vehicle type on the cost is accepted as $\vartheta 1$ provided that the road safety coefficient Si_j from the starting point i to the target point j, the distance di_j, the number of different types of vehicles used for transfer is l. The cost function is rewritten as presented in Eq 10.

$$c_{ij} = d_{ij}(1 / S_{ij}\vartheta_1) \tag{10}$$

Using the cost function presented in Eq.10, the objective function of the vehicle routing problem for biomass is written as Eq.11.

$$\min z \sum_{k=1}^{K} \sum_{ij \in A}^{A} \sum_{l=1}^{L} x_{ij} d_{ij} (1 / S_{ij}\vartheta_l) \tag{11}$$

The sum of the biomass supply amount to be visited by each vehicle should not exceed the capacity of the type l vehicle (VN_l). The travel time from supply point i to supply point j is c_{ij}, x_{ijk}=1 otherwise x_{ijk}=0 if vehicle k moves from customer i to customer j. The constraint presented in Eq.2 is used if each customer is visited by only one vehicle. The constraint determines that each vehicle sent from the starting point is sent to only one supply point (see Eq.3.) If a transfer vehicle's vehicle is visiting a supply point, Eq.4 is used for the constraint to move at that supply point simultaneously, Eq.5 is used for the constraint of only one node to connect to the warehouse at the end of the routes. To amount at the supply points to exceed the capacity of the l vehicle type, Eq.6 is revised and used in Eq.12.

$$\sum_{i\in\Delta^-(n+1)}^{i\in\Delta^- j} q_i \sum_{i\in\Delta^+ i} X_{ijk} \leq C_l \; \forall k\in K, \forall l\in L \tag{12}$$

Positive variables in the routing problem The constraint presented in Eq.7 is written. Eq.8 is written for the assignment constraint [0,1], which takes the value of 1 if the k vehicles move from customer i to customer j. Eq.9 is written for the maximum distance constraint that each vehicle assigned to the determined routes can travel. The revised model, which contains objective function and constraint for biomass collection presented in Eq.13.

Objective function: (13)

$$\min \sum_{k=1}^{K} \sum_{ij\in A}^{A} \sum_{l=1}^{L} x_{ij} d_{ij} \left(\frac{1}{S_{ij}\vartheta_l}\right)$$

Constraints:

$$\sum_{k=1}^{K} \sum_{j\in\Delta^+ i} X_{ijk} = 1$$

$$\sum_{j\in(0)} X_{0jk} = 1$$

$$\sum_{i\in(0)} X_{0ik} = 1$$

$$\sum_{i\in\Delta^- j} X_{ijk} - \sum_{i\in\Delta^+ j} X_{ij} = 0,\ \forall k \in K,\ i \in N$$

$$\sum_{i\in\Delta^-(n+1)}^{i\in\Delta^- j} X_{i,n+1,k} = 1,\ \forall k \in K\ X_{ijk} \geq 0,\ \forall k \in K,\ (i,j) \in A$$

$$X_{ijk} \in (0,1),\ \forall k \in K,\ (i,j) \in A$$

$$\sum_{i=0}\sum_{j=0} c_{ijk} X_{ijk} \leq T$$

Simulated Annealing

The simulated annealing (SA) aims to find the optimal solution with the initial and randomly generated neighbor solutions (Kirkpatrick et al., 1983). The advantage of panning simulation compared to swarm-based optimization techniques is that SA does not get stuck on local goodness by sacrificing the improvement in the objective function and can obtain better solutions in the subsequent iterations (Brooks & Morgan, 1995). The function of the annealing process is to start from a solution at a sufficiently high temperature and gradually decrease the temperature to cycle between good and bad solutions and finally arrive at the best solution (Romeo & Sangiovanni-Vincentelli, 1991). To show the objective function value of any solution in a minimization problem z, the steps of simulated annealing are as follows. The steps of Simulated Annealing are presented in Table 1.

The number of iterations of the annealing simulation is controlled by the cooling coefficient and the initial temperature. Temperature T is a parameter that controls the probability of the algorithm accepting bad solutions. High temperatures can cause a completely random walk, increased resolution time, or poor performance. On the other hand, low temperatures can lead to the path to the best being overlooked. Therefore, the temperature should be kept high at the beginning and gradually reduced as the research progresses. The flow algorithm of the simulated annealing is shown in Figure 3. Biomass vehicle routing has some differences from the classical vehicle routing problem. Vehicles depart from the same biomass storage center. They visit supply points. Biomass loading is done at each supply point. The total biomass loaded on the vehicle cannot be greater than the vehicle's capacity. However, the distance that the transport vehicle can travel is limited. This intent can vary from business to business. While the constraint is the shift time in the companies that collect on a triple shift basis, it is the fuel tank capacity for the transport vehicles that cannot refuel on the route.

Table 1. The steps of simulated annealing (SA)

Step 1: Randomly generate an initial solution that meets the constraints *S* Step 2: Calculate objective function value *Z* Step 3: The current solution is S_c=S and the best solution is S_B=S. The urrent solution objective function value is Z_c=Z The est solution objective function value is Z_B=Z. Step 4: Set algorithm parameters Detemine initial temperature T Assin final temperature T_e Assin cooling rate C_r Step 5: Repeat the loop $T<T_e$ Create a neighbor solution S_N Calculate neighbor solution objective function value Z_N Calculate $\Delta Z=Z_S-Z_N$ If $\Delta Z>0$ $S_C=S_N$, $Z_S=Z_N$ Else Generate a random number RN in the range [0,1] If $RN \geq e^{-\Delta Z/T}$ then $S_C=S_N$, $Z_S=Z_N$ T=TxC$_r$ If $C_B-C_S>0$ then $S_B=S_C$, $Z_B=Z_C$ Step 6: Output S_B and Z_B

The initial solution is randomly generated. The number of elements of the initial solution (I+J-1) is the permutation matrix, provided that the number of supply points (*I*) and the number of routes (*J*) in the initial solution are. The order of the elements with the permutation matrix elements greater than *I* is found, and the matrix is divided into the number of routes. The type of truck used on each route is randomly generated as many as *J* routes. The vehicle capacity assigned to the route is compared with the total waste amount at the supply points on the route. If the vehicle capacity is insufficient, the permutation matrix is created again, and the process is repeated. The route is a permutation of supply points. When generating a neighboring solution, the previous solution is used to order the supply points. The supply points at two randomly selected positions in the permutation matrix are interchanged, and the same solution is obtained. The graphical representation of obtaining the neighbor solution is shown in Figure 4. The pseudo-code for creating the initial solution is presented in Table 2.

Figure 3. The simulated annealing flowchart

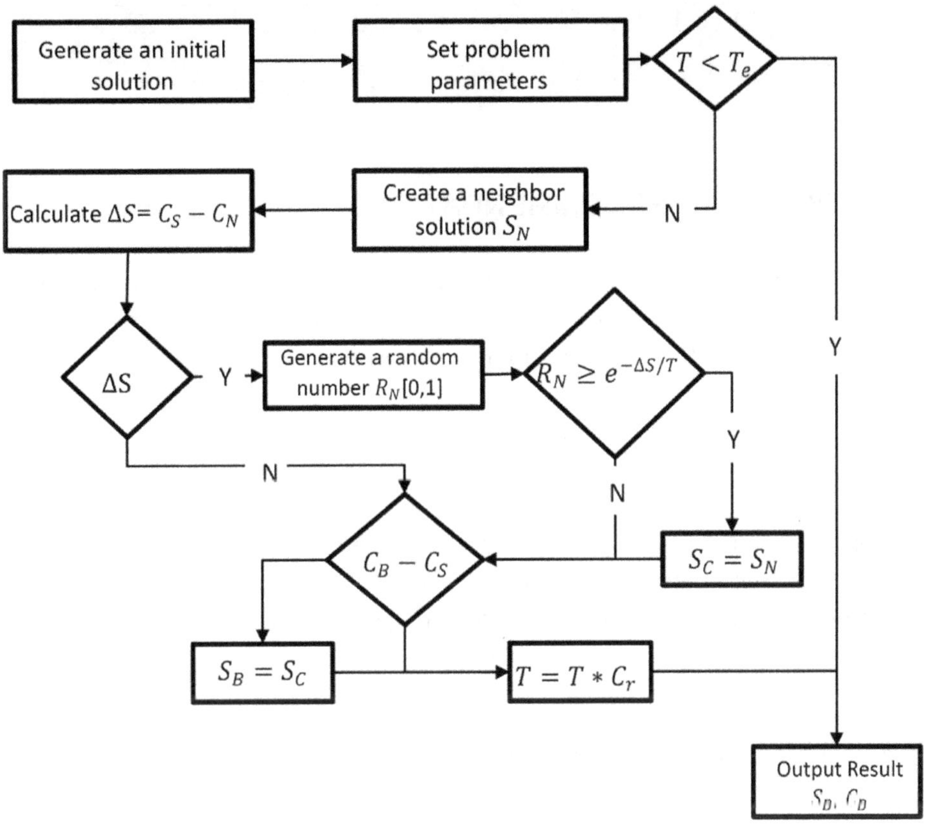

Table 2. Initial solution pseudocode

```
GET supply points number (I)
GET routes number (J)
  Loop solution.feasible =TRUE
    Randperm(I+J–1);
    q=find(Sol>I);
    Split q to J piece and get Routes
    Sum supply quantity of each Route  ∑_{j=1}^{J} ∑_{i=1}^{VN} r_ij
      Assign a vehicle type to each route VRA=randi(VN,1,J) (VN: available vehicle number)
      if VC(j) > ∑_{j=1}^{J} ∑_{i=1}^{VN} r_ij
        then solution.feasible= TRUE
        else solution.feasible=FALSE
Output solution
```

Figure 4. Generating the neighboring solution

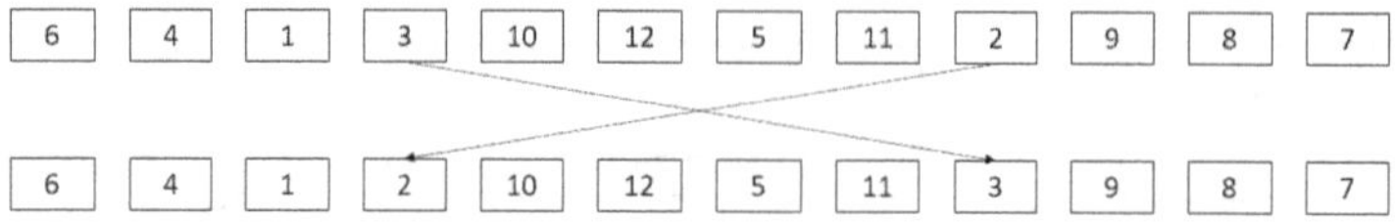

SOLUTIONS AND RECOMMENDATIONS

This study focuses on reducing transportation costs in biomass collection. The multi-capacity vehicle routing problem is modeled to minimize transportation costs. The model is solved by using a randomly generated dataset annealing simulation.

Data

The data used in this study were randomly generated. It is assumed that there are three different routes on the planning route. It is accepted that transportation operations are carried out with five different types of vehicles. Vehicle types and carrying capacities are presented in Table 3.

Table 3. Vehicle types, impact on costs and carrying capacities

Vehicle type	1	2	3	4	5
Cost Impact	1	2	4	6	8
Vehicle Capacity (tons)	4	6	10	20	30

It is assumed that there are ten different supply points on the planning route. The amount of bio-waste supplied at each supply point was randomly generated and presented in Table 4.

Table 4. Supply points and amount of bio-waste

Supply point	1	2	3	4	5	6	7	8	9	10
Bio waste quantity	2	5	1	2	1	3	5	7	2	7

The distance of the supply points to each other and the biopower plant was randomly distributed between 0 and 200. The distances are presented in Table 5.

Table 5. Distances

Destination	PC	SP 1	SP 2	SP 3	SP 4	SP 5	SP 6	SP 7	SP 8	SP 9	SP 10
PC	0	28	30	78	50	20	80	108	80	98	120
SP 1	28	0	32	45	45	48	85	120	130	110	80
SP 2	30	32	0	91.83	15	50	65	80	1113	137	110
SP 3	78	45	91.83	0	140	98	140	180	90	57	43
SP 4	50	45	15	140	0	66	26	50	71	160	165
SP 5	20	48	50	98	66	0	85	89.4	55	96	180
SP 6	80	85	65	140	26	85	0	58	46	130	206
SP 7	108	120	80	180	50	89.4	58.6	0	33	117	210
SP 8	80	130	1113	90	71	55	46	33	0	70	157
SP 9	98	110	137	57	160	96	130	117	70.2	0	84
SP 10	120	80	110	43	165	180	206	210	157	84	0

It is assumed that there are three different road types between the supply points and the biopower plant installation site. These road types are double lane asphalt road (DLAR), single-lane asphalt road (SLAR), and single-lane dirt road (SLDR), and their S_{ij} values are taken as 1, 0.95, 0.9, respectively. The road state cost coefficients are presented in Table 6.

Table 6. Road state cost coefficients

Destination	PC	SP 1	SP 2	SP 3	SP 4	SP 5	SP 6	SP 7	SP 8	SP 9	SP 10
PC	0	1	0.95	1	0.9	1	0.9	0.9	1	1	1
SP 1	1	0	0.95	1	0.9	1	0.9	0.9	1	1	1
SP 2	0.95	0.95	0	0.95	0.9	0.95	0.9	0.9	0.95	0.95	0.95
SP 3	1	1	0.95	0	0.9	1	0.9	0.9	1	1	0.9
SP 4	0.9	0.9	0.9	0.9	0	0.9	0.9	0.9	0.9	0.9	1
SP 5	1	1	0.95	1	0.9	0	0.9	0.9	1	1	0.9
SP 6	0.9	0.9	0.9	0.9	0.9	0.9	0	0.9	0.9	0.9	1
SP 7	0.9	0.9	0.9	0.9	0.9	0.9	0.9	0	0.9	0.9	0.9
SP 8	1	1	0.95	1	0.9	1	0.9	0.9	0	1	0.9
SP 9	1	1	0.95	1	0.9	1	0.9	0.9	1	0	1
SP 10	1	1	0.95	0.9	1	0.9	1	0.9	0.9	1	0

RESULTS AND FINDINGS

The simulated annealing algorithm developed for the vehicle routing problem for biomass collection was run in Matlab 2018. The cooling rate (Cr) was taken as 0.99. The initial temperature (T0) is assumed as 1000 degrees. The proposed objective function equation (see Eq.11) for the vehicle routing problem in biomass collection was used. Thus, more realistic decision-making has been enabled by including the state of the road between the supply points and the center and the cost-effectiveness ratio of the vehicles used for transportation into the objective function. The algorithm was run on Win10 64 bit operating system installed on an I5 processor and 8 GB memory.

As a result of running the algorithm, three different routes were determined. Vehicle number 4 is planned for the first route, number 3 for the second route, and number 5 for the third. The capacities of the transport vehicles are 20, 10 and 30 tons, respectively.

The order of visits of the transport vehicles of the routes and supply points is presented in Table 7. It has been proposed to have supply points 4, 7, 8, and 3 on Route 1. Route 1' vehicle number 4 is planned. The load capacity of vehicle number 4 is 20 tons, and the maximum amount of biowaste collected from the supply points on the route is 15 tons. Vehicle number 1 is planned for Route 2. The maximum amount of waste collected from route two supply points is 1 ton, and vehicle number 1 with a capacity of 4 tons is planned. Route three consisted of supply points 2, 1, 9,6, and 10.

Table 7. Route and solution

	Supply points				
Route 1	4	7	8	3	
Route 2	5				
Route 3	2	1	9	6	10

The maximum amount of waste collected from Route 2 supply points is 19 tons, and vehicle number 4, with a capacity of 20 tons, is planned. Vehicle capacity constraint is provided for all three routes. The graphical representation of the solution is defined in a 200x100 pixel area. The x and y coordinate points used for the graphical representation of the solution are presented in Table 8. The supply points coordinates, and biopower plant center coordinates are SP (in), PC respectively. The connections of the solution between supply points and bio plant are shown in Figure 5.

Table 8. Coordinate matrix

	PC	SP 1	SP 2	SP 3	SP 4	SP 5	SP 6	SP 7	SP 8	SP 9	SP 10
X	110	113.1	141.3	54.5	164.5	123.3	152.9	196.5	153.5	81.7	12
Y	56	85.3	75.1	60.3	56.7	34.9	37.1	19.5	10.3	13.3	41.5

Figure 5. Connections between supply points and biopower plant

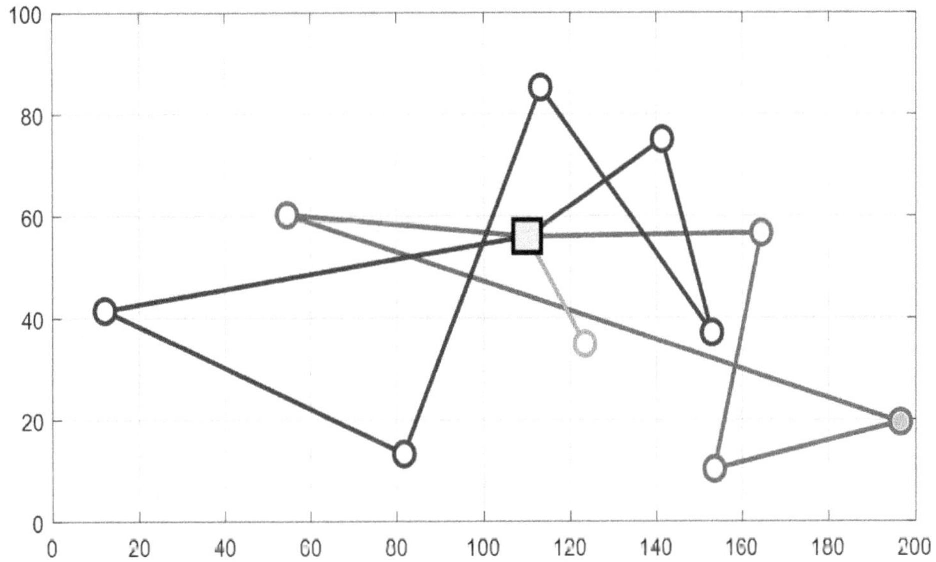

The hundredth iteration of the algorithm reached the best objective function value. The objective function value was determined as 1270.187. As a result of running the algorithm, the best objective function & iteration graph is presented in Figure 6.

FUTURE RESEARCH DIRECTIONS

Energy is one of the resources that societies and states need. The energy needs must be met uninterruptedly and safely. The preferred energy source should be sustainable, environmentally, and cost-effective. The solution to the energy problem is in a structure that requires cooperation and coordination. The choice of energy source is a strategic decision, and therefore, it is of great importance to determine a practical decision-making approach. Evaluating energy resources to determine the most suitable one for energy production requires simultaneous evaluation of many decision criteria in terms of decision making.

Figure 6. Best objective function and iteration graph

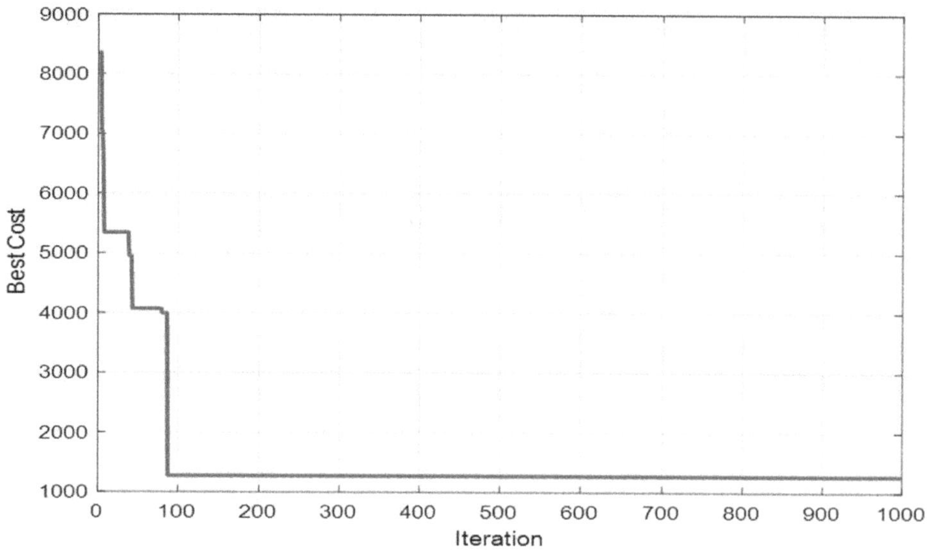

The environmental, socio-cultural, and economic effects of renewable energy resources can be examined. Thus, the environmental effects can be revealed, and the national park areas that need to be protected can be determined, or conservation programs can be developed. For power plants that produce energy from biomass, studies can be carried out in river, river, and lake areas, considering the distance to water areas. The most critical environmental problem that may occur in wind turbines is noise. When choosing the installation site of wind power plants, an examination can be made on the choice decision where there is no settlement, or the noise is felt very little due to altitude differences. Multi-criteria decisions and techniques can be used for potential site compatibility with renewable energy sources.

The cost of alternative routes can be inspected by taking a fixed value for the number of routes and supply points. Other techniques in the literature can be used to generate neighbor solutions in annealing simulation. The data used in the workshop can be used instead of randomly generated.

CONCLUSION

One of the biggest problems of our age is global climate change. As the pressure of global climate change on society, state, economy, and the gas emission effect of fossil resources increases, the use of environmentally friendly energy resources has gained momentum. To meet the increasing energy need, research on renewable energy sources, reducing costs, and increasing efficiency gain importance.

In this study, vehicle route optimization and vehicle planning on routes were inspected to reduce the cost of biomass collection in energy generation using biomass wastes. Unlike the literature, the state of the road between the supply points and the center is modeled for the classical vehicle routing problem. A constraint has been added to meet the amount of biomass collected for the transport vehicle capacity planned on the route. The data used in the multi-capacity vehicle routing for biomass solution are randomly generated.

A vehicle routing model developed simulated annealing algorithm for biomass collection was used to solve the problem. The modeled problem has been solved by determining the supply points that the vehicles will visit in three different routes. The three different routes were determined, the most suitable vehicle type was assigned to routes to minimize transportation costs.

The model and solution technique we have developed will significantly reduce the costs and increase the efficiency of enterprises producing energy from biomass.

ACKNOWLEDGMENT

This research received no specific grant from any funding agency in the public, commercial, or not-for-profit sectors.

REFERENCES

Ajansı, S. K. (2016). *Project of Research*. Planning and Feasibility Studies on Small-scale Biogas Plants.

Alniak, M. O. (2011). *Advanced Technologies Workshop (IATW 2020) Proceedings Book*. Academic Press.

Archetti, C., Bertazzi, L., Laporte, G., & Speranza, M. G. (2007). A Branch-and-Cut Algorithm for a Vendor-Managed Inventory-Routing Problem. *Transportation Science*, *41*(3), 382–391. doi:10.1287/trsc.1060.0188

Aworanti, O. A., Agarry, S. E., & Ogunleye, O. O. (2017). Biomethanization of Cattle Manure, Pig Manure and Poultry Manure Mixture in Co-digestion with Waste of Pineapple Fruit and Content of Chicken-Gizzard- Part I: Kinetic and Thermodynamic Modelling Studies. *The Open Biotechnology Journal*, *11*(1), 36–53. doi:10.2174/1874070701711010036

Azevedo, S. G., Sequeira, T., Santos, M., & Mendes, L. (2019). Biomass-related sustainability: A review of the literature and interpretive structural modeling. *Energy*, *171*, 1107–1125. doi:10.1016/j.energy.2019.01.068

Bhushan, S., Rana, M. S., Bhandari, M., Sharma, A. K., Simsek, H., & Prajapati, S. K. (2021). Enzymatic pretreatment of algal biomass has different optimal conditions for biogas and bioethanol routes. *Chemosphere*, *284*(131264), 1–10. doi:10.1016/j.chemosphere.2021.131264 PMID:34216928

Binswanger, M. (2001). Technological progress and sustainable development: What about the rebound effect? *Ecological Economics*, *36*(1), 119–132. doi:10.1016/S0921-8009(00)00214-7

Brooks, S. P., & Morgan, B. J. T. (1995). Optimization Using Simulated Annealing. *The Statistician*, *44*(2), 241–257. doi:10.2307/2348448

Cao, J. X., Wang, X., & Gao, J. (2021). A two-echelon location-routing problem for biomass logistics systems. *Biosystems Engineering*, *202*, 106–118. doi:10.1016/j.biosystemseng.2020.12.007

Cao, J. X., Zhang, Z., & Zhou, Y. (2021). A location-routing problem for biomass supply chains. *Computers & Industrial Engineering*, *152*(107017), 1–11. doi:10.1016/j.cie.2020.107017

Çepelioğullar, Ö., & Pütün, A. E. (2014). A pyrolysis study for the thermal and kinetic characteristics of an agricultural waste with two different plastic wastes. *Waste Management & Research*, *32*(10), 971–979. doi:10.1177/0734242X14542684 PMID:25062939

Chen, X. (2016). Economic potential of biomass supply from crop residues in China. *Applied Energy*, *166*, 141–149. doi:10.1016/j.apenergy.2016.01.034

Dincer, I. (2000). Renewable energy and sustainable development: A crucial review. *Renewable & Sustainable Energy Reviews*, *4*(2), 157–175. doi:10.1016/S1364-0321(99)00011-8

Gendreau, M., Guertin, F., Potvin, J.-Y., & Taillard, É. (1999). Parallel Tabu Search for Real-Time Vehicle Routing and Dispatching. *Transportation Science*, *33*(4), 381–390. doi:10.1287/trsc.33.4.381

Hajinajaf, N., Mehrabadi, A., & Tavakoli, O. (2021). Practical strategies to improve harvestable biomass energy yield in microalgal culture: A review. *Biomass and Bioenergy*, *145*(105941), 1–11. doi:10.1016/j.biombioe.2020.105941

Haugen, H. H., Halvorsen, B. M., & Eikeland, M. S. (2015).. . *Simulation of Gasification of Livestock Manure with Aspen Plus.*, *119*, 271–277. doi:10.3384/ecp15119271

Hosseinabadi, A. A. R., Rostami, N. S. H., Kardgar, M., Mirkamali, S., & Abraham, A. (2017). A new efficient approach for solving the capacitated Vehicle Routing Problem using the Gravitational Emulation Local Search Algorithm. *Applied Mathematical Modelling*, *49*, 663–679. doi:10.1016/j.apm.2017.02.042

Iasimone, F., Seira, J., Panico, A., De Felice, V., Pirozzi, F., & Steyer, J.-P. (2021). Insights into bioflocculation of filamentous cyanobacteria, microalgae and their mixture for a low-cost biomass harvesting system. *Environmental Research*, *199*(111359), 1–10. doi:10.1016/j.envres.2021.111359 PMID:34022232

Jacob, D. J., & Winner, D. A. (2009). Effect of climate change on air quality. *Atmospheric Environment*, *43*(1), 51–63. doi:10.1016/j.atmosenv.2008.09.051

Jayathilakan, K., Sultana, K., Radhakrishna, K., & Bawa, A. S. (2012). Utilization of byproducts and waste materials from meat, poultry and fish processing industries: A review. *Journal of Food Science and Technology*, *49*(3), 278–293. doi:10.100713197-011-0290-7 PMID:23729848

Jin, W., Pastor-Pérez, L., Yu, J., Odriozola, J. A., Gu, S., & Reina, T. R. (2020). Cost-effective routes for catalytic biomass upgrading. *Current Opinion in Green and Sustainable Chemistry*, *23*, 1–9. doi:10.1016/j.cogsc.2019.12.008

Johnson, L., Lippke, B., & Oneil, E. (2012). Modeling Biomass Collection and Woods Processing Life-Cycle Analysis*. *Forest Products Journal*, *62*(4), 258–272. doi:10.13073/FPJ-D-12-00019.1

Kara, I., & Bektas, T. (2006). Integer linear programming formulations of multiple salesman problems and its variations. *European Journal of Operational Research*, *174*(3), 1449–1458. doi:10.1016/j.ejor.2005.03.008

Karayilmazlar, S. S., Cabuk, Y.; & Kurt, R. (2011). Biyokütlenin Türkiye'de Enerji Üretiminde Degerlendirilmesi. *Bartin Orman Fakültesi Dergisi*, *13*(19), 63–75.

Karl, T. R., & Trenberth, K. E. (2003). Modern Global Climate Change. *Science*, *302*(5651), 1719–1723. doi:10.1126cience.1090228 PMID:14657489

Kingwell, R., & Abadi, A. (2014). Cereal straw for bioenergy production in an Australian region affected by climate change. *Biomass and Bioenergy*, *61*, 58–65. doi:10.1016/j.biombioe.2013.11.026

Kirkpatrick, S., Gelatt, C. D. Jr, & Vecchi, M. P. (1983). Optimization by Simulated Annealing. *Science*, *220*(4598), 671–680. doi:10.1126cience.220.4598.671 PMID:17813860

Kohlheb, N., & Krausmann, F. (2009). Land use change, biomass production and HANPP: The case of Hungary 1961–2005. *Ecological Economics*, *69*(2), 292–300. doi:10.1016/j.ecolecon.2009.07.010

Kumar, A., & Jayanti, S. (2021). A land-use-constrained, generation–transmission model for electricity generation through solar photovoltaic technology: A case study of south India. *Clean Technologies and Environmental Policy*, *23*(9), 2757–2774. doi:10.100710098-021-02202-z

Li, J.-M., Li, A.-H., Varbanov, P. S., & Liu, Z.-Y. (2017). Distance potential concept and its applications to the design of regional biomass supply chains and solving vehicle routing problems. *Journal of Cleaner Production*, *144*, 426–436. doi:10.1016/j.jclepro.2016.12.166

Li, S., Wang, Z., Wang, X., Zhang, D., & Liu, Y. (2019). Integrated optimization model of a biomass feedstock delivery problem with carbon emissions constraints and split loads. *Computers & Industrial Engineering*, *137*(106013), 1–18. doi:10.1016/j.cie.2019.106013

Liserre, M., Sauter, T., & Hung, J. Y. (2010). Future Energy Systems: Integrating Renewable Energy Sources into the Smart Power Grid Through Industrial Electronics. *IEEE Industrial Electronics Magazine*, *4*(1), 18–37. https://doi.org/10.1109/MIE.2010.935861

Liu, L., Wang, J., Wang, F., & Yang, X. (2021). The impact of the planting of forest biomass energy plants under the embedded Internet of Things technology on the biodiversity of the local environmental ecology. *Environmental Technology & Innovation, 24*, 2-16. doi:10.1016/j.eti.2021.101894

Malladi, K. T., & Sowlati, T. (2018). Biomass logistics: A review of important features, optimization modeling and the new trends. *Renewable & Sustainable Energy Reviews*, *94*, 587–599. https://doi.org/10.1016/j.rser.2018.06.052

Mazaheri, N., Akbarzadeh, A. H., Madadian, E., & Lefsrud, M. (2019). Systematic review of research guidelines for numerical simulation of biomass gasification for bioenergy production. *Energy Conversion and Management*, *183*, 671–688. https://doi.org/10.1016/j.enconman.2018.12.097

McKendry, P. (2002). Energy production from biomass (part 1): Overview of biomass. *Bioresource Technology*, *83*(1), 37–46. https://doi.org/10.1016/S0960-8524(01)00118-3

Misra, A. K. (2014). Climate change and challenges of water and food security. *International Journal of Sustainable Built Environment*, *3*(1), 153–165. https://doi.org/10.1016/j.ijsbe.2014.04.006

Mutlu, N. T., Karaca, C., & Öztürk, H. H. (2020). Developments in Biomass Gasification Technology. *Journal of Agricultural Machinery. Science*, *15*(2), 53–59.

Nalan, Ç. B., Murat, Ö., & Nuri, Ö. (2009). Renewable energy market conditions and barriers in Turkey. *Renewable & Sustainable Energy Reviews*, *13*(6), 1428–1436. https://doi.org/10.1016/j.rser.2008.09.001

Naughton-Treves, L., Kammen, D. M., & Chapman, C. (2007). Burning biodiversity: Woody biomass use by commercial and subsistence groups in western Uganda's forests. *Biological Conservation*, *134*(2), 232–241. https://doi.org/10.1016/j.biocon.2006.08.020

Niu, Y., Zhang, Y., Cao, Z., Gao, K., Xiao, J., Song, W., & Zhang, F. (2021). MIMOA: A membrane-inspired multi-objective algorithm for green vehicle routing problem with stochastic demands. *Swarm and Evolutionary Computation*, *60*(100767), 1–12. https://doi.org/10.1016/j.swevo.2020.100767

Ocal, F. (2013). *Biogas Energy Production and Application for Eskişehir* [Master of Science Thesis]. Department of Mechanical Engineering.

Pahla, G., Mamvura, T. A., Ntuli, F., & Muzenda, E. (2017). Energy densification of animal waste lignocellulose biomass and raw biomass. *South African Journal of Chemical Engineering*, *24*, 168–175. https://doi.org/10.1016/j.sajce.2017.10.004

Piscioneri, I., Sharma, N., Baviello, G., & Orlandini, S. (2000). Promising industrial energy crop, Cynara cardunculus: A potential source for biomass production and alternative energy. *Energy Conversion and Management*, *41*(10), 1091–1105. https://doi.org/10.1016/S0196-8904(99)00135-1

Polat, M. (2021). Türkiye'nin Tarımsal Atık Biyokütle Enerji Potansiyelindeki Degisim. *Toprak Su Dergisi,* 19-24. doi:10.21657/topraksu.692275

Polprasert, C. (2007). *Organic Waste Recycling: Technology and Management* (3rd ed.). IWA Publishing.

Rabbani, M., Akbarian-Saravi, N., Ansari, M., & Musavi, M. (2020). A Bi-Objective Vehicle-Routing Problem for Optimization of a Bioenergy Supply Chain by Using NSGA-II Algorithm. *Journal of Quality Engineering and Production Optimization*, *5*(1), 87–102. https://doi.org/10.22070/jqepo.2020.3650.1079

Röder, M., Mohr, A., & Liu, Y. (2020). Sustainable bioenergy solutions to enable development in low- and middle-income countries beyond technology and energy access. *Biomass and Bioenergy*, *143*(105876), 1–8. https://doi.org/10.1016/j.biombioe.2020.105876

Romeo, F., & Sangiovanni-Vincentelli, A. (1991). A theoretical framework for simulated annealing. *Algorithmica*, *6*(1), 302–345. https://doi.org/10.1007/BF01759049

Salihoglu, N. K., Teksoy, A., & Altan, K. (2019). Determination of Biogas Production Potential From Cattle And Sheep Wastes: Balikesir Case Study. *Ömer Halisdemir Üniversitesi Mühendislik Bilimleri Dergisi*, *8*(1), 31-47. doi:10.28948/ngumuh.516798

Schievano, A., D'Imporzano, G., & Adani, F. (2009). Substituting energy crops with organic wastes and agro-industrial residues for biogas production. *Journal of Environmental Management*, *90*(8), 2537–2541. https://doi.org/10.1016/j.jenvman.2009.01.013

Shabani, N., & Sowlati, T. (2013). A mixed integer non-linear programming model for tactical value chain optimization of a wood biomass power plant. *Applied Energy*, *104*, 353–361. https://doi.org/10.1016/j.apenergy.2012.11.013

Singh, J., Panesar, B. S., & Sharma, S. K. (2010). A mathematical model for transporting the biomass to biomass based power plant. *Biomass and Bioenergy*, *34*(4), 483–488. https://doi.org/10.1016/j.biombioe.2009.12.012

Spinelli, R., & Magagnotti, N. (2010). Comparison of two harvesting systems for the production of forest biomass from the thinning of Picea abies plantations. *Scandinavian Journal of Forest Research*, *25*(1), 69–77. https://doi.org/10.1080/02827580903505194

Toklu, E. (2017). Biomass energy potential and utilization in Turkey. *Renewable Energy*, *107*, 235–244. https://doi.org/10.1016/j.renene.2017.02.008

Trivedi, J. K., Sareen, H., & Dhyani, M. (2008). Rapid urbanization - Its impact on mental health: A South Asian perspective. *Indian Journal of Psychiatry*, *50*(3), 161–165. https://doi.org/10.4103/0019-5545.43623

van Dyken, S., Bakken, B. H., & Skjelbred, H. I. (2010). Linear mixed-integer models for biomass supply chains with transport, storage and processing. *Energy, 35*(3), 1338–1350. https://doi.org/10.1016/j.energy.2009.11.017

Vida, E., & Tedesco, D. E. A. (2017). The carbon footprint of integrated milk production and renewable energy systems – A case study. *The Science of the Total Environment, 609*, 1286–1294. https://doi.org/10.1016/j.scitotenv.2017.07.271

Vitali, F., Parmigiani, S., Vaccari, M., & Collivignarelli, C. (2013). Agricultural waste as household fuel: Techno-economic assessment of a new rice-husk cookstove for developing countries. *Waste Management, 33*(12), 2762-2770. doi:10.1016/j.wasman.2013.08.026

Voivontas, D., Assimacopoulos, D., & Koukios, E. G. (2001). Aessessment of biomass potential for power production: A GIS based method. *Biomass and Bioenergy, 20*(2), 101–112. https://doi.org/10.1016/S0961-9534(00)00070-2

Wzorek, M., Junga, R., Yilmaz, E., & Niemiec, P. (2021). Combustion behavior and mechanical properties of pellets derived from blends of animal manure and lignocellulosic biomass. *Journal of Environmental Management, 290*(112487), 1–8. https://doi.org/10.1016/j.jenvman.2021.112487

Yergin, D. (2006). Ensuring Energy Security. *Foreign Affairs, 85*(2), 69–82. https://doi.org/10.2307/20031912

Zhang, L., & Hu, G. (2013). Supply chain design and operational planning models for biomass to drop-in fuel production. *Biomass and Bioenergy, 58*, 238–250. https://doi.org/10.1016/j.biombioe.2013.08.016

Zhu, L., Li, Z., & Hiltunen, E. (2018). Microalgae Chlorella vulgaris biomass harvesting by natural flocculant: effects on biomass sedimentation, spent medium recycling and lipid extraction. *Biotechnology for Biofuels, 11*(1), 1-10. doi:10.1186/s13068-018-1183-z

ADDITIONAL READING

Atashbar, N. Z., Labadie, N., & Prins, C. (2016). Modeling and optimization of biomass supply chains: A review and a critical look. *IFAC-PapersOnLine, 49*(12), 604–615. doi:10.1016/j.ifacol.2016.07.742

Kumar, A., Sokhansanj, S., & Flynn, P. C. (2006). Development of a multicriteria assessment model for ranking biomass feedstock collection and transportation systems. *Applied Biochemistry and Biotechnology*, *129*(1), 71–87. doi:10.1385/ABAB:129:1:71 PMID:16915632

Nunes, L. J. R., Causer, T. P., & Ciolkosz, D. (2020). Biomass for energy: A review on supply chain management models. *Renewable & Sustainable Energy Reviews*, *120*, 109658. doi:10.1016/j.rser.2019.109658

Perpiñá, C., Alfonso, D., Pérez-Navarro, A., Peñalvo, E., Vargas, C., & Cárdenas, R. (2009). Methodology based on Geographic Information Systems for biomass logistics and transport optimisation. *Renewable Energy*, *34*(3), 555–565. doi:10.1016/j.renene.2008.05.047

Ruiz, J. A., Juárez, M. C., Morales, M. P., Muñoz, P., & Mendívil, M. A. (2013). Biomass logistics: Financial & environmental costs. Case study: 2 MW electrical power plants. *Biomass and Bioenergy*, *56*, 260–267. doi:10.1016/j.biombioe.2013.05.014

Sokhansanj, S., & Hess, J. R. (2009). Biomass Supply Logistics and Infrastructure. In J. R. Mielenz (Ed.), *Biofuels: Methods and Protocols* (pp. 1–25). Humana Press. doi:10.1007/978-1-60761-214-8_1

Sokhansanj, S., Kumar, A., & Turhollow, A. F. (2006). Development and implementation of integrated biomass supply analysis and logistics model (IBSAL). *Biomass and Bioenergy*, *30*(10), 838–847. doi:10.1016/j.biombioe.2006.04.004

Tatsiopoulos, I. P., & Tolis, A. J. (2003). Economic aspects of the cotton-stalk biomass logistics and comparison of supply chain methods. *Biomass and Bioenergy*, *24*(3), 199–214. doi:10.1016/S0961-9534(02)00115-0

Velazquez-Marti, B., & Annevelink, E. (2009). GIS Application to Define Biomass Collection Points as Sources for Linear Programming of Delivery Networks. *Transactions of the ASABE*, *52*(4), 1069–1078. doi:10.13031/2013.27776

KEY TERMS AND DEFINITIONS

Constraint: The optimization expresses the limits of the investigated problem.

Cost Function: Function that minimizes cost and maximizes production efficiency.

Meta Heuristic: Unified version of heuristic methods. It aims to find the most suitable solution in the solution space by using search algorithms.

Multi-Capacity Vehicle Routing: A special version of the vehicle routing problem in which the capacities of the planning vehicles are taken into account.

Objective Function: Defines the purpose of optimization. It is expressed as profit maximization or cost minimization.

Simulated Annealing: A probabilistic approximation technique that allows finding the global optimum value of a given function solution.

Supply Chain: The management of the activities of businesses, individuals and institutions in delivering natural resources to the end user as a final product.

Vehicle Routing Problem: An integer programming model that allows the determination of the route that allows the vehicles that are the basis of planning to visit a certain set of customers.

Chapter 8
Multi-Criteria Decision-Making Methods for Biomass Energy Systems:
A Review

Meral Güldeş
https://orcid.org/0000-0002-1256-2751
Sivas Cumhuriyet University, Turkey

Ömer Faruk Gürcan
Sivas Cumhuriyet University, Turkey

ABSTRACT

Global climate change is one of the most challenging problems of today's world and its effects have become more noticeable day after day. The magnitude of climate change is closely related to our carbon footprint, so replacing the resources such as petroleum, coal, nuclear energy by which humankind generates their energy requirements with new ones is essential. The usage of renewable energy resources is one of the effective ways to decrease CO2 emissions and environmental pollution. Biomass energy is one of the promising future energies as a renewable resource. Therefore, many requirements should be considered and evaluated carefully to produce and sustain a successful biomass energy system. This chapter presents a review of academic research attempting to face the biomass energy sector's problems using multi-criteria decision-making (MCDM) methods. Related articles in the international journals from 2010 to 2021 are collected and reviewed to answer the following questions: (1) Which methods are mainly used? (2) Which problems attract the most attention?

DOI: 10.4018/978-1-6684-2472-8.ch008

INTRODUCTION

Industrialization and motorization are increasing all over the world. Energy demand is increasing in proportion to these developments, and 80% of this need is met from fossil fuels. Despite the rapidly increasing energy demand, the effort to meet these demands with limited fossil fuels has accelerated the searching to introduce new energy sources in the coming years. One of the main motivations in these studies is that the environmental damage caused by fossil fuels is becoming challenging to reverse. Increasing energy consumption causes an increase in GHG emissions; therefore, environmental damage increases. In order to meet the energy demand in the coming years, the world has accelerated the search for environmentally friendly, renewable, efficient, and cost-effective alternative energy sources instead of fossil fuels, which are gradually decreasing and causing an increase in GHG emissions. While countries are creating their energy policies, they are taking initiatives to supply their energy demands from cheap, reliable, and sustainable alternative sources. Alternative energy is a concept that expresses the substitution of other energy sources for the need for crude oil and thus reducing the environmental damage brought about by climate change. Alternative energy sources include solar, wind, geothermal, and biomass energy. When considering renewable energy sources, solar and wind energies are the first ones that come to mind. However, from the first human to this day, many people have used wood, fertilizer, and coal as fuel while fulfilling their daily routines such as cooking. Considering the increasing population, mostly living in rural areas, especially in developing countries, the use of these resources is undeniable.

Fuelwood consumption has increased by 250% since 1960. This fact shows us that this energy, which we call biomass, is an important element of renewable energy sources. Today, biomass energy has a share of 14% in primary energy sources. Every year, several million tons of agricultural waste are destroyed in different ways, such as incineration, land applications, and landfilling. As a result, high potential bio-renewable energy sources cannot be used. Biomass energy meets 35% of the energy needs of developing countries. It is predicted that by 2050, 90% of the world population will live in developing countries. Therefore, biomass energy is of great importance for these countries where agriculture is an important source of livelihood (Pathak, Chaudhari, & Fulekar, 2013).

The world population continues to increase rapidly. Today's world population is twice that of 1960 and is expected to approach 9 billion by 2050 (Perea-Moreno, Samerón-Manzano, & Perea-Moreno, 2019). The increase in the world population brings an increase in energy demand. Studies show that this increase in energy need has an annual acceleration of 2% (Zahid, Tahir, Khan, & Naeem, 2021). Developing countries account for 99% of this population growth, and this increase is observed as 50% in urban areas. While the share of global energy in cities was less than half of the energy produced in 1990, today, this rate constitutes two-thirds of the total energy. Fossil fuels are still used as the primary energy source in cities, and they constitute 70% of CO2 emissions, which is one of the causes of global warming. Air pollution in urban areas reaches significant levels. According to the World Health Organization (WHO), 90% of those living in urban areas are exposed to levels of environmental pollution that exceed recommended limits. 90% of all urban areas are located on the coastline. Rising sea levels make some cities vulnerable to danger in developing countries. (Perea-Moreno et al., 2019)

Different forms of biomass such as plant products, vegetation, and animal manure are made usable in the form of biogas and liquid fuels by utilizing thermochemical conversion technologies. These technologies have significantly contributed to reducing the need for fossil fuels. It also helps these regions develop economically by reducing the damage to the environment and enabling new job opportunities for people living in rural areas. Especially in developing countries, the emergence of different livelihoods helps reduce migration from rural areas to cities and prevents agricultural land from being idle. Due to political uncertainties, fluctuations in oil prices also play an essential role in developing biofuels that can replace fossil fuels. Many countries in the world meet their crude oil needs through export. In addition to environmental concerns, foreign dependency on energy causes countries to explore their energy potential. Efficient and cost-effective access to energy has a vital role in the sustainable development of countries.

Energy generation from renewable energy sources includes long-term investments from designing the necessary systems to implementing these systems. The decisions to be taken in this process include many different stakeholders. Each of these stakeholders has multifaceted decisions to make. Stakeholders should also consider essential criteria for themselves when making these decisions. In this study, multi-criteria decision-making studies on biomass energy, one of the renewable energy sources, were examined between 2010-2021 years. The popularity of the methods used was evaluated, and the problems that these studies mostly addressed were investigated by examining the literature.

Biomass

Biomass is a renewable organic material. Recently, biomass has been the most common form of renewable energy, and its importance has increased because of the impacts of fossil fuels. There are high potentials of producing biofuels for heating, electricity, and transportation. Released CO_2 from combustion and biomass utilization does not increase atmospheric CO_2 because of the biogenic origin of biomass. Biomass usage enables faster transfer of CO_2 into the atmosphere, which plants will adsorb to produce biomass again. Biomass is generally categorized based on biomass type in nature and application and use of biomass as feedstock. Nature-based classification is the most preferred one (Tursi, 2019).

A general classification of biomass as a solid fuel resource is given in Table 1 (Vassilev et al., 2012).

Table 1. A general classification of biomass as a resource of solid fuel

Biomass Class	Examples
Wood-based	barks, stems, twigs, foliage, bushes, pellets, sawdust,
Herbaceous and agricultural	grasses, flowers, straws (corn, bean, sunflower, others), fibers (flax, palm, coconut coir, others), stalks (corn, bean, kenaf, tobacco, others), shells and husks (coconut, cotton, olive, rice, coffee, others), pits (olive, plum, peach, others), other residues (seeds, cobs, food, fruits, pips, marc, cakes, others)
Aquatic	microalgae or macroalgae, freshwater or marine, unicellular or multicellular species (blue or red algae, duckweed, kelp, duckweed, water hyacinth, others)
Animal and human biomass wastes	various manures, sponges, bones, others
Contaminated and industrial biomass wastes	solid waste from municipal areas, hospital waste, waste papers, demolition wood, tannery waste, fibreboard, chipboard, others

Biomass is converted to energy through different processes: direct combustion (burning) for heating; thermochemical conversion, which generates liquid, solid and gaseous fuels; chemical conversion for producing liquid fuels; biological conversion for producing gaseous and liquid fuels.

Direct combustion is the most widely used method for producing energy from biomass. Pruduced energy can be used in heating water or buildings, steam turbines, or industrial process heat.

Thermochemical conversion includes gasification and pyrolysis. Both processes use thermal decomposition, where materials are heated at gasifiers (pressurized vessels) at very high temperatures (400-900° C). The difference between these two processes is due to the amount of O_2 present and temperature in processes.

The chemical conversion process is called transesterification, in which biomass is converted into fatty acid methyl esters (FAME). FAME is used in biodiesel production.

Biological conversion involves fermentation and anaerobic digestion processes that produce ethanol and renewable natural gas.

Recently, researchers have tried to improve productivity in current processes and also to explore other ways in converting and using more biomass

Multi-Criteria Decision Making

Multi-criteria decision-making (MCDM) is a branch of operational research. Its development began in 1971. MCDM methods are applied to rank, compare, classify, or select multiple alternatives, including incommensurate attributes. MCDM methods help decision-makers find robust and consistent solutions for multi-criteria problems in various industries or science (Zare et al., 2016).

A typical process of a MCDM method has seven steps: problem formulating, defining of objectives, criteria selecting, formation of alternatives, defining scores/weights to criteria, and selecting proper methods (Gebre et al., 2021).

MCDM methods place decision-makers in the center of the decision-making process. These methods are not automatable, which gives the same solution for each decision-maker. They are comprised of subjective (preference) information. There are many MCDM methods in the literature. Each of them has some advantages and disadvantages in related to application areas. The publications about multi-criteria decision analysis (MCDA) have recently increased steadily. The software availability and efficiency of researchers contribute to growth in use (Ishizaka and Nemery, 2013).

Ishizaka and Nemery (2013) listed MCDM methods relative to the type of problems. Such as AHP, ANP, MAUT/UTA, PTOMETHEE, TOPSIS, DEA, MACBETH are used in both ranking and choice problems. AHPSort, ELECTRE-Tri, UTADIS is used in sorting problems. GAIA, FS-Gaia are used in description problems.

In the literature, there are various classifications for MCDM methods. The methods are classified based on criteria, data, number of decision-makers, alternatives.

One of the classifications is based on a number of alternatives: discrete MCDM or Multi-Attribute Decision Making (MADM) against continuous Multi-Objective Decision Making (MODM) methods. The MADM methods deal with the cases where decision alternatives are defined explicitly by a finite list of alternatives. The aim is to make a rational selection from the finite number of alternatives or assess and rank the finite number of alternatives (Zavadskas et al., 2014).

MADM methods make preference decisions (prioritize, select, rank, classify, screen) over the finite number of alternatives where problems are characterized by attributes (conflicting, multiple, weighted and incommensurable) (Yoon and Hwang, 1995).

On the other hand, MODM methods deal with problems in which problems are non-predetermined, and the number of alternatives is continuous (infinite). The aim is to design the optimal alternative with well-defined constraints and measurable objectives (Zavadskas et al., 2014). The classification is presented in Figure x. Some example methods are given for each category. Additionally, Yalcin et al. (2022) presented MODM methods in four categories: no-preference methods such as Global Criterion; a priori methods such as Goal Programming; a posteriori methods such as

Epsilon Constraint, multi-objective particle swarm optimization; interactive methods such as STEM, Geoffrion, Interactive Compromise Programming.

Figure 1. A classification for MCDM methods

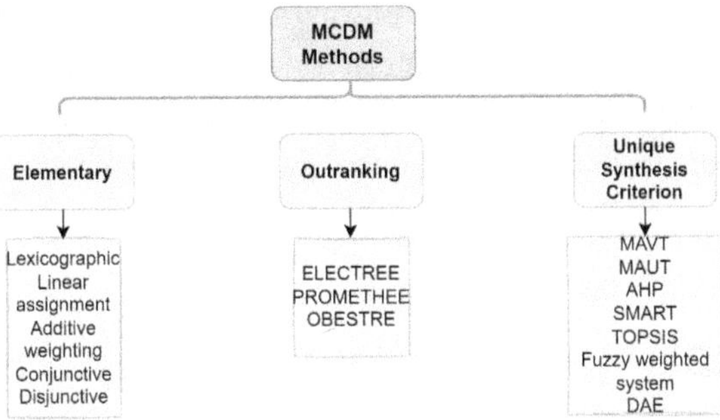

Zardari et al. (2015) presented another classification for MCDM methods, shown in Figure 2. There are three classes: elementary, unique synthesis, and outranking methods. Unique synthesis criterion methods combine different views into a unique function that will be optimized. In outranking methods, a relationship is developed based on the preferences of decision-makers, and then the relationship is explored to enable the solution of problems.

Figure 2. A classification of MCDM methods (Zarderi et al., 2015)

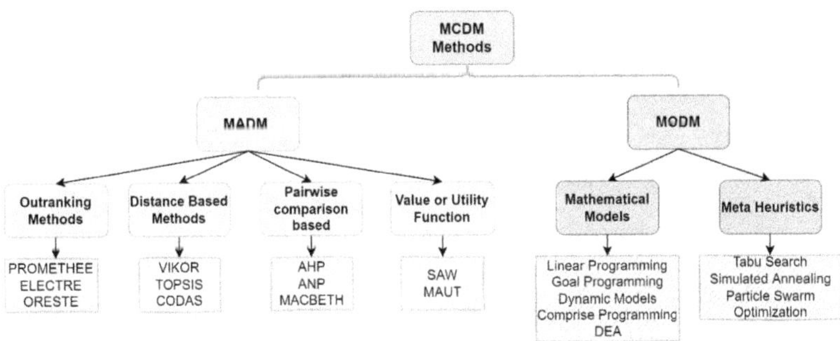

Arslan (2018) is defined hybrid methods in addition to basic methods (such as goal programming, integer, or linear programming) and single analytic methods (such as AHP, SMART, DAE). Some examples of Hybrid MCDM are AHP-VIKOR, fuzzy-TOPSIS.

Mutikanga (2012) summarized the weaknesses and strengths of MCDM methods. Different methods can give different results for the same problem. A multi-criteria problem is uncertain or ill-defined. Such that an action can be better than another action regarding one criterion and worse as regards another. In the process, subjective information is used. Information loss happens such as alternatives are generally defined as a single abstract value can reason that. During the problem structuring, adding or disregarding some options or criteria may manipulate results. The scaling of variables and requirements for additional information should be considered carefully. It is robust to represent and quantify the performance of options using a single value (Zarderi et al., 2015).

BACKGROUND

Bioenergy project developers need to make decisions that include many criteria to implement a project. Some of these decisions are how the biomass raw material will be stored and transported, which technologies will be used, the capacity and financing of the project to be made, etc. Using the ScienceDirect, Scopus, ProQuest, and Google scholar databases, studies conducted between 2010 and 2021 years were reviewed. While searching the databases, keywords considered necessary for the subject were used. First, the keyword "biomass" was used. Then, in order to keep the study broad, studies on byproducts such as biodiesel, biofuel, and biogas produced from biomass were also considered. The used search terms are presented in Table 2.

Table 2. Used search terms

	Search terms used for literature	Search terms not used
Methods	MCDM, multi-criteria decision making	Decision analysis, decision making
Application	Biomass, biofuel, biogas, biodiesel, bioenergy	Renewable energy, waste, RES

A pool containing 2678 articles is created, and 150 of these articles, in which multi-criteria decision-making methods are used, are selected for literature review. These studies are tabulated by considering the topics they cover and the methods they use. The literature survey on biomass and MCDM is given in Table 3.

Table 3. Summary of studies made between 2010-2021 years

# of the Study	Name of the Study	Subject	Used MCDM in the Study
1.	Wang et al., 2021	Selecting biomass furnace suppliers.	Fuzzy Analytic Hierarchy Process (FAHP)-Combined Compromise Solution (CoCoSo) algorithm
2.	Unay et al., 2021	Selecting the most suitable growth and harvesting method for microalgal biomass.	TOPSIS (Technique for Order Preference by Similarity to Ideal Solution)
3.	Tazzit et al., 2021	Selecting a biomass pelleting processing depot.	Preference Index Selection Method (PSI)-Grey relational analysis (GRA),
4.	Sunil Chaitanya et al., 2021	Analyze and rank the different blends of biodiesel from the engine performance data.	AHP-TOPSIS
5.	Shahraki Shahdabadi et al., 2021	Selecting the best biomass power plant location in Iran.	SAW, TOPSIS, ELECTRE
6.	Rani et al., 2021	Selecting appropriate biomass technology for agricultural residues.	The Pythagorean Fuzzy Set (PFS)-Weighted Discrimination-based Approximation (WDBA)
7.	Ossei-Bremang & Kemausuor, 2021	Selection of appropriate biomass resources for bioenergy production.	AHP-Fuzzy TOPSIS
8.	Narwane et al., 2021	Ranking of challenges in biofuel industry in India.	Interpretive Structural Modelling (ISM)-DEMATEL, MICMAC analysis
9.	Nantasaksiri et al., 2021	Selecting of biogas power plant location in Thailand.	AHP
10.	Najafi et al., 2021	Selection of biodiesel production location in Iran.	Additive Ratio Assessment Method (ARAS)
11.	Mukeshimana et al., 2021	Ranking of the barriers for biogas dissemination in Rwanda	AHP
12.	Liano et al., 2021	Assessing biogas plants in Reykjavik, Iceland.	Weighted Summation Method (WSM)
13.	Li et al., 2021	Assessing biomass gasification.	The Linear Programming Technique for Multidimensional Analysis of Preference (LINMAP)
14.	Kokkinos et al., 2021	Optimization of microalgal biomass feedstock selection for biofuel production.	FAHP, Fuzzy TOPSIS, Fuzzy Cognitive Mapping (FCM)
15.	Kheybari et al., 2021	Measuring the importance of the selection criteria for biofuel production technology.	Best–Worst Method (BWM)
16.	Kheybari et al., 2021	Location selection for bioethanol production.	BWM-PROMETHEE II
17.	Júnior et al., 2021	Selecting biomass feedstock to produce biogas.	AHP-TOPSIS
18.	Guler, et al., 2021	Selecting the location of biomass facility in Turkey.	BWM
19.	González-Cruz et al., 2021	Assessment of various biorefinery eco-design alternatives to select the optimal pathway to produce biodiesel.	TOPSIS, AHP, M-TOPSIS, Fuzzy AHP

continues on following page

Table 3. Continued

# of the Study	Name of the Study	Subject	Used MCDM in the Study
20.	George et al., 2021	Selection of biomass feedstock for gasification technology.	AHP-TOPSIS
21.	Galang et al., 2021	Analyzing sustainability of GIS-based biomass energy in Cebu.	AHP
22.	Firouzi et al., 2021	Selection of appropriate biomass resources for biofuel production.	TOPSIS-ARAS-WASPAS
23.	Boafo-Mensah et al., 2021	Assessing the performance potential of biomass cookstoves.	Entropy-TOPSIS
24.	Barry et al., 2021	Analyzing the barriers to biomass gasification in Burkina Faso.	AHP
25.	Abdel-Basset et al., 2021	Selecting the most suitable sustainable bioenergy technology.	DEMATEL-EDAS-TNNs
26.	Zhou et al., 2020	Ranking risk for biomass power generation projects.	Picture Fuzzy Set (PFS)-Entropy Weight Method (EWM),VIKOR
27.	Yeo et al., 2020	Prioritizing parameters for multi-objective optimization of the biomass supply network.	Principal Component Analysis (PCA) and Analytic Network Process (ANP)
28.	Shin & Zul, 2020	Selecting the most feasible biomass resource for bioethanol generation.	AHP
29.	Wannasiri, 2020	Analyzing land suitability for a biomass power plant in Thailand.	Fuzzy AHP
30.	Vlachokostas et al., 2020	Selecting biowaste treatment for producing bioenergy.	ELECTRE III
31.	Su et al., 2020	Evaluating the performance of four primary crop straw energy utilization methods in the eastern Chinese province of Jiangsu.	Fuzzy AHP
32.	Sivaraja et al., 2020	Selecting the optimum bio-diesel fuel blend.	Fuzzy TOPSIS-Fuzzy VIKOR
33.	Rasheed et al., 2020	Evaluating the sustainability of bioenergy projects in South Asia.	Simple Multi-Attribute Rating Techniqu (SMART)
34.	Rahemi et al., 2020	Designing bioethanol supply chain network.	PROMETHEE II
35.	Pehlken et al., 2020	Selection of more sustainable bioenergy.	PROMETHEE II
36.	Omrani et al., 2020	Selection of pretreatment process in a biofuel production line.	ANP
37.	Naeini et al., 2020	Assessing biodiesel production from biomass feedstock.	SWARA- Fuzzy Cognitive Map (FCM)-WASPAS
38.	Mostafaeipour et al., 2020	Evaluating location for bioethanol production.	VIKOR, TOPSIS, SAW
39.	Mojaver et al., 2020	Evaluating biomass fueled power generation system.	Entropy method-VIKOR
40.	Mishra et al., 2020	Assessment of the bioenergy production process.	SWARA-COPRAS
41.	Madhu et al., 2020	Selection of a suitable biomass material for maximum bio-oil yield during pyrolysis.	FAHP, TOPSIS, VIKOR, EDAS, PROMETHEE II
42.	Hong & Mwakalonge, 2020	Assessing biofuel logistics network.	DEA

continues on following page

Table 3. Continued

# of the Study	Name of the Study	Subject	Used MCDM in the Study
43.	Dhanalakshmi et al., 2020	Selecting pyrolysis material.	FAHP-TOPSIS-EDAS (Distance from Average Solution)
44.	Costa et al., 2020	Assessing the location of bioenergy plants.	AHP
45.	Wu et al., 2019	Assessing location decision for biomass cogeneration project.	MABAC, SAW, TOPSIS
46.	Wang et al., 2019	Selecting the location of biomass energy power plants.	FAHP-TOPSIS
47.	Schröder et al., 2019	Assessing bioenergy pathways.	PROMETHEE
48.	Rodrigues et al., 2019	Assessing the location of a biogas plant.	AHP
49.	Rentizelas et al., 2019	Assessing biomass supply chain pathways.	DEA
50.	Nahand, 2019	Selecting appropriate biomass products for biofuel generation.	BWM-PROMETHEE
51.	Martinkus et al., 2019	Biorefinery siting.	Weighted Sum Method (WSM)
52.	Lu et al., 2019	Supplier selection for the biomass industry.	Fuzzy AHP
53.	Lerche et al., 2019	Providing decision support for bioenergy projects.	PROMETHEE
54.	Siamak Kheybari et al., 2019	Assessing energy production technologies.	AHP
55.	Ghose et al., 2019	Selecting a suitable biomass facility location in Sikkim.	AHP
56.	Curiel-Esparza et al., 2019	Selecting the biogas desulfurization technique.	Fuzzy set theory VIKOR
57.	Bastan et al., 2019	Selecting biomass product for biofuel generation.	BWM-PROMETHEE
58.	Arabi et al., 2019	Assessing biomass product supply chain for biofuel generation.	DEA
59.	Ali & Waewsak, 2019	Selecting the location of a biomass power plant.	AHP
60.	Zhang et al., 2018	Assessing performances of biomass technologies.	PROMETHEE
61.	Xiang et al., 2018	Assessing energy performances biomass feedstock for bioenergy production.	Entropy method-AHP-GRA
62.	Woo et al., 2018	Assessing the location of biomass energy facilities.	AHP
63.	Wheeler et al., 2018	Designing biomass supply chain network.	SMART, SWING, AHP, TRADE OF
64.	Sanaei et al., 2018	Assessing biorefinery strategies.	Multi attribute utility theory (MAUT)
65.	Memona et al., 2018	Analyzing power generation from biomass.	AHP
66.	Meidiana et al., 2018	Assessing the location of anaerobic digester location.	AHP
67.	Martinkus et al., 2018	Assessing facility siting biorefinery supply chain method.	Weighted Sum Method (WSM)
68.	Madhu et al., 2018	Selecting biomass feedstock for bio-oil.	Fuzzy AHP-TOPSIS
69.	Jeong, 2018	Biomass feedstock for biomass power facilities.	F-DEMATEL-SAW

continues on following page

Table 3. Continued

# of the Study	Name of the Study	Subject	Used MCDM in the Study
70.	Ioannou et al., 2018	Assessing the location of biomass energy production.	Fuzzy system-AHP
71.	Geri et al., 2018	Evaluation of bioenergy planning.	AHP
72.	Gandhi et al., 2018	Biogas production from waste.	VIKOR
73.	Davtalab & Alesheikh, 2018	Assessing the biomass power plant location.	Fuzzy ANP- Weighted Linear Combination (WLC)
74.	Andiloro et al., 2018	Assessing alternative feeds for biogas plants.	AHP-SAW
75.	Schillo et al., 2017	Assessing biofuels policies.	QFD- AHP
76.	Rodríguez et al., 2017	Assessing the allocation of bioenergy plants.	Fuzzy AHP
77.	Quinta-Nova et al., 2017	Assessing forest biomass energy potential.	AHP
78.	Priyanka & Rajneesh, 2017	Optimal biomass usage.	Fuzzy VIKOR
79.	Moulogianni & Bournaris, 2017	Ranking of agro-energy regions.	ELECTRE III
80.	Mostafaeipour et al., 2017	Assessing the location of plant planning for bioethanol.	DEA
81.	Khishtandar et al., 2017	Assessing the technologies for bioenergy production.	Multi Actor Multi Criteria Outranking Method (MAMCA)
82.	Jeong & Ramírez-Gómez, 2017	Optimizing biomass facility sites.	Fuzzy logic
83.	Hamid et al., 2017	Assessing biorefinery feedstock for biomass energy.	AHP
84.	den Herder et al., 2017	Evaluating land usage for bioenergy targets.	Simple Multi-Attribute Rating Technique (SMART)
85.	de Clercq et al., 2017	Evaluating performances of different waste for biogas projects.	SWING
86.	Babazadeh et al., 2017	Designing biodiesel supply chain network.	DEA
87.	Anish Kumar et al., 2017	Assessing biodiesel production method.	Fuzzy AHP
88.	Ubando et al., 2016	Selecting appropriate biofuel feedstock.	AHP
89.	Ubando et al., 2016	Evaluating biofuel production feedstock.	AHP
90.	Sacchelli & Cipollaro, 2016	Evaluating bioenergy chain perception.	Compromise Programming (CP)
91.	Rupf et al., 2016	Assessing biogas system design.	TOPSIS
92.	Pezdevšek Malovrh et al., 2016	Assessing forest biomass.	SMART
93.	Khang et al., 2016	Selecting appropriate biodiesel feedstock.	AHP
94.	Gautam & LeBel, 2016	Locating a terminal in bioenergy supply chains.	AHP
95.	Djaković et al., 2016	Evaluating the use of biomass.	SWOT-AHP, SWOT-ANP
96.	Cutz et al., 2016	Assessing biomass technologies and sources.	Fuzzy Multi-Actor Multi-Criteria Decision-Making (FMAMCDM)

continues on following page

Table 3. Continued

# of the Study	Name of the Study	Subject	Used MCDM in the Study
97.	Cebi et al., 2016	Selecting suitable locations for biomass power plants.	Fuzzy logic
98.	Billig & Thrän, 2016	Evaluation of appropriate biomethane technologies.	AHP
99.	Baltazar et al., 2016	Selecting suitable locations for biomass power plants.	AHP
100.	Yadav et al., 2015	Selecting appropriate biomass energy sources.	AHP
101.	Sakthivel et al., 2015	Selecting appropriate biodiesel blend.	ANP-TOPSIS, ANP-VIKOR
102.	Ren et al., 2015	Assessing bioethanol production pathways.	AHP-VIKOR
103.	Parajuli et al., 2015	Selecting feedstock for biorefineries.	PROMETHEE-AHP
104.	Nwokoagbara et al., 2015	Selecting bioenergy feedstock for biodiesel.	AHP, WSM, WPM, DCP, TOPSIS
105.	Lewis et al., 2015	Selecting bioenergy feedstock.	Fuzzy logic
106.	Galvez et al., 2015	Designing reverse logistics network for a biogas plant.	AHP
107.	Franco et al., 2015	Selecting suitable locations for biogas plants.	AHP
108.	Delivand et al., 2015	Selecting the optimal location of bioenergy facilities in south Italy.	AHP
109.	Cobuloglu & Büyüktahtakin, 2015	Selecting sustainable biomass crop.	AHP
110.	Arce, 2015	Determining influences of parameter variability on biomass selection.	GRA
111.	Ahmad et al., 2015	Selecting microorganisms for the production of oils for biodiesel production.	AHP-PROMETHEE
112.	Ziolkowska, 2014	Optimizing biofuels production.	Fuzzy logic-PROMETHEE
113.	Yan & Tao, 2014	Evaluating the efficiency of a biomass power generation industry.	DEA
114.	von Doderer & Kleynhans, 2014	Assessing sustainable bioenergy system.	AHP
115.	Ubando et al., 2014	Evaluating the suitable cultivation system for sustainable production of algal biofuels.	AHP
116.	Silva et al., 2014	Selecting the location of a biogas plant in a Portuguese region.	ELECTRE
117.	Sakthivel et al., 2014	Selecting biodiesel blend selection.	GRA-TOPSIS
118.	Saelee et al., 2014	Selecting biomass type for boilers.	TOPSIS
119.	Recanatesi et al., 2014	Selecting the location of a biomass power plant.	Fuzzy logic
120.	Rao et al., 2014	Analyzing alternative biogas technologies.	AHP
121.	Pastare et al., 2014	Using sustainable macro-algae for biogas production.	TOPSIS
122.	Okello et al., 2014	Appraising bioenergy alternatives in Uganda.	AHP
123.	Madugu & Collu, 2014	Analyzing microalgae cultivation for biofuels.	TOPSIS
124.	Kigozi et al., 2014	Selecting of biogas digesters technology.	SMART

continues on following page

Table 3. Continued

# of the Study	Name of the Study	Subject	Used MCDM in the Study
125.	Durairaj et al., 2014	Selection of biodiesel for a power generator.	FAHP-GRA-TOPSIS
126.	Cobuloglu & Büyüktahtakin, 2014	Determining criteria for decision-makers and farmers to select the crop type in biomass production.	Fuzzy AHP
127.	Ziolkowska, 2013	Evaluating the sustainability of biofuels feedstock.	PROMETHEE
128.	Šiškova, 2013	Risk analyzing in a biogas plant.	PROMETHEE
129.	Scott et al., 2013	Supplier selection for biomass schemes.	QFD–AHP
130.	Sakthivel et al., 2013	Selection of best biodiesel blend for engines.	FAHP-TOPSIS FAHP-VIKOR
131.	Ren et al., 2013	Assessing biomass-based technologies for hydrogen production.	Fuzzy Multi-actor Multi-criteria Decision Making (FMAMCDM)
132.	Perpiña et al., 2013	Selecting the location of biomass plants.	Simple Additive Weighting (SAW), Ideal Point Method (IPM)
133.	Myllyviita et al., 2013	Assessing sustainability wood-based bioenergy production in eastern Finland.	AHP
134.	Kurka, 2013	Evaluating the regional sustainability of bioenergy developments.	AHP
135.	de Carlo & Schiraldi, 2013	Selecting the location of a biomass plant in Tuscany.	ANP
136.	Darshini et al., 2013	Capturing stakeholder views in biofuel and biomass utilization in Malaysia.	SWOT-AHP
137.	Balezentiene et al., 2013	Selection of sustainable energy crop.	Fuzzy MULTIMOORA
138.	Zubaryeva et al., 2012	Assessment of local biomass availability for distributed biogas production.	AHP
139.	Zhou et al., 2012	Selecting the location of a biofuel refinery.	Fuzzy TOPSIS
140.	van Dael et al., 2012	Determine potential locations for biomass valorization in a specified region.	AHP, Multi-Attribute Value Theory (MAVT)
141.	Sultana & Kumar, 2012	Ranking of biomass pellets.	PROMETHEE
142.	Myllyviita et al., 2012	Process of assessing environmental impacts of two alternative raw materials in the biomass supply chain.	Simple Multi-Attribute Rating Technique (SMART)
143.	Turcksin et al., 2011	Assessing several biofuel options for Belgium.	Multi-Actor Multi-Criteria Analysis (MAMCA)
144.	Smyth et al., 2011	Assessing the potential of a grass biogas/ biomethane industry are identified and analyzed.	Weighted Sum Model (WSM)
145.	Rao & Baral, 2011	Selecting of feedstock for biogas production.	TOPSIS
146.	Halog, 2011	Analyzing biofuel systems in pursuit of sustainable large-scale production.	AHP
147.	Barin et al., 2011	Selecting hybrid energy systems fueled by biogas.	Fuzzy logic
148.	Volkova et al., 2010	Selection of the location of wood fuel-based cogeneration plants in Estonia.	AHP

continues on following page

Table 3. Continued

# of the Study	Name of the Study	Subject	Used MCDM in the Study
149.	Vindiš et al., 2010	Assessing energy crops for biogas production.	AHP
150.	Herrera-Seara et al., 2010	Selecting biomass power plant location.	AHP

According to the literature survey, AHP and Fuzzy AHP are the most applied methods in the biomass literature survey; 70 out of 150 studies that are presented in Table 3 applied AHP and Fuzzy AHP in decision making for biomass applications. These methods are followed by TOPSIS, Fuzzy TOPSIS, PROMETHEE I & II, VIKOR, Fuzzy VIKOR, ANP, Fuzzy ANP, DEA, SMART, GRA, WSM. The number of studies and frequency information on related methods is given in Table 4.

Table 4. The mostly applied MCDM methods in the biomass literature survey

Method	# of studies	%
AHP, Fuzzy AHP	70	46,6
TOPSIS, Fuzzy TOPSIS	27	18
PROMETHEE I, II	15	10
VIKOR, Fuzzy VIKOR	11	7,3
ANP, Fuzzy ANP	6	4
DEA	6	4
SMART	6	4
GRA	5	3,3
WSM	5	3,3

Summary of application areas based on literature survey is presented in Table 5. The most applied areas of MCDM methods in biomass are technical issues, location selection, technology selection, respectively. A brief explanation of application areas is given as follow:

Political or legal issues: Any studies that consider political or legal issues in biomass energy processes in related areas.

Technical issues: Any studies try to enhance the conversion or process efficiency of biomass conversion technologies.

Economic issues: Any studies aim to solve market difficulties, economic issues or increase the cost-effectiveness of biomass processes through financial models.

Environmental issues: Any studies that focus on decreasing the environmental impact of biomass processes.

Social issues: Any studies that consider the positive and negative effects of biomass processes in terms of social issues.

Operational issues: Any studies that evaluate problems of biomass operations and organization on a day to-day basis or planning stage. These problems include the flow of materials and logistic processes.

Location: Any studies that try to decide on location alternatives to build biomass-related facilities.

Technology selection: Any studies that consider the proper technology selection.

Capacity: Any studies that concentrate on capacity-related problems.

Table 5. Summary of application areas based on biomass literature survey

Category of application	# of studies	%
Political or legal issues	26	17,3
Technical issues	66	44
Economic issues	44	29,3
Environmental issues	34	22,7
Social issues	8	5,3
Operational	44	29,3
Location selection	53	35,3
Technology selection	45	30
Capacity	16	10,7

FUTURE RESEARCH DIRECTIONS

Literature studies are important in terms of shedding light on future studies. In this study, bioenergy studies and methods used in the last ten years are tabulated. As can be seen from the table, one of the MCDM methods or hybrid approaches was used in the studies. In future studies, different hybrid methods (such as Machine Learning or Stochastic Algorithms that can be used with MCDM methods in decision making) can be used to fill the gap in the literature in accordance with the study subject. In the studies examined, it was seen that the studies on the choice of place and technology were dominant. Biomass energy production takes place in a large supply chain that includes the production, collection, transportation and conversion of the biomass source to be used. There are many stakeholders interacting within this supply chain. These stakeholders make decisions that need to be addressed from multiple perspectives. MCDMs make important contributions to these decisions.

CONCLUSION

Biomass is currently the most common source of renewable energy and its utilization is further increasing because of the negative impacts of fossil fuel consumption. This study presents a literature review based on the examination of articles related to biomass energy, which is one of the renewable energy types and using MCDM methods. A total of 150 papers are included in the study and reviewed papers published between 2010-2021. The literature survey reveals the mostly used MCDM methods and application areas, which helps decision-makers and researchers to keep up with trend in biomass.

REFERENCES

Abdel-Basset, M., Gamal, A., Chakrabortty, R. K., & Ryan, M. (2021). Development of a hybrid multi-criteria decision-making approach for sustainability evaluation of bioenergy production technologies: A case study. *Journal of Cleaner Production*, *290*, 125805. doi:10.1016/j.jclepro.2021.125805

Ahmad, F. B., Zhang, Z., Doherty, W. O. S., & O'Hara, I. M. (2015). A multi-criteria analysis approach for ranking and selection of microorganisms for the production of oils for biodiesel production. *Bioresource Technology*, *190*, 264–273. doi:10.1016/j.biortech.2015.04.083 PMID:25958151

Ali, S., & Waewsak, J. (2019). *GIS-MCDM approach to scrutinize the suitable sites for a biomass power plant in southernmost provinces of Thailand*. Academic Press.

Andiloro, S., Romeo, G., Marciano, C., Zimbone, S. M., & Zema, D. A. (2018). A local multi-criteria assessment of alternative feeds for biogas plants in Calabria (Southern Italy). *Chemical Engineering Transactions*, *65*, 859–864. doi:10.3303/CET1865144

Anish Kumar, K., Senthil Kumar, P., Madhusudanan, S., Pasupathy, V., Vignesh, P. R., & Sankaranarayanan, A. R. (2017). A simplified model for evaluating best biodiesel production method: Fuzzy analytic hierarchy process approach. *Sustainable Materials and Technologies*, *12*, 18–22. doi:10.1016/j.susmat.2017.03.002

Arabi, M., Yaghoubi, S., & Tajik, J. (2019). Algal biofuel supply chain network design with variable demand under alternative fuel price uncertainty: A case study. *Computers & Chemical Engineering*, *130*, 106528. doi:10.1016/j.compchemeng.2019.106528

Arce, M. E. (2015). The influence of parameters variability on biomass selection for energy use. *Energy, Sustainability and Society*, *5*(1), 1–10. doi:10.118613705-015-0042-z

Babazadeh, R., Razmi, J., Rabbani, M., & Pishvaee, M. S. (2017). An integrated data envelopment analysis–mathematical programming approach to strategic biodiesel supply chain network design problem. *Journal of Cleaner Production*, *147*, 694–707. doi:10.1016/j.jclepro.2015.09.038

Balezentiene, L., Streimikiene, D., & Balezentis, T. (2013). Fuzzy decision support methodology for sustainable energy crop selection. *Renewable & Sustainable Energy Reviews*, *17*, 83–93. doi:10.1016/j.rser.2012.09.016

Baltazar, B. M., Remolador, M., Sevilla, K. H., Saladaga, I., Ang, M. R. C. O., & Inocencio, L. C. V. (2016). *Locating potential site for biomass power plant development in Central Luzon, Philippines using LANDSAT-based suitability map*. Academic Press.

Barin, A., Canha, L. N., Magnago, K. M., Matos, M. A., & Wottrich, B. (2011). A novel fuzzy-based methodology for biogas fuelled hybrid energy systems decision making. In K. Gopalakrishnan, S. K. Khaitan, & S. Kalogirou (Eds.), Studies in Fuzziness and Soft Computing (Vol. 269, pp. 183–198). Academic Press.

Barry, F., Sawadogo, M., Bologo, M., Ouédraogo, I. W. K., & Dogot, T. (2021). Key barriers to the adoption of biomass gasification in Burkina Faso. *Sustainability (Switzerland)*, *13*(13). Advance online publication. doi:10.3390u13137324

Bastan, M., Nahand, P. K., Korlou, S., & Hamid, M. (2019). *Selection of a biomass product using a hybrid approach of BW-PROMETHEE*. Academic Press.

Billig, E., & Thrän, D. (2016). Evaluation of biomethane technologies in Europe – Technical concepts under the scope of a Delphi-Survey embedded in a multi-criteria analysis. *Energy*, *114*, 1176–1186. doi:10.1016/j.energy.2016.08.084

Boafo-Mensah, G., Neba, F. A., Tornyeviadzi, H. M., Seidu, R., Darkwa, K. M., & Kemausuor, F. (2021). Modelling the performance potential of forced and natural-draft biomass cookstoves using a hybrid Entropy-TOPSIS approach. *Biomass and Bioenergy*, *150*, 106106. doi:10.1016/j.biombioe.2021.106106

Cebi, S., Ilbahar, E., & Atasoy, A. (2016). A fuzzy information axiom based method to determine the optimal location for a biomass power plant: A case study in Aegean Region of Turkey. *Energy*, *116*, 894–907. doi:10.1016/j.energy.2016.10.024

Cobuloglu, H. I., & Büyüktahtakin, I. E. (2014). *A multi-criteria approach for biomass crop selection under fuzzy environment*. Academic Press.

Cobuloglu, H. I., & Büyüktahtakin, I. E. (2015). A stochastic multi-criteria decision analysis for sustainable biomass crop selection. *Expert Systems with Applications*, *42*(15-16), 6065–6074. doi:10.1016/j.eswa.2015.04.006

Costa, F. R., Ribeiro, C. A. A. S., Marcatti, G. E., Lorenzon, A. S., Teixeira, T. R., Domingues, G. F., Castro, N. L. M., Santos, A. R., Soares, V. P., Menezes, S. J. M. C., Mota, P. H. S., Telles, L. A. A., & Carvalho, J. R. D. (2020). GIS applied to location of bioenergy plants in tropical agricultural areas. *Renewable Energy*, *153*, 911–918. doi:10.1016/j.renene.2020.01.050

Curiel-Esparza, J., Reyes-Medina, M., Martin-Utrillas, M., Martinez-Garcia, M. P., & Canto-Perello, J. (2019). Collaborative elicitation to select a sustainable biogas desulfurization technique for landfills. *Journal of Cleaner Production*, *212*, 1334–1344. doi:10.1016/j.jclepro.2018.12.095

Cutz, L., Haro, P., Santana, D., & Johnsson, F. (2016). Assessment of biomass energy sources and technologies: The case of Central America. *Renewable & Sustainable Energy Reviews*, *58*, 1411–1431. doi:10.1016/j.rser.2015.12.322

Darshini, D., Dwivedi, P., & Glenk, K. (2013). Capturing stakeholders´ views on oil palm-based biofuel and biomass utilisation in Malaysia. *Energy Policy*, *62*, 1128–1137. doi:10.1016/j.enpol.2013.07.017

Davtalab, M., & Alesheikh, A. A. (2018). Spatial optimization of biomass power plant site using fuzzy analytic network process. *Clean Technologies and Environmental Policy*, *20*(5), 1033–1046. doi:10.100710098-018-1531-5

de Carlo, F., & Schiraldi, M. M. (2013). *Sustainable choice of the location of a biomass plant: An application in Tuscany*. Academic Press.

de Clercq, D., Wen, Z., & Fan, F. (2017). Performance evaluation of restaurant food waste and biowaste to biogas pilot projects in China and implications for national policy. *Journal of Environmental Management*, *189*, 115–124. doi:10.1016/j.jenvman.2016.12.030 PMID:28012386

Delivand, M. K., Cammerino, A. R. B., Garofalo, P., & Monteleone, M. (2015). Optimal locations of bioenergy facilities, biomass spatial availability, logistics costs and GHG (greenhouse gas) emissions: A case study on electricity productions in South Italy. *Journal of Cleaner Production*, *99*, 129–139. doi:10.1016/j.jclepro.2015.03.018

den Herder, M., Kurttila, M., Leskinen, P., Lindner, M., Haatanen, A., Sironen, S., Salminen, O., Juusti, V., & Holma, A. (2017). Is enhanced biodiversity protection conflicting with ambitious bioenergy targets in eastern Finland? *Journal of Environmental Management*, *187*, 54–62. doi:10.1016/j.jenvman.2016.10.065 PMID:27883939

Dhanalakshmi, C. S., Madhu, P., Karthick, A., Mathew, M., & Vignesh Kumar, R. (2020). A comprehensive MCDM-based approach using TOPSIS and EDAS as an auxiliary tool for pyrolysis material selection and its application. *Biomass Conversion and Biorefinery*. Advance online publication. doi:10.100713399-020-01009-0

Djaković, D. D., Gvozdenac-Urošević, B. D., & Vasić, G. M. (2016). Multi-criteria analysis as a support for national energy policy regarding the use of biomass: Case study of Serbia. *Thermal Science*, *20*(2), 371–380. doi:10.2298/TSCI150602190D

Durairaj, S., Sathiya Sekar, K., & Ilangkumaran, M., RamManohar, M., Thyalan, B., Yuvaraj, E., & Ramesh, S. (2014). Multi-criteria decision model for biodiesel selection in an electrical power generator based on FAHP-GRA-TOPSIS. *International Journal of Research in Engineering and Technology*, *3*(23), 226–233. doi:10.15623/ijret.2014.0323050

Firouzi, S., Allahyari, M. S., Isazadeh, M., Nikkhah, A., & Van Haute, S. (2021). Hybrid multi-criteria decision-making approach to select appropriate biomass resources for biofuel production. *The Science of the Total Environment*, *770*, 144449. doi:10.1016/j.scitotenv.2020.144449 PMID:33513499

Franco, C., Bojesen, M., Hougaard, J. L., & Nielsen, K. (2015). A fuzzy approach to a multiple criteria and Geographical Information System for decision support on suitable locations for biogas plants. *Applied Energy*, *140*, 304–315. doi:10.1016/j.apenergy.2014.11.060

Galang, W. N., Tabañag, I. D., & Loretero, M. (2021). GIS-based biomass energy sustainability analysis using analytical hierarchy process: A case study in Medellin, Cebu. *International Journal of Renewable Energy Development*, *10*(3), 551–561. doi:10.14710/ijred.0.33260

Galvez, D., Rakotondranaivo, A., Morel, L., Camargo, M., & Fick, M. (2015). Reverse logistics network design for a biogas plant: An approach based on MILP optimization and Analytical Hierarchical Process (AHP). *Journal of Manufacturing Systems*, *37*, 616–623. doi:10.1016/j.jmsy.2014.12.005

Gandhi, P., Kunwar, P., Pareek, N., Mathur, S., Lizasoain, J., Gronauer, A., ... Vivekanand, V. (2018). Multicriteria Decision Model and Thermal Pretreatment of Hotel Food Waste for Robust Output to Biogas: Case Study from City of Jaipur, India. *BioMed Research International*, *13*. Advance online publication. doi:10.1155/2018/9416249 PMID:30306090

Gautam, S., & LeBel, L. (2016). *A multi-criteria decision making approach to locate a terminal in bioenergy supply chains*. Canadian Institute of Forestry.

George, J., Arun, P., & Muraleedharan, C. (2021). Region-specific biomass feedstock selection for gasification using multi-attribute decision-making techniques. *International Journal of Sustainable Engineering*, *14*(5), 1101–1109. doi:10.1080/19397038.2020.1790058

Geri, F., Sacchelli, S., Bernetti, I., & Ciolli, M. (2018). Urban-rural bioenergy planning as a strategy for the sustainable development of inner areas: A GIS-based method to chance the forest chain. *Green Energy Technology*, *0*, 539–550.

Ghose, D., Naskar, S., & Uddin, S. (2019). *Q-GIS-MCDA based approach to identify suitable biomass facility location in Sikkim*. doi:10.1109/ICACCP.2019.8882978

González-Cruz, L. A., Morales-Mendoza, L. F., Aguilar-Lasserre, A. A., Azzaro-Pantel, C., Martínez-Isidro, P., & Meza-Palacios, R. (2021). Optimal ecodesign selection for biodiesel production in biorefineries through multicriteria decision making. *Clean Technologies and Environmental Policy*, *23*(8), 2337–2356. Advance online publication. doi:10.100710098-021-02141-9

Guler, D., Charisoulis, G., Buttenfield, B. P., & Yomralioglu, T. (2021). Suitability modeling and sensitivity analysis for biomass energy facilities in Turkey. *Clean Technologies and Environmental Policy*, *23*(7), 2183–2199. doi:10.100710098-021-02126-8

Halog, A. (2011). Sustainable development of bioenergy sector: An integrated methodological framework. *International Journal of Multicriteria Decision Making*, *1*(3), 338–361. doi:10.1504/IJMCDM.2011.041193

Hamid, N. N. A., Martinez-Hernandez, E., & Lim, J. S. (2017). Technological screening of algae-based biorefinery for sustainable biofuels production using analytic hierarchy process (AHP) with feature scaling normalisation. *Chemical Engineering Transactions*, *61*, 1369–1374. doi:10.3303/CET1761226

Herrera-Seara, M. A., Aznar Dols, F., Zamorano, M., & Alameda-Hernández, E. (2010). Optimal location of a biomass power plant in the province of granada analyzed by multi-criteria evaluation using appropriate geographic information system according to the analytic hierarchy process. *Renewable Energy and Power Quality Journal*, *1*(8), 813–818. doi:10.24084/repqj08.484

Hong, J. D., & Mwakalonge, J. L. (2020). Biofuel logistics network scheme design with combined data envelopment analysis approach. *Energy*, *209*, 118342. Advance online publication. doi:10.1016/j.energy.2020.118342

Ioannou, K., Tsantopoulos, G., Arabatzis, G., Andreopoulou, Z., & Zafeiriou, E. (2018). A spatial decision support system framework for the evaluation of biomass energy production locations: Case study in the regional unit of drama, greece. *Sustainability*, *10*(2), 531. doi:10.3390u10020531

Jeong, J. S. (2018). Biomass feedstock and climate change in agroforestry systems: Participatory location and integration scenario analysis of biomass power facilities. *Energies*, *11*(6), 1404. Advance online publication. doi:10.3390/en11061404

Jeong, J. S., & Ramírez-Gómez, A. (2017). A multi-criteria GIS-based assessment to optimize biomass facility sites with parallel environment - A case study in Spain. *Energies*, *10*(12), 2095. Advance online publication. doi:10.3390/en10122095

Júnior, E. S., Colmenero, J. C., & Junior, A. B. (2021). Biomass selection method to produce biogas with a multi-criteria approach. *Waste and Biomass Valorization*, *12*(6), 3169–3177. doi:10.100712649-020-01231-x

Khang, D. S., Promentilla, M. A. B., Tan, R. R., Abe, N., Tuan, P. D., & Razon, L. F. (2016). Multi-criteria approach to assess stakeholders preferences for selection of biodiesel feedstock in Vietnam. *International Journal of Business and Systems Research*, *10*(2-4), 306–331. doi:10.1504/IJBSR.2016.075738

Kheybari, S., Javdanmehr, M., Rezaie, F. M., & Rezaei, J. (2021). Corn cultivation location selection for bioethanol production: An application of BWM and extended PROMETHEE II. *Energy*, *228*, 120593. doi:10.1016/j.energy.2021.120593

Kheybari, S., Mahdi Rezaie, F., & Rezaei, J. (2021). Measuring the importance of decision-making criteria in biofuel production technology selection. *IEEE Transactions on Engineering Management*, *68*(2), 483–497. doi:10.1109/TEM.2019.2908037

Kheybari, S., Rezaie, F. M., Naji, S. A., & Najafi, F. (2019). Evaluation of energy production technologies from biomass using analytical hierarchy process: The case of Iran. *Journal of Cleaner Production*, *232*, 257–265. doi:10.1016/j.jclepro.2019.05.357

Khishtandar, S., Zandieh, M., & Dorri, B. (2017). A multi criteria decision making framework for sustainability assessment of bioenergy production technologies with hesitant fuzzy linguistic term sets: The case of Iran. *Renewable & Sustainable Energy Reviews*, *77*, 1130–1145. doi:10.1016/j.rser.2016.11.212

Kigozi, R., Aboyade, A. O., & Muzenda, E. (2014). *Technology selection of biogas digesters for OFMSW via multi-criteria decision analysis*. Academic Press.

Kokkinos, K., Karayannis, V., & Moustakas, K. (2021). Optimizing microalgal biomass feedstock selection for nanocatalytic conversion into biofuel clean energy, using fuzzy multi-criteria decision making processes. *Frontiers in Energy Research*, *8*, 622210. Advance online publication. doi:10.3389/fenrg.2020.622210

Kurka, T. (2013). Application of the analytic hierarchy process to evaluate the regional sustainability of bioenergy developments. *Energy*, *62*, 393–402. doi:10.1016/j.energy.2013.09.053

Leiche, N., Wilkens, I., Schmehl, M., Eigner-Thiel, S., & Geldermann, J. (2019). Using methods of Multi-Criteria Decision Making to provide decision support concerning local bioenergy projects. *Socio-Economic Planning Sciences*, *68*, 100594. doi:10.1016/j.seps.2017.08.002

Lewis, S. M., Gross, S., Visel, A., Kelly, M., & Morrow, W. (2015). Fuzzy GIS-based multi-criteria evaluation for US Agave production as a bioenergy feedstock. *Global Change Biology. Bioenergy*, *7*(1), 84–99. doi:10.1111/gcbb.12116

Li, X., Chen, J., Sun, X., Zhao, Y., Chong, C., Dai, Y., & Wang, C. H. (2021). Multi-criteria decision making of biomass gasification-based cogeneration systems with heat storage and solid dehumidification of desiccant coated heat exchangers. *Energy*, *233*, 121122. Advance online publication. doi:10.1016/j.energy.2021.121122

Liano, T., Dosal, E., Lindorfer, J., & Finger, D. C. (2021). Application of multi-criteria decision-making tools for assessing biogas plants: A case study in reykjavik, Iceland. *Water (Basel)*, *13*(16), 2150. doi:10.3390/w13162150

Lu, Z., Sun, X., Wang, Y., & Xu, C. (2019). Green supplier selection in straw biomass industry based on cloud model and possibility degree. *Journal of Cleaner Production, 209*, 995–1005. doi:10.1016/j.jclepro.2018.10.130

Madhu, P., Dhanalakshmi, C. S., & Mathew, M. (2020). Multi-criteria decision-making in the selection of a suitable biomass material for maximum bio-oil yield during pyrolysis. *Fuel, 277*, 118109. doi:10.1016/j.fuel.2020.118109

Madhu, P., Nithiyesh Kumar, C., Anojkumar, L., & Matheswaran, M. (2018). Selection of biomass materials for bio-oil yield: A hybrid multi-criteria decision making approach. *Clean Technologies and Environmental Policy, 20*(6), 1377–1384. doi:10.100710098-018-1545-z

Madugu, F., & Collu, M. (2014). *Techno-economic modelling analysis of microalgae cultivation for biofuels and co-products*. doi:10.2495/EQ141022

Martinkus, N., Latta, G., Brandt, K., & Wolcott, M. (2018). A multi-criteria decision analysis approach to facility siting in a wood-based depot-and-biorefinery supply chain model. *Frontiers in Energy Research, 6*(NOV), 124. Advance online publication. doi:10.3389/fenrg.2018.00124

Martinkus, N., Latta, G., Rijkhoff, S. A. M., Mueller, D., Hoard, S., Sasatani, D., Pierobon, F., & Wolcott, M. (2019). A multi-criteria decision support tool for biorefinery siting: Using economic, environmental, and social metrics for a refined siting analysis. *Biomass and Bioenergy, 128*, 105330. doi:10.1016/j.biombioe.2019.105330

Meidiana, C., Nurfitriya, I. D., & Sari, K. E. (2018). Multi-criteria evaluation for determination of anaerobic di-gester location in rural area. *International Journal of Recent Technology and Engineering, 7*(4), 153–157.

Memona, L. R., Harijana, K., Mirjata, N. H., & Nixonb, J. D. (2018). *A multi-criteria analysis of options for power generation from biomass in Pakistan*. Academic Press.

Mishra, A. R., Rani, P., Pandey, K., Mardani, A., Streimikis, J., Streimikiene, D., & Alrasheedi, M. (2020). Novel multi-criteria intuitionistic fuzzy SWARA-COPRAS approach for sustainability evaluation of the bioenergy production process. *Sustainability (Switzerland), 12*(10), 4155. Advance online publication. doi:10.3390u12104155

Mojaver, P., Khalilarya, S., & Chitsaz, A. (2020). Multi-objective optimization and decision analysis of a system based on biomass fueled SOFC using couple method of entropy/VIKOR. *Energy Conversion and Management, 203*, 112260. doi:10.1016/j.enconman.2019.112260

Mostafaeipour, A., Sarikhani, S., Sedaghat, A., & Arabnia, H. R. (2017). *Location planning of bioethanol plants from agricultural crop residues for fuel cells using DEA*. Academic Press.

Mostafaeipour, A., Sedaghat, A., Hedayatpour, M., & Jahangiri, M. (2020). Location planning for production of bioethanol fuel from agricultural residues in the south of Caspian Sea. *Environmental Development*, *33*, 100500. doi:10.1016/j.envdev.2020.100500

Moulogianni, C., & Bournaris, T. (2017). Biomass production from crops residues: Ranking of agro-energy regions. *Energies*, *10*(7), 1061. Advance online publication. doi:10.3390/en10071061

Mukeshimana, M. C., Zhao, Z. Y., Ahmad, M., & Irfan, M. (2021). Analysis on barriers to biogas dissemination in Rwanda: AHP approach. *Renewable Energy*, *163*, 1127–1137. doi:10.1016/j.renene.2020.09.051

Myllyviita, T., Holma, A., Antikainen, R., Lähtinen, K., & Leskinen, P. (2012). Assessing environmental impacts of biomass production chains-Application of life cycle assessment (LCA) and multi-criteria decision analysis (MCDA). *Journal of Cleaner Production*, *29-30*, 238–245. doi:10.1016/j.jclepro.2012.01.019

Myllyviita, T., Leskinen, P., Lähtinen, K., Pasanen, K., Sironen, S., Kähkönen, T., & Sikanen, L. (2013). Sustainability assessment of wood-based bioenergy - A methodological framework and a case-study. *Biomass and Bioenergy*, *59*, 293–299. doi:10.1016/j.biombioe.2013.07.010

Naeini, M. A., Zandieh, M., Najafi, S. E., & Sajadi, S. M. (2020). Analyzing the development of the third-generation biodiesel production from microalgae by a novel hybrid decision-making method: The case of Iran. *Energy*, *195*, 116895. doi:10.1016/j.energy.2020.116895

Nahand, P. K. (2019). *Selection of a biomass product using a Hybrid Approach of BW-PROMETHEE*. Academic Press.

Najafi, F., Sedaghat, A., Mostafaeipour, A., & Issakhov, A. (2021). Location assessment for producing biodiesel fuel from Jatropha Curcas in Iran. *Energy*, *236*, 121446. doi:10.1016/j.energy.2021.121446

Nantasaksiri, K., Charoen-amornkitt, P., & Machimura, T. (2021). Integration of multi-criteria decision analysis and geographic information system for site suitability assessment of Napier grass-based biogas power plant in southern Thailand. *Renewable and Sustainable Energy Transition*, *100011*, 100011. Advance online publication. doi:10.1016/j.rset.2021.100011

Narwane, V. S., Yadav, V. S., Raut, R. D., Narkhede, B. E., & Gardas, B. B. (2021). Sustainable development challenges of the biofuel industry in India based on integrated MCDM approach. *Renewable Energy, 164*, 298–309. doi:10.1016/j.renene.2020.09.077

Nwokoagbara, E., Olaleye, A. K., & Wang, M. (2015). Biodiesel from microalgae: The use of multi-criteria decision analysis for strain selection. *Fuel, 159*, 241–249. doi:10.1016/j.fuel.2015.06.074

Okello, C., Pindozzi, S., Faugno, S., & Boccia, L. (2014). Appraising bioenergy alternatives in Uganda using strengths, weaknesses, opportunities and threats (SWOT)-Analytical hierarchy process (AHP) and a desirability functions approach. *Energies, 7*(3), 1171–1192. doi:10.3390/en7031171

Omrani, K., Safaei, A. S., Paydar, M. M., & Nikzad, M. (2020). Pretreatment process selection in a biofuel production line. *International Journal of Industrial Engineering and Production Research, 31*(1), 51–61. doi:10.22068/ijiepr.31.1.51

Ossei-Bremang, R. N., & Kemausuor, F. (2021). A decision support system for the selection of sustainable biomass resources for bioenergy production. *Environment Systems & Decisions, 41*(3), 437–454. doi:10.100710669-021-09810-6

Parajuli, R., Knudsen, M. T., & Dalgaard, T. (2015). Multi-criteria assessment of yellow, green, and woody biomasses: Pre-screening of potential biomasses as feedstocks for biorefineries. *Biofuels, Bioproducts & Biorefining, 9*(5), 545–566. doi:10.1002/bbb.1567

Pastare, L., Romagnoli, F., Lauka, D., Dzene, I., & Kuznecova, T. (2014). Sustainable use of Macro-Algae for biogas production in Latvian conditions: A preliminary study through an integrated MCA and LCA approach. *Environmental and Climate Technologies, 13*(1), 44–56. doi:10.2478/rtuect-2014-0006

Pathak, B., Chaudhari, S., & Fulekar, M. H. (2013). Biomass-resource for sustainable development. *Int J Adv Res Technol, 2*(6), 271–287.

Pehlken, A., Wulf, K., Grecksch, K., Klenke, T., & Tsydenova, N. (2020). More sustainable bioenergy by making use of regional alternative biomass? *Sustainability (Switzerland), 12*(19), 7849. Advance online publication. doi:10.3390u12197849

Perea-Moreno, M. A., Samerón-Manzano, E., & Perea-Moreno, A. J. (2019). Biomass as renewable energy: Worldwide research trends. *Sustainability, 11*(3), 863. doi:10.3390u11030863

Perpiña, C., Martínez-Llario, J. C., & Pérez-Navarro, Á. (2013). Multicriteria assessment in GIS environments for siting biomass plants. *Land Use Policy*, *31*, 326–335. doi:10.1016/j.landusepol.2012.07.014

Pezdevšek Malovrh, Š., Kurttila, M., Hujala, T., Kärkkäinen, L., Leban, V., Lindstad, B. H., Peters, D. M., Rhodius, R., Solberg, B., Wirth, K., Zadnik Stirn, L., & Krč, J. (2016). Decision support framework for evaluating the operational environment of forest bioenergy production and use: Case of four European countries. *Journal of Environmental Management*, *180*, 68–81. doi:10.1016/j.jenvman.2016.05.021 PMID:27208996

Priyanka & Rajneesh. (2017). *A fuzzy VIKOR model for selection of optimal Biomass usage in India*. Academic Press.

Quinta-Nova, L., Fernandez, P., & Pedro, N. (2017). *GIS-based suitability model for assessment of forest biomass energy potential in a region of Portugal*. Academic Press.

Rahemi, H., Torabi, S. A., Avami, A., & Jolai, F. (2020). Bioethanol supply chain network design considering land characteristics. *Renewable & Sustainable Energy Reviews*, *119*, 109517. Advance online publication. doi:10.1016/j.rser.2019.109517

Rani, P., Mishra, A. R., Saha, A., & Pamucar, D. (2021). Pythagorean fuzzy weighted discrimination-based approximation approach to the assessment of sustainable bioenergy technologies for agricultural residues. *International Journal of Intelligent Systems*, *36*(6), 2964–2990. Advance online publication. doi:10.1002/int.22408

Rao, B., Mane, A., Rao, A. B., & Sardeshpande, V. (2014). *Multi-criteria analysis of alternative biogas technologies*. Academic Press.

Rao, P. V., & Baral, S. S. (2011). Attribute based specification, comparison and selection of feed stock for anaerobic digestion using MADM approach. *Journal of Hazardous Materials*, *186*(2-3), 2009–2016. doi:10.1016/j.jhazmat.2010.12.108 PMID:21247688

Rasheed, R., Javed, H., Rizwan, A., Yasar, A., Tabinda, A. B., Mahfooz, Y., Wang, Y., & Su, Y. H. (2020). Sustainability and CDM potential analysis of a novel vs conventional bioenergy projects in South Asia by multi-criteria decision-making method. *Environmental Science and Pollution Research International*, *27*(18), 23081–23093. doi:10.100711356-020-08862-6 PMID:32333350

Recanatesi, F., Tolli, M., & Lord, R. (2014). Multi criteria analysis to evaluate the best location of plants for renewable energy by forest biomass: A case study in central Italy. *Applied Mathematical Sciences*, *8*(129), 6447–6458. doi:10.12988/ams.2014.46451

Ren, J., Fedele, A., Mason, M., Manzardo, A., & Scipioni, A. (2013). Fuzzy multi-actor multi-criteria decision making for sustainability assessment of biomass-based technologies for hydrogen production. *International Journal of Hydrogen Energy*, *38*(22), 9111–9120. doi:10.1016/j.ijhydene.2013.05.074

Ren, J., Manzardo, A., Mazzi, A., Zuliani, F., & Scipioni, A. (2015). Prioritization of bioethanol production pathways in China based on life cycle sustainability assessment and multi-criteria decision-making. *The International Journal of Life Cycle Assessment*, *20*(6), 842–853. doi:10.100711367-015-0877-8

Rentizelas, A., Melo, I. C., Alves, P. N. Junior, Campoli, J. S., & Aparecida do Nascimento Rebelatto, D. (2019). Multi-criteria efficiency assessment of international biomass supply chain pathways using data envelopment analysis. *Journal of Cleaner Production*, *237*, 117690. Advance online publication. doi:10.1016/j.jclepro.2019.117690

Rodrigues, C., Rodrigues, A. C., Vilarinho, C., Alves, M., & Alonso, J. M. (2019). Spatial multi-criteria gis-based analysis to anaerobic biogas plant location for dairy waste and wastewater treatment and energy recovery (Barcelos, NW Portugal). Springer Verlag.

Rodríguez, R., Gauthier-Maradei, P., & Escalante, H. (2017). Fuzzy spatial decision tool to rank suitable sites for allocation of bioenergy plants based on crop residue. *Biomass and Bioenergy*, *100*, 17–30. doi:10.1016/j.biombioe.2017.03.007

Rupf, G. V., Bahri, P. A., de Boer, K., & McHenry, M. P. (2016). Development of a model for identifying the optimal biogas system design in Sub-Saharan Africa. In Z. Kravanja & M. Bogataj (Eds.), *Computer Aided Chemical Engineering* (Vol. 38, pp. 1533–1538). Elsevier.

Sacchelli, S., & Cipollaro, M. (2016). Public perception of bioenergy chain: An integrated evaluation based on semantic differential approach and multi-criteria analysis. *Chemical Engineering Transactions*, *50*, 427–432. doi:10.3303/CET1650072

Saelee, S., Paweewan, B., Tongpool, R., Witoon, T., Takada, J., & Manusboonpurmpool, K. (2014). Biomass type selection for boilers using TOPSIS multi-criteria model. *International Journal of Environmental Sciences and Development*, *5*(2), 181–186. doi:10.7763/IJESD.2014.V5.474

Sakthivel, G., Ilangkumaran, M., & Gaikwad, A. (2015). A hybrid multi-criteria decision modeling approach for the best biodiesel blend selection based on ANP-TOPSIS analysis. *Ain Shams Engineering Journal*, *6*(1), 239–256. doi:10.1016/j.asej.2014.08.003

Sakthivel, G., Ilangkumaran, M., Nagarajan, G., Priyadharshini, G. V., Dinesh Kumar, S., Satish Kumar, S., Suresh, K. S., Thirumalai Selvan, G., & Thilakavel, T. (2014). Multi-criteria decision modelling approach for biodiesel blend selection based on GRA-TOPSIS analysis. *International Journal of Ambient Energy*, *35*(3), 139–154. doi:10.1080/01430750.2013.789984

Sakthivel, G., Ilangkumaran, M., Nagarajan, G., & Shanmugam, P. (2013). Selection of best biodiesel blend for IC engines: An integrated approach with FAHP-TOPSIS and FAHP-VIKOR. *International Journal of Oil, Gas and Coal Technology*, *6*(5), 581–612. doi:10.1504/IJOGCT.2013.056153

Sanaei, S., Chambost, V., & Stuart, P. R. (2018). Systematic assessment of triticale-based biorefinery strategies: Sustainability assessment using multi-criteria decision-making (MCDM). *Biofuels, Bioproducts & Biorefining*, *12*, S73–S86. doi:10.1002/bbb.1482

Schillo, R. S., Isabelle, D. A., & Shakiba, A. (2017). Linking advanced biofuels policies with stakeholder interests: A method building on Quality Function Deployment. *Energy Policy*, *100*, 126–137. doi:10.1016/j.enpol.2016.09.056

Schröder, T., Lauven, L. P., Beyer, B., Lerche, N., & Geldermann, J. (2019). Using PROMETHEE to assess bioenergy pathways. *Central European Journal of Operations Research*, *27*(2), 287–309. doi:10.100710100-018-0590-3

Scott, J. A., Ho, W., & Dey, P. K. (2013). Strategic sourcing in the UK bioenergy industry. *International Journal of Production Economics*, *146*(2), 478–490. doi:10.1016/j.ijpe.2013.01.027

Shahraki Shahdabadi, R., Maleki, A., Haghighat, S., & Ghalandari, M. (2021). Using multi-criteria decision-making methods to select the best location for the construction of a biomass power plant in Iran. *Journal of Thermal Analysis and Calorimetry*, *145*(4), 2105–2122. doi:10.100710973-020-10281-1

Shin, Y. O., & Zul, I. (2020). *Energy priority estimation model for quantitative analysis of potential bioethanol feedstock*. Academic Press.

Silva, S., Alçada-Almeida, L., & Dias, L. C. (2014). Biogas plants site selection integrating Multicriteria Decision Aid methods and GIS techniques: A case study in a Portuguese region. *Biomass and Bioenergy*, *71*, 58–68. doi:10.1016/j.biombioe.2014.10.025

Šiškova, J. (2013). Multi-criterion analysis of the risks involved in a biogas plant in relation to the structure and sources of biomass and its application in agricultural companies. *Acta Universitatis Agriculturae et Silviculturae Mendelianae Brunensis*, *61*(7), 2843–2850. doi:10.11118/actaun201361072843

Sivaraja, C. M., Sakthivel, G., & Jegadeeshwaran, R. (2020). Selection of optimum bio-diesel fuel blend using fuzzy TOPSIS and fuzzy VIKOR approaches. *International Journal of Oil, Gas and Coal Technology*, *23*(2), 261–291. doi:10.1504/IJOGCT.2020.105455

Smyth, B. M., Smyth, H., & Murphy, J. D. (2011). Determining the regional potential for a grass biomethane industry. *Applied Energy*, *88*(6), 2037–2049. doi:10.1016/j.apenergy.2010.12.069

Su, S., Yu, Z., Zhu, W., & Chang, W. (2020). A comprehensive evaluation and optimal utilization structure of crop straw-based energy production in eastern China. *BioResources*, *15*(2), 2850–2868. doi:10.15376/biores.15.2.2850-2868

Sultana, A., & Kumar, A. (2012). Ranking of biomass pellets by integration of economic, environmental and technical factors. *Biomass and Bioenergy*, *39*, 344–355. doi:10.1016/j.biombioe.2012.01.027

Sunil Chaitanya, G., Raj Kumar, M., & Deivanathan, R. (2021). *Multi criteria decision making approach for selection of biodiesel blend using AHP-TOPSIS analysis*. Academic Press.

Tazzit, S., Ibne Hossain, N. U., Nur, F., Elakramine, F., Jaradat, R., & El Amrani, S. (2021). Selecting a biomass pelleting processing depot using a data driven decision-making approach. *Systems*, *9*(2), 32. Advance online publication. doi:10.3390ystems9020032

Turcksin, L., Macharis, C., Lebeau, K., Boureima, F., Van Mierlo, J., Bram, S., De Ruyck, J., Mertens, L., Jossart, J.-M., Gorissen, L., & Pelkmans, L. (2011). A multi-actor multi-criteria framework to assess the stakeholder support for different biofuel options: The case of Belgium. *Energy Policy*, *39*(1), 200–214. doi:10.1016/j.enpol.2010.09.033

Ubando, A. T., Cuello, J. L., Culaba, A. B., Promentilla, M. A. B., & Tan, R. R. (2014). Multi-criterion evaluation of cultivation systems for sustainable algal biofuel production using Analytic Hierarchy Process and Monte Carlo simulation. *Energy Procedia*, *61*, 389–392. doi:10.1016/j.egypro.2014.11.1132

Ubando, A. T., Cuello, J. L., El-Halwagi, M. M., Culaba, A. B., Promentilla, M. A. B., & Tan, R. R. (2016). Application of stochastic analytic hierarchy process for evaluating algal cultivation systems for sustainable biofuel production. *Clean Technologies and Environmental Policy*, *18*(5), 1281–1294. doi:10.100710098-015-1073-z

Ubando, A. T., Promentilla, M. A. B., Culaba, A. B., & Tan, R. R. (2016). *Application of spatial analytic hierarchy process in the selection of algal cultivation site for biofuel production: A case study in the Philippines*. Academic Press.

Unay, E., Ozkaya, B., & Yoruklu, H. C. (2021). A multi-criteria decision analysis for the evaluation of microalgal growth and harvesting. *Chemosphere*, *279*, 130561. doi:10.1016/j.chemosphere.2021.130561 PMID:33892454

van Dael, M., Van Passel, S., Pelkmans, L., Guisson, R., Swinnen, G., & Schreurs, E. (2012). Determining potential locations for biomass valorization using a macro screening approach. *Biomass and Bioenergy*, *45*, 175–186. doi:10.1016/j.biombioe.2012.06.001

Vindiš, P., Muršec, B., Rozman, Č., & Čus, F. (2010). A Multi-Criteria Assessment of Energy Crops for Biogas Production. Strojniski Vestnik. *Jixie Gongcheng Xuebao*, *56*(1).

Vlachokostas, C., Achillas, C., Agnantiaris, I., Michailidou, A. V., Pallas, C., Feleki, E., & Moussiopoulos, N. (2020). Decision support system to implement units of alternative biowaste treatment for producing bioenergy and boosting local bioeconomy. *Energies*, *13*(9), 2306. Advance online publication. doi:10.3390/en13092306

Volkova, A., Latosov, E., & Siirde, A. (2010). Selection of the most appropriate regions for wood fuel based cogeneration plants using multi criteria decision analysis methods. *International Journal of Exergy*, *4*(2).

von Doderer, C. C. C., & Kleynhans, T. E. (2014). Determining the most sustainable lignocellulosic bioenergy system following a case study approach. *Biomass and Bioenergy*, *70*, 273–286. doi:10.1016/j.biombioe.2014.08.014

Wang, C. N., Fu, H. P., Hsu, H. P., Nguyen, V. T., Nguyen, V. T., & Ahmar, A. S. (2021). A model for selecting a biomass furnace supplier based on qualitative and quantitative factors. Computers, Materials, &. *Continua*, *69*(2), 2339–2353. doi:10.32604/cmc.2021.016284

Wang, C. N., Tsai, T. T., & Huang, Y. F. (2019). A model for optimizing location selection for biomass energy power plants. *Processes (Basel, Switzerland)*, *7*(6), 353. Advance online publication. doi:10.3390/pr7060353

Wannasiri, W. (2020). The potential of biomass fuel and land suitability for biomass power plant based on gis spatial analysis in the Nakhon Ratchasima Province, Thailand. *Chemical Engineering Transactions*, *78*, 325–330. doi:10.3303/CET2078055

Wheeler, J., Páez, M. A., Guillén-Gosálbez, G., & Mele, F. D. (2018). Combining multi-attribute decision-making methods with multi-objective optimization in the design of biomass supply chains. *Computers & Chemical Engineering*, *113*, 11–31. doi:10.1016/j.compchemeng.2018.02.010

Woo, H., Acuna, M., Moroni, M., Taskhiri, M. S., & Turner, P. (2018). Optimizing the location of biomass energy facilities by integrating multi-criteria analysis (MCA) and geographical information systems (GIS). *Forests*, *9*(10), 585. Advance online publication. doi:10.3390/f9100585

Wu, Y., Yan, Y., Wang, S., Liu, F., Xu, C., & Zhang, T. (2019). Study on location decision framework of agroforestry biomass cogeneration project: A case of China. *Biomass and Bioenergy*, *127*, 105289. Advance online publication. doi:10.1016/j.biombioe.2019.105289

Xiang, W., Xue, S., Qin, S., Xiao, L., Liu, F., & Yi, Z. (2018). Development of a multi-criteria decision making model for evaluating the energy potential of Miscanthus germplasms for bioenergy production. *Industrial Crops and Products*, *125*, 602–615. doi:10.1016/j.indcrop.2018.09.050

Yadav, S., Srivatava, A. K., & Singh, R. S. (2015). Selection and ranking of multifaceted criteria for the prioritization of most appropriate conversion technology for biomass to biofuel in Indian perspective using analytic hierarchy process. *International Journal of Advanced Technology in Engineering and Science*, *3*, 869–881.

Yan, Q., & Tao, J. (2014). Biomass power generation industry efficiency evaluation in China. *Sustainability*, *6*(12), 8720–8735. doi:10.3390u6128720

Yeo, S. Z., How, B. S., Ngan, S. L., Ng, W. P. Q., Leong, W. D., Lim, C. H., & Lam, H. L. (2020). An integrated approach to prioritise parameters for multi-objective optimisation: A case study of biomass network. *Journal of Cleaner Production*, *274*, 123053. doi:10.1016/j.jclepro.2020.123053

Zahid, F., Tahir, A., Khan, H. U., & Naeem, M. A. (2021). Wind farms selection using geospatial technologies and energy generation capacity in Gwadar. *Energy Reports*, *7*, 5857–5870.

Zhang, W., Wang, C., Zhang, L., Xu, Y., Cui, Y., Lu, Z., & Streets, D. G. (2018). Evaluation of the performance of distributed and centralized biomass technologies in rural China. *Renewable Energy*, *125*, 445–455. doi:10.1016/j.renene.2018.02.109

Zhou, S., Zhang, Y., & Bao, X. (2012). *Methodology of location selection for biofuel refinery based on fuzzy TOPSIS*. Academic Press.

Zhou, X. Y., Wang, X. K., Wang, J. Q., Li, J. B., & Li, L. (2020). Decision support framework for the risk ranking of agroforestry biomass power generation projects with picture fuzzy information. *Journal of Intelligent & Fuzzy Systems*, •••, 1–20.

Ziolkowska, J. R. (2013). Evaluating sustainability of biofuels feedstocks: A multi-objective framework for supporting decision making. *Biomass and Bioenergy*, *59*, 425–440. doi:10.1016/j.biombioe.2013.09.008

Ziolkowska, J. R. (2014). Optimizing biofuels production in an uncertain decision environment: Conventional vs. advanced technologies. *Applied Energy*, *114*, 366–376. doi:10.1016/j.apenergy.2013.09.060

Zubaryeva, A., Zaccarelli, N., Del Giudice, C., & Zurlini, G. (2012). Spatially explicit assessment of local biomass availability for distributed biogas production via anaerobic co-digestion – Mediterranean case study. *Renewable Energy*, *39*(1), 261–270. doi:10.1016/j.renene.2011.08.021

ADDITIONAL READING

Estévez, R. A., Espinoza, V., Ponce Oliva, R. D., Vásquez-Lavín, F., & Gelcich, S. (2021). Multi-Criteria Decision Analysis for Renewable Energies: Research Trends, Gaps and the Challenge of Improving Participation. *Sustainability*, *13*(6), 3515. doi:10.3390u13063515

Hobbs, B. F., & Meier, P. (2012). *Energy decisions and the environment: a guide to the use of multicriteria methods* (Vol. 28). Springer Science & Business Media.

Mardani, A., Zavadskas, E. K., Khalifah, Z., Zakuan, N., Jusoh, A., Nor, K. M., & Khoshnoudi, M. (2017). A review of multi-criteria decision-making applications to solve energy management problems: Two decades from 1995 to 2015. *Renewable & Sustainable Energy Reviews*, *71*, 216–256. doi:10.1016/j.rser.2016.12.053

San Cristóbal, J. R. (2012). *Multi criteria analysis in the renewable energy industry*. Springer Science & Business Media. doi:10.1007/978-1-4471-2346-0

Scott, J. A., Ho, W., & Dey, P. K. (2012). A review of multi-criteria decision-making methods for bioenergy systems. *Energy*, *42*(1), 146–156. doi:10.1016/j.energy.2012.03.074

KEY TERMS AND DEFINITIONS

AHP: Developed by Thomas L. Saaty in 1970, it is an approach that uses a system based on mathematics and psychology in the complex decision-making process.

Bioenergy: It is one of the renewable energy types obtained from organic materials. It can be used in a wide range from transportation to electricity production.

Biomass: It is the material obtained from plants and animals used to generate electricity or heat.

Multi-Criteria Decision Making: It is a sub-discipline of operations research that can be used in a wide variety of sectors and used in decision-making processes by considering conflicting criteria in these sectors.

PROMETHEE: It is a method used to make decisions based on mathematics and sociology developed in the early 1980s. This method points to the alternative that fits the understanding and goals of the decision-makers rather than the right decision.

Renewable Energy: It is an energy type. Its primary resources are renewable resources such as sunlight, wind, rain, tides, waves, and geothermal heat.

SWOT Analysis: It is a strategic planning and management technique. It helps individuals or institutions to identify strengths, weaknesses, opportunities, and threats related to business competition or project planning.

TOPSIS: It is one of the techniques used to evaluate alternatives in decision-making. First, weight is determined for each criterion. Next, the score of each criterion is then normalized. Finally, the geometric distance between each alternative and the ideal alternative is calculated.

Chapter 9
The Review of Multi-Criteria Decision Making in the Renewable Energy Industry of Turkey

Ayse Topal
Nigde Omer Halisdemir University, Turkey

ABSTRACT

Renewable energy resources have become popular in energy policies as sustainable development in the energy field requires the transition to clean or renewable energy resources such as solar, wind, and hydro to mitigate global warming. Renewable resources play a more significant role in the energy future of Turkey. However, despite renewable energy resources being cleaner and causing fewer environmental problems, the renewable energy selection problem is a complex task due to the involvement of various conflicting factors and uncertainty. Therefore, multi-criteria decision-making methods are commonly used to handle this complexity successfully. In this chapter, the studies focused on renewable energy resource selection problem in Turkey with multi-criteria decision-making methods were reviewed. Findings suggest that the number of studies increased due to the growing importance of renewables. Also, AHP, TOPSIS, and ANP have risen to the top of the literature as the most extensively used approaches.

DOI: 10.4018/978-1-6684-2472-8.ch009

INTRODUCTION

Environmental issues such as global warming was not a new problem, but it has never been as evident as it has been in recent years which we have had worldwide fires, floods, and pandemics. Energy is a significant input for economic growth, but it is also a significant issue for the environment because most energy resources are based on fossil fuels. Fossil fuels release a significant amount of carbon into the atmosphere, which is the primary cause of greenhouse gases that contribute to global warming. Sustainable growth in the energy area necessitates the shift to clean or renewable energy resources such as solar, wind, and hydro.

Renewable energy resources are not only significant from environmental perspective but also from economic perspective. Turkey is an energy-poor country in terms of fossil fuels thus relies significantly on imports which causes a deficit in the national budget (TCMB, 2021). Therefore, energy production, consumption, imports, dependency, and the current account deficit are all crucial economic elements in determining Turkey's energy policy. The rise in oil prices in the 1970s coincided with an increase in Turkey's energy dependency. Energy consumption in Turkey expanded quickly after 1980, as the population and industrialisation of the country accelerated. Energy began to be used more often with the increase of economic growth and the need for fossil fuels such as oil, natural gas, and coal has increased as a result (Efeoglu and Pehlivan, 2018). As a result, renewable energy sources become important in Turkey's energy policies. The main objectives of Turkey's current energy policy in terms of energy supply security are to increase the share of renewable clean energy sources in energy supply, increase energy efficiency, follow an environmentally friendly energy policy by reducing the use of fossil fuels, and increase the use of national natural resources by reducing energy dependency on foreign sources (MFA, 2015).

Although renewable energy resources are cleaner and create fewer environmental problems than fossil fuels, choosing a renewable energy source is a difficult task as it involves multiple criteria and alternatives in addition to confliction and uncertainty. Multi-Criteria Decision Making (MCDM) methodologies are successfully used to handle decision problems. MCDM allows choosing the best option from various options by weighing them against several criteria. Renewable energy selection is likewise an MCDM process that needs consideration of various criteria, including technical, economic, environmental, technical, and socio-political factors. MCDM techniques are viewed as effective methods for assessing all elements of decision-making situations and obtaining a satisfying answer for decision-makers (Sengul et al., 2015). This chapter thoroughly examines the studies that apply classic MCDM approaches to handle Turkey's renewable energy selection problem. The studies were categorized based on document type, year, journal, MCDM techniques, and energy resources. The number of publications is also looked at to see how energy decision-making issues have changed over time.

The remainder of this study is structured as follows: The current state of renewable energy in the world and Turkey is briefly given in section 2. In Section 3, novel MCDM approaches were discussed in general. Section 4 provides a comprehensive review of the literature on novel MCDM studies conducted in Turkey about renewable energy decision-making. In Section 5, the acquired results and future study proposals were discussed.

CURRENT STATE OF RENEWABLE ENERGY RESOURCES

Renewable resources are known as local, clean, and limitless energy supplies. Biomass, hydropower, geothermal, solar, wind, and marine energies are all examples of renewable energy sources. Hydropower, solar, and wind are the most commonly used renewable resources globally. They accounted for 12% of the global energy supply in 2020, and it is projected to be 26% by 2030. The ratio of renewable energy in electricity production is 17%, and it is expected to be 46% by 2030 (IEA, 2021). Figure 1 shows the global renewable energy outlook for 2050.

Figure 1. Renewable electricity share development
Source: IRENA (2020)

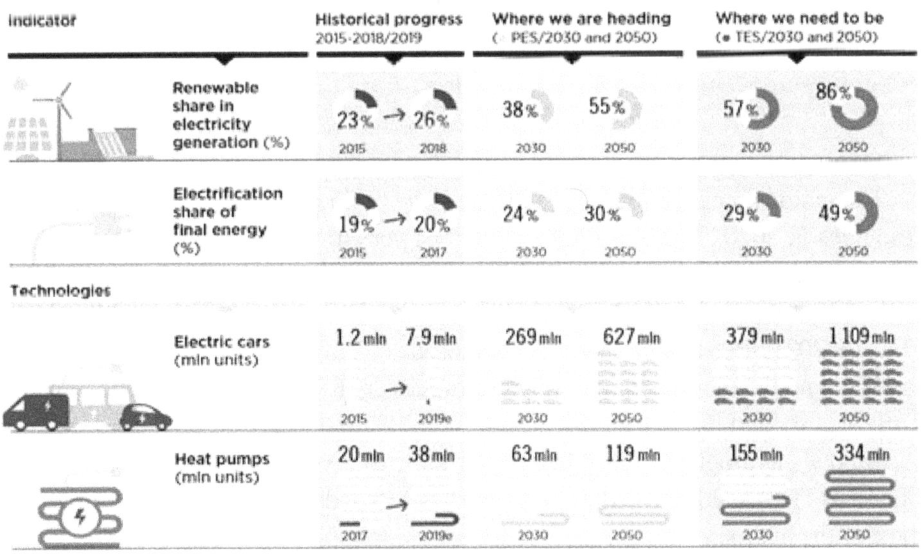

As the world is in an energy transformation trend towards renewables, Turkey has also followed this trend. Turkey is an energy poor country as it has low reserves in fossil fuels such as oil, coal, and natural gas. While the potential for renewable energy resources is high in Turkey due to the geographical and climate advantages, electricity generation is still heavily dependent on imported resources. The share of imported resources in electricity generation has reached to 43.6 percent in 2020 (Figure 2). Therefore, Turkey's national energy strategy focuses on minimizing the country's reliance on imported resources until 2023, while also maximizing the use of local resources by lowering the share of fossil fuels in the country's energy mix.

Figure 2. The share of local and imported resources in electricity generation of Turkey (2002-2020)
Source: TEIAS (2021)

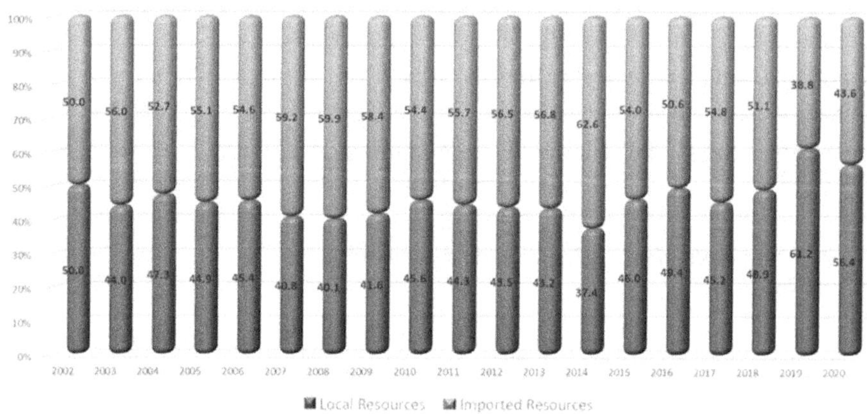

The national energy strategy led to an increase in the investment of renewable resources in 2020. Turkey's total installed power climbed by 4.623 MW in 2020, bringing the installed power to 95.890 MW by the end of the year. Most of the installed power added in 2020 came from renewable resources with 4493 MW (TEIAS, 2021). Renewable energy accounted for 51% of primary energy consumption in 2020, and the most used renewable resources are hydropower, solar and wind in Turkey as seen in the Figures 3.

Turkey's hydroelectric installed power grew to 30,983 MW in 2020, as seen in Figure 4 (TEIAS, 2021). Turkey currently has 685 hydroelectric power stations and most of the hydroelectric energy in Turkey is supplied from the Atatürk, Karakaya, Keban and Altınkaya Dams (Energy Atlas, 2021).

Figure 3. The share of installed capacity ratio by primary energy sources in 2020
Source: TEIAS (2021)

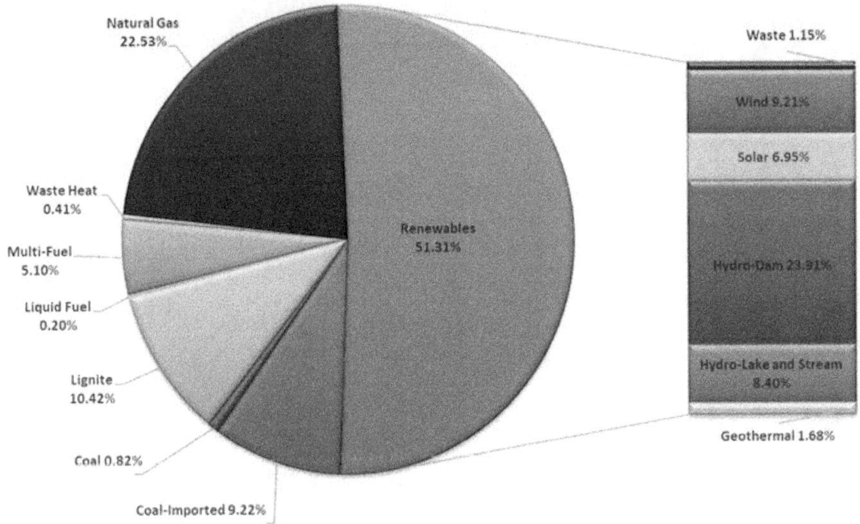

Figure 4. Hydro power capacity between 2004 and 2020
Source: TEIAS (2021)

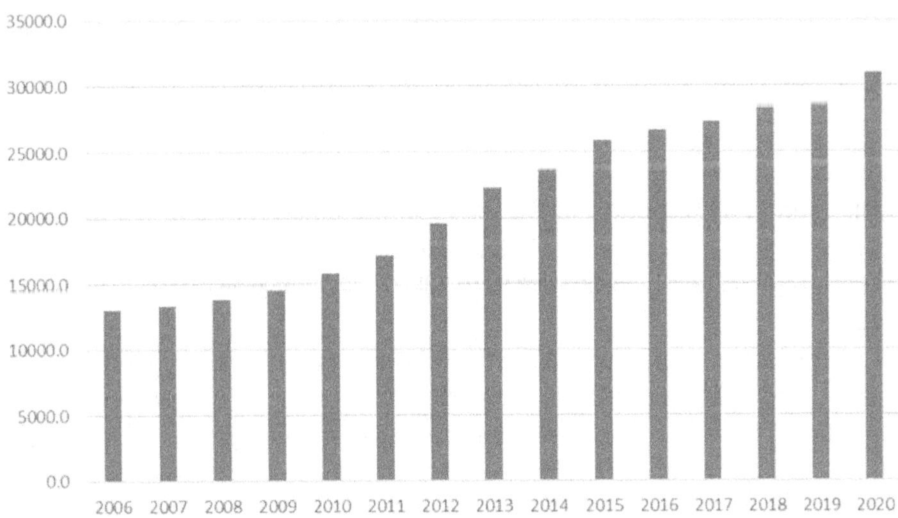

Turkey's average annual sunshine duration is 2640 hours, which is equivalent to a total of 7.2 hours per day. The region that receives the most sunshine in Turkey is the Southeastern Anatolia Region, followed by the Mediterranean Region. The capacity for solar power in Turkey has grown dramatically between 2014 and 2020 and reached to 6,668 megawatts as seen in Figure 5 (ETKB, 2021a).

Figure 5. Solar power capacity between 2014 and 2020
Source: ETKB (2021a)

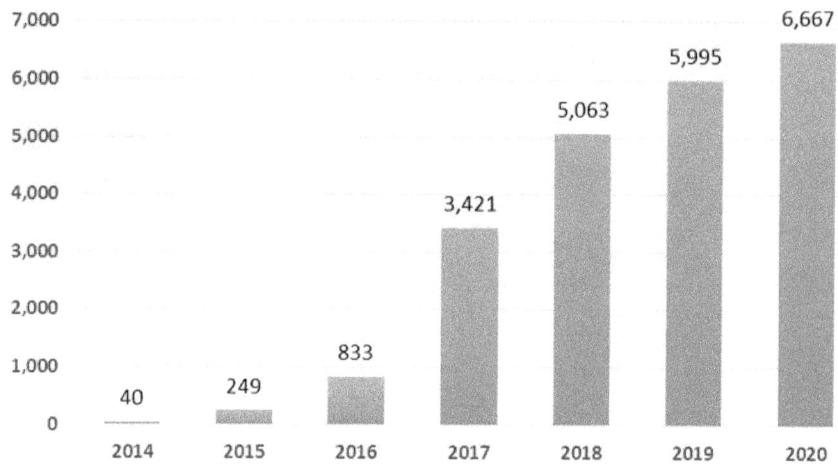

The potential of wind power is high in the country's coastal regions. Aegean Region accounts for around 38 percent (2868 MW) of Turkey's installed wind energy capacity. Aegean Region is followed by Marmara Region, which has 34 percent (2603 MW) installed electricity, and Mediterranean Region, which has 13 percent (996 MW) installed power. The most wind power stations are placed in the provinces of Izmir, Balikesir, Manisa, and Canakkale (TUREB, 2019). Wind power capacity increased from 19 MW in 2004 to 8832 MW in 2020, as presented in Figure 6 (ETKB, 2021b).

Turkey ranks first in Europe and fourth worldwide in geothermal, coming after US, Indonesia, and Philippines. 90% of our geothermal resources are at low and medium temperatures. The lower end of temperatures makes most of the use of geothermal for direct application such as heating therefore a small portion (approximately 10%) has been used for electricity generation. Turkey's geothermal potential is projected to be 2000 MW, which would provide 31500 MWh of electricity (ETKB, 2021c). Installed geothermal energy capacity had a significant increase in recent years. It increased from 81.9 MW in 2006 to 1613.2 MW in 2020 (TEIAS, 2021).

Biomass has been a steady resource for heating. With the technological developments and the expansion of use, it has begun to be used in biofuels and electricity generation. It was projected that Turkey's biomass waste potential is 8.6 million tons of oil equivalent (MTEP), with a biogas production capacity of 1.5-2 MTEP (ETKB, 2021d). The capacity of biomass in electricity generation has increased in the last decades. In 2006, the installed biomass energy generation capacity in Turkey was 41.3 MW. It increased to 1502.8 MW by 2020 (TEIAS, 2021).

Figure 6. Wind power capacity between 2004 and 2020
Source: ETKB (2021b)

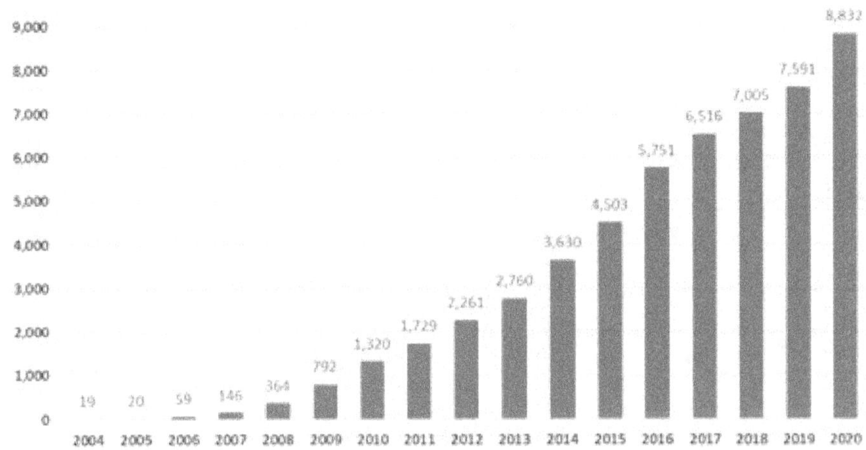

THE REVIEW OF MCDM METHODS IN TURKEY FOR RENEWABLE ENERGY SELECTION

Extensive attempts have been made in the last decades to increase renewable energy, diversify power generating techniques, do multi-dimensional assessments and develop approaches in decision making of renewable energy. MCDM approaches have been commonly used in the renewable energy literature to choose the most appropriate renewable energy resource among multiple options. It consists of various decision-making techniques which make it possible to compare several alternatives based on several conflicting criteria. Because they comprise a range of criteria and sub-criteria that should be evaluated, MCDM approaches are effective for providing an efficient solution to renewable energy selection decision problem. Therefore, the aim of this study is to conduct a literature review to identify which MCDM approaches are used to handle renewable energy selection problem in Turkey. The focus of this study is on the whole country. For this reason, the studies assessing renewable energy resources for a specific region were eliminated.

The following databases were reviewed for common MCDM techniques in title, summary, and keywords: Web of Science, Google Scholar and TR-Dizin to find papers that supplied the most useful information about renewable energy in Turkey. Then, these studies were narrowed down to papers that focused on MCDM techniques used in Turkey for renewable energy selection decision problem as shown in Table 1.

Table 1. The details of MCDM studies about renewable energy selection in Turkey

Authors	Date	Technique	Results (Resources)
Kahraman & Kaya	2010	Fuzzy AHP	Wind
Kahraman et al.	2010	Fuzzy Axiomatic Design	Wind
Kaya & Kahraman	2011	Fuzzy TOPSIS	Biomass
Boran et al.	2012	Intuitionistic fuzzy TOPSIS	Hydro
Demirtas	2013	AHP	Biomass
Ertay et al.	2013	MACBETH, fuzzy AHP	Wind
Yakici Ayan & Pabuccu	2013	AHP	Hydro
Buyukozkan & Guleryuz	2014	Fuzzy AHP, fuzzy TOPSIS	Wind
Kabak & Dagdeviren	2014	BOCR (Benefits, Opportunities, Costs and Risks) and ANP	Hydro
Kuleli Pak et al.	2015	ANP, TOPSIS	Hydro
Sengül et al.	2015	Fuzzy TOPSIS	Geothermal and wind
Buyukozkan & Guleryuz	2016	DEMATEL-ANP	Solar/Wind
Celikbilek & Tuysuz	2016	Grey DEMATEL-ANP-VIKOR	Solar
Balin & Baracli	2017	Fuzzy AHP, interval type-2 TOPSIS	Wind
Buyukozkan & Guleryuz	2017	Fuzzy DEMATEL, ANP, TOPSIS	Geothermal
Colak & Kaya	2017	Interval type-2 fuzzy AHP, Hesitant fuzzy TOPSIS	Wind
Damgaci et al.	2017	Fuzzy TOPSIS	Biomass
Ozcan et al.	2017	ANP, TOPSIS	Wind
Ozkale et al.	2017	SWOT, PROMETHEE	Hydro
Tuysuz	2017	Grey relational analysis (GRA), Monte Carlo	Hydro
Baysal & Cetin	2018	Fuzzy AHP, Mixed-Integer Linear Programming (MILP)	Solar
Boran	2018	Intuitionistic fuzzy VIKOR	Wind
Buyukozkan et al.	2018a	Hesitant fuzzy linguistic term sets, SAW, TOPSIS	Hydro

Tabe 1. Continued

Authors	Date	Technique	Results (Resources)
Dogan & Uludag	2018	AHP, Fuzzy Grey Relational Analysis	Solar
Karaca & Ulutas	2018	WASPAS	Hydro
Toklu & Taskin	2018	Fuzzy AHP, Fuzzy TOPSIS	Wind
Tolga & Turgut	2018	Fuzzy TODIM	Solar
Buyukozkan et al.	2018b	HFL-AHP, HFL-COPRAS	Solar
Aksoy	2019	AHP, Mixed-Integer Linear Programming (MILP)	Wind/Solar
Karakas & Yildiran	2019	Fuzzy AHP	Wind/Solar
Alkan & Albayrak	2020	Fuzzy COPRAS, fuzzy MULTIMOORA	Hydro
Derse & Yontar	2020	SWARA, TOPSIS	Hydro
Deveci et al.	2020	Intuitionistic fuzzy CODAS	Wind
Eroglu & Sahin	2020	Extended neutrosophic VIKOR	Solar
Karaaslan & Aydin	2020	AHP, COPRAS, MULTIMOORA	Hydro
Karasan & Kahraman	2020	Interval-valued neutrosophic ELECTRE I	Wind
Kayahan Karakul	2020	Fuzzy AHP	Solar
Karatop et al.	2021	Fuzzy AHP, EDAS and Fuzzy FMEA	Hydro
Yürek et al.	2021	Pythagorean fuzzy AHP-TOPSIS	Hybrids: Hydro-solar Wind-solar
Ezbakhe & Pérez-Foguet	2021	ELECTRE III	Wind
Ecer et al.	2021	Interval Rough Number (IRN) extension of CODAS	Hydro

The review is conducted based on publication date, techniques, and results (resources). Journal papers, focusing mostly on the fields of operations research and management science were reviewed. The period of this study has been limited to the years between 2010 and 2021 to identify trends, tendencies, and gaps in the last decade. There is also a classification of the studies depending on their MCDM techniques and results.

Initially, the studies were reviewed according to their publication dates. As demonstrated in Figure 7, the number of studies is increasing by year. Because of finite fossil energy resources and their effects on environment, the importance of renewable energy use has grown by time. As a result, in recent years, there has been a general trend towards using MCDM approaches to make decisions for renewable energy selection.

Figure 7. The number of renewable energy selection studies in Turkey between 2010 and 2021

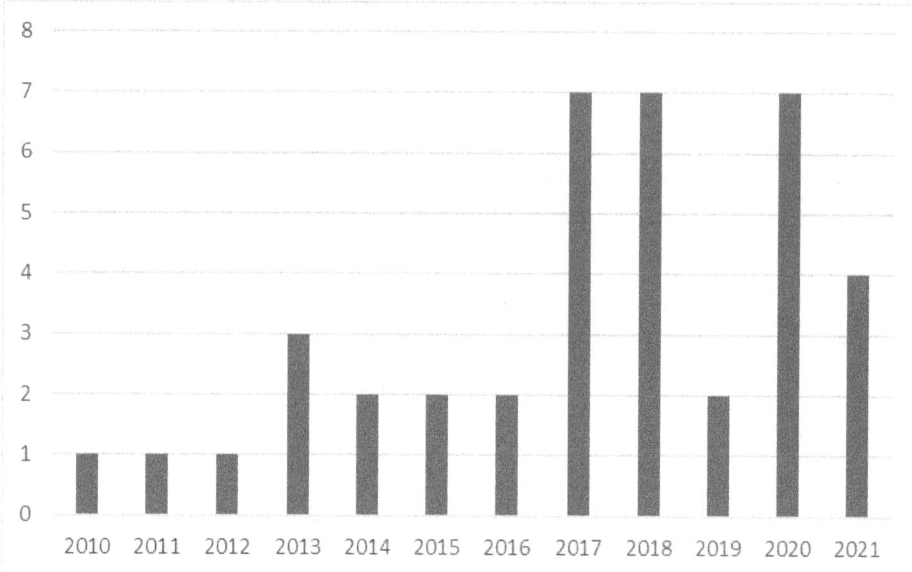

It was found that the most used techniques are AHP and TOPSIS as presented in Table 2 and Figure 8. This is due to AHP's straightforward structure and the analyst's ability to discuss outcomes until consistency is obtained and a near-unanimous judgment is reached (Abu-Taha, 2011). TOPSIS has been widely used as it has a simple and easy algorithm (Bottani & Rizzi, 2006) its procedure for calculation is consistent regardless of the number of alternatives (Shih et al., 2007).

In terms of energy resources, MCDM techniques have been used in a variety of contexts throughout the literature. Some studies study only one of the renewable energy sources, but others evaluate many energy sources to pick amongst them or establish the best mix of sources. Studies selected hydro, solar and wind energy have risen throughout time, as can be shown in Table 1. Geothermal and biomass, on the other hand, appear to have received less attention over time. The percentages of energy resources selected for Turkey as the most optimal energy resource by studies are shown in Figure 9. Wind is selected as the most optimal option by 34 percent of studies, hydro 27 percent, solar 25 percent, biomass 6 percent, geothermal 4 percent, and hybrid options 4 percent.

Table 2. The distribution of MCDM techniques used for renewable energy selection

MCDM Techniques	Year	No. (#)	MCDM Techniques	Year	No. (#)
ELECTRE	1966	2	VIKOR	1998	3
MAUT	1967	-	MOORA	2006	-
SMART	1971	-	ARAS	2010	-
DEMATEL	1972	3	MULTIMOORA	2010	2
ORESTE	1979	-	PSI	2010	-
AHP	1980	17	MACBETH	2012	1
EVAMIX	1982	-	MOOSRA	2012	-
PROMETHEE	1982	1	WASPAS	2014	1
GRA	1989	2	EDAS	2015	1
Axiomatic Design	1990	1	MABAC	2015	-
TODIM	1991	1	CODAS	2016	2
TOPSIS	1992	14	MAIRCA	2017	-
ROV	1993	-	PIPRECIA	2017	-
COPRAS	1994	3	PIV	2018	-
OCRA	1994	-	COCOSO	2019	-
ANP	1996	6	MARCOS	2020	-

Figure 8. The percentage of MCDM techniques used for renewable energy selection

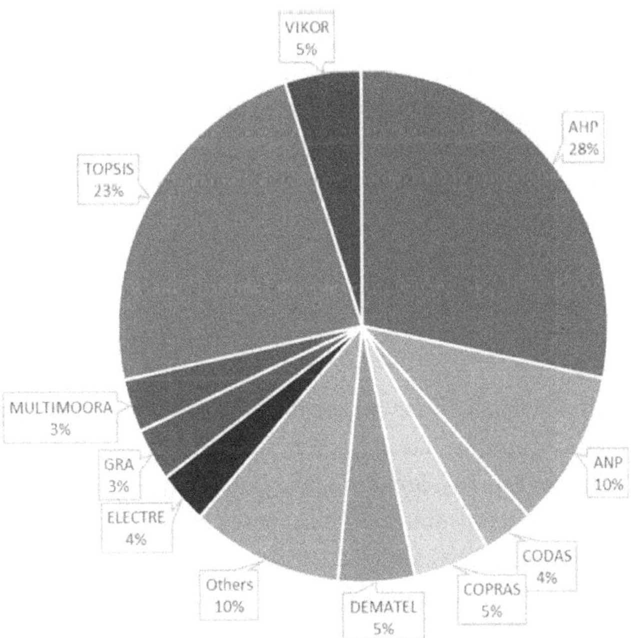

Figure 9. The percentage of selected renewable energy resources for Turkey

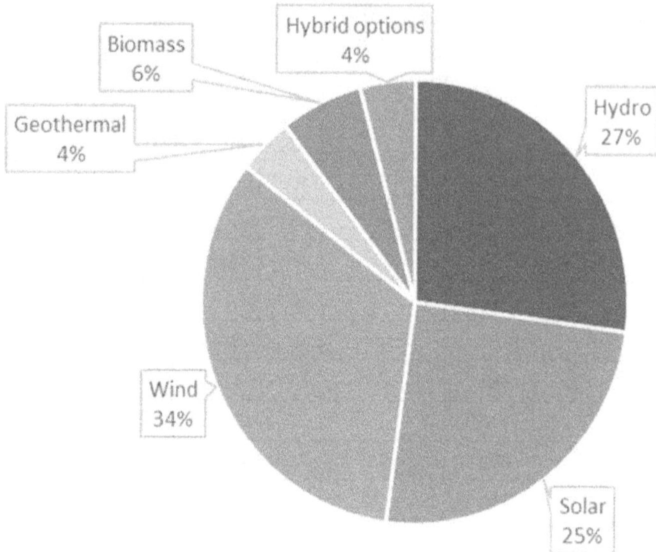

FUTURE RESEARCH DIRECTIONS

While this study thoroughly examined the publications mentioned, it also includes a number of limitations. For example, only papers written in English and Turkish were evaluated in the study. Also, only studies published in journals were reviewed. Book chapters, conference proceedings, thesis, notes, reports, and other items were eliminated. These can be included in future studies. In addition, the studies assessing renewable energy resources for a specific region were eliminated. Region specific reviews can be conducted in the future. Furthermore, the publications were found using the Web of Science, Google Scholar and TR-Dizin databases. Other databases can be used in future studies.

CONCLUSION

In order to mitigate the negative environmental and economic effects of fossil fuel consumption in Turkey, appropriate use of renewable energy sources is critical. This study provides a thorough and systematic evaluation of papers that have used MCDM techniques to address concerns of renewable energy selection in Turkey. Thirty-six studies published between 2010 and 2021 using a systematic selection procedure were examined. In this study, extensive evaluations of the most widely discussed renewable energy solutions, as well as the most favoured MCDM approaches for evaluating them, are offered.

The extensive review led to the following findings. MCDM techniques have been commonly used in the literature to select the best option from various renewable resources, categorized in this paper as solar, wind, hydro, geothermal and biomass. The number of studies applying MCDM techniques for renewable energy selection has risen over time. AHP and TOPSIS have emerged as the most extensively used techniques in the literature. Furthermore, fuzzy sets have been widely used in the literature. The results also presented that the number of papers favouring hydro, wind and solar has risen over time. A few studies in the literature find geothermal and biomass the most optimal renewable resource for Turkey.

REFERENCES

Abu-Taha, R. (2011, July). Multi-criteria applications in renewable energy analysis: A literature review. In *2011 Proceedings of PICMET'11: Technology Management in the Energy Smart World (PICMET)* (pp. 1-8). IEEE.

Aksoy, A. (2019). Integrated model for renewable energy planning in Turkey. *International Journal of Green Energy*, *16*(1), 34–48. doi:10.1080/15435075.2018.1531872

Alkan, O., & Albayrak, O. K. (2020). Ranking of renewable energy sources for regions in Turkey by fuzzy entropy based fuzzy COPRAS and fuzzy MULTIMOORA. *Renewable Energy*, *162*, 712–726. doi:10.1016/j.renene.2020.08.062

Balin, A., & Baracli, H. (2017). A fuzzy multi-citeria decision making methadology based upon the interval Type-2 fuzzy sets for evaluating renewable energy alternatives in Turkey. *Technological and Economic Development of Economy*, *23*(5), 742–763. doi:10.3846/20294913.2015.1056276

Baysal, M. E., & Cetin, N. C. (2018). Priority ranking for energy resources in Turkey and investment planning for renewable energy resources. *Complex & Intelligent Systems*, *4*(4), 261–269. doi:10.100740747-018-0075-y

Boran, F. E. (2018). A new approach for evaluation of renewable energy resources: A case of Turkey. *Energy Sources. Part B, Economics, Planning, and Policy*, *13*(3), 196–204. doi:10.1080/15567249.2017.1423414

Boran, F. E., Boran, K., & Menlik, T. (2012). The evaluation of renewable energy technologies for electricity generation in Turkey using intuitionistic fuzzy TOPSIS. *Energy Sources. Part B, Economics, Planning, and Policy*, *7*(1), 81–90. doi:10.1080/15567240903047483

Bottani, E., & Rizzi, A. (2006). A fuzzy TOPSIS methodology to support outsourcing of logistics services. *Supply Chain Management*, *11*(4), 294–308. doi:10.1108/13598540610671743

Buyukozkan, G., & Guleryuz, S. (2014). A new GDM based AHP framework with linguistic interval fuzzy preference relations for renewable energy planning. *Journal of Intelligent & Fuzzy Systems*, *27*(6), 3181–3195. doi:10.3233/IFS-141275

Buyukozkan, G., & Guleryuz, S. (2016). An integrated DEMATEL-ANP approach for renewable energy resources selection in Turkey. *International Journal of Production Economics*, *182*, 435–448. doi:10.1016/j.ijpe.2016.09.015

Buyukozkan, G., & Guleryuz, S. (2017). Evaluation of renewable energy resources in Turkey using an integrated MCDM approach with linguistic interval fuzzy preference relations. *Energy*, *123*, 149–163. doi:10.1016/j.energy.2017.01.137

Buyukozkan, G., Karabulut, Y., & Guler, M. (2018a). Strategic renewable energy source selection for turkey with hesitant fuzzy MCDM method. In *Energy Management—Collective and Computational Intelligence with Theory and Applications* (pp. 229–250). Springer.

Buyukozkan, G., Karabulut, Y., & Mukul, E. (2018b). A novel renewable energy selection model for United Nations' sustainable development goals. *Energy*, *165*, 290–302. doi:10.1016/j.energy.2018.08.215

Celikbilek, Y., & Tuysuz, F. (2016). An integrated grey based multi-criteria decision making approach for the evaluation of renewable energy sources. *Energy*, *115*, 1246–1258. doi:10.1016/j.energy.2016.09.091

Colak, M., & Kaya, I. (2017). Prioritization of renewable energy alternatives by using an integrated fuzzy MCDM model: A real case application for Turkey. *Renewable & Sustainable Energy Reviews*, *80*, 840–853. doi:10.1016/j.rser.2017.05.194

Damgaci, E., Boran, K., & Boran, F. E. (2017). Evaluation of Turkey's renewable energy using intuitionistic fuzzy TOPSIS method. *Journal of Polytechnic*, *20*(3), 629–637.

Demirtas, O. (2013). Evaluating the best renewable energy technology for sustainable energy planning. *International Journal of Energy Economics and Policy*, *3*, 23–33.

Derse, O., & Yontar, E. (2020). Determination of the most appropriate renewable energy source by SWARA-TOPSIS Method. *Journal of Industrial Engineering*, *31*(3), 389–419.

Deveci, K., Cin, R., & Kagızman, A. (2020). A modified interval valued intuitionistic fuzzy CODAS method and its application to multi-criteria selection among renewable energy alternatives in Turkey. *Applied Soft Computing*, *96*, 106660. doi:10.1016/j.asoc.2020.106660

Dogan, H., & Uludag, A. S. (2018). Evaluation of renewable energy alternatives and selection of suitable facility location: A study in Turkey. *The International Journal of Economic and Social Research*, *14*(2), 157–180.

Ecer, F., Pamucar, D., Mardani, A., & Alrasheedi, M. (2021). Assessment of renewable energy resources using new interval rough number extension of the level based weight assessment and combinative distance-based assessment. *Renewable Energy*, *170*, 1156–1177. doi:10.1016/j.renene.2021.02.004

Efeoglu, R., & Pehlivan, C. (2018). The effects of energy consumption and current deficit on economic growth in Turkey. *Journal of Political Economic Theory*, *2*(1), 104–105.

Energy Atlas. (2021). *Hidroelecticity power plants*. https://www.enerjiatlasi.com/hidroclcktrik/

Eroglu, H., & Sahin, R. (2020). A neutrosophic VIKOR method based decision-making with an improved distance measure and score function: Case study of selection for rencwablc cnergy alternatives. *Cognitive Computation*, *12*(6), 1338–1355. doi:10.100712559-020-09765-x

Ertay, T., Kahraman, C., & Kaya, I. (2013). Evaluation of renewable energy alternatives using MACBETH and fuzzy AHP multicriteria methods: The case of Turkey. *Technological and Economic Development of Economy*, *19*(1), 38–62. doi:10.3846/20294913.2012.762950

ETKB. (2021a). *Solar*. https://enerji.gov.tr/eigm-yenilenebilir-enerji-kaynaklar-gunes

ETKB. (2021b). *Wind energy potential atlas*. https://repa.enerji.gov.tr/REPA/bolgeler/TURKIYE-GENELI.pdf

ETKB. (2021c). *Geothermal*. https://enerji.gov.tr/bilgi-merkezi-enerji-

ETKB. (2021d). *Biomass*. https://enerji.gov.tr/bilgi-merkezi-enerji-biyokutle

Ezbakhe, F., & Pérez-Foguet, A. (2021). Decision analysis for sustainable development: The case of renewable energy planning under uncertainty. *European Journal of Operational Research*, *291*(2), 601–613. doi:10.1016/j.ejor.2020.02.037

IEA. (2021). *World Energy Outlook 2021.* https://iea.blob.core.windows.net/assets/888004cf-1a38-4716-9e0c-3b0e3fdbf609/WorldEnergyOutlook2021.pdf

IRENA. (2020). *Global renewables outlook 2050.* https://irena.org/-/media/Files/IRENA/Agency/Publication/2020/Apr/IRENA_GRO_Summary_2020.pdf?la=en&hash=1F18E445B56228AF8C4893CAEF147ED0163A0E47

jeotermal

Kabak, M., & Dagdeviren, M. (2014). Prioritization of renewable energy sources for Turkey by using a hybrid MCDM methodology. *Energy Conversion and Management*, *79*, 25–33. doi:10.1016/j.enconman.2013.11.036

Kahraman, C., Cebi, S., & Kaya, I. (2010). Selection among renewable energy alternatives using fuzzy axiomatic design: The case of Turkey. *Journal of Universal Computer Science*, *16*(1), 82–102.

Kahraman, C., & Kaya, I. (2010). A fuzzy multicriteria methodology for selection among energy alternatives. *Expert Systems with Applications*, *37*(9), 6270–6281. doi:10.1016/j.eswa.2010.02.095

Karaaslan, A., & Aydin, S. (2020). Evaluation of renewable energy resources with multi criteria decision making techniques: Evidence from Turkey. *Ataturk University Journal of Economics and Administrative Sciences*, *34*(4), 1351–1375.

Karaca, C., & Ulutas, A. (2018). The selection of appropriate renewable energy source for Turkey by using Entropy and WASPAS methods. *Ege Academic Review*, *18*(3), 483–494.

Karakas, E., & Yildiran, O. V. (2019). Evaluation of renewable energy alternatives for Turkey via modified fuzzy AHP. *International Journal of Energy Economics and Policy*, *9*(2), 31–39. doi:10.32479/ijeep.7349

Karasan, A., & Kahraman, C. (2020). Selection of the most appropriate renewable energy alternatives by using a novel interval-valued neutrosophic ELECTRE I method. *Informatica (Vilnius)*, *31*(2), 225–248. doi:10.15388/20-INFOR388

Karatop, B., Taskan, B., Adar, E., & Kubat, C. (2021). Decision analysis related to the renewable energy investments in Turkey based on a fuzzy AHP-EDAS-fuzzy FMEA approach. *Computers & Industrial Engineering*, *151*, 106958. doi:10.1016/j.cie.2020.106958

Kaya, T., & Kahraman, C. (2011). Multicriteria decision making in energy planning using a modified fuzzy TOPSIS methodology. *Expert Systems with Applications*, *38*(6), 6577–6585. doi:10.1016/j.eswa.2010.11.081

Kayahan Karakul, A. (2020). Selection of renewable energy source using fuzzy AHP Method. *Journal of Social Sciences Institute*, *10*(19), 127–150.

Kuleli Pak, B., Albayrak, Y. E., & Erensal, Y. C. (2015). Renewable energy perspective for Turkey using sustainability indicators. *International Journal of Computational Intelligence Systems*, *8*(1), 187–197.

MFA. (2015). *Turkey's energy profile and strategy*. https://www.mfa.gov.tr/turkeys-energy-strategy.en.mfa

Ozcan, E. C., Unlusoy, S., & Tamer, E. (2017). Evaluation of the renewable energy investments in Turkey using ANP and TOPSIS methods. *Selcuk University Journal of Engineering. Science and Technology*, *5*(2), 204–219.

Ozkale, C., Celik, C., Turkmen, A. C., & Cakmaz, E. S. (2017). Decision analysis application intended for selection of a power plant running on renewable energy sources. *Renewable & Sustainable Energy Reviews*, *70*, 1011–1021. doi:10.1016/j.rser.2016.12.006

Sengul, U., Eren, M., Shiraz, S. E., Gezder, V., & Sengul, A. B. (2015). Fuzzy TOPSIS method for ranking renewable energy supply systems in Turkey. *Renewable Energy*, *75*, 617–625. doi:10.1016/j.renene.2014.10.045

Shih, H. S., Shyur, H. J., & Lee, E. S. (2007). An extension of TOPSIS for group decision making. *Mathematical and Computer Modelling*, *45*(7-8), 801–813. doi:10.1016/j.mcm.2006.03.023

TCMB. (2021). *Electronic data delivery system*. https://evds2.tcmb.gov.tr/

TEIAS. (2021). *Turkey electricity generation-transmission statistics for 2020*. https://www.teias.gov.tr/tr-TR/turkiye-elektrik-uretim-iletim-istatistikleri

Toklu, M. C., & Taskin, H. (2018). A fuzzy hybrid decision model for renewable energy sources selection. *International Journal of Computational and Experimental Science and Engineering*, *4*(1), 6–10. doi:10.22399/ijcesen.399976

Tolga, A. C., & Turgut, Z. K. (2018). Sustainable and renewable energy power plants evaluation by fuzzy TODIM technique. *Alphanumeric Journal*, *6*(1), 49–68. doi:10.17093/alphanumeric.371754

TUREB. (2019). *Turkey wind energy statistics report.* https://tureb.com.tr//lib/uploads/4e77501b714739a9.pdf

Tuysuz, F. (2017). A hybrid multi-criteria analysis approach for the assessment of renewable energy resources under uncertainty. *Alphanumeric Journal, 5*(2), 317-328.

Yakici Ayan, T., & Pabuccu, H. (2013). Evaluation of the renewable energy investment project with Analytic Hierarchy Process method. *Suleyman Demirel University The Journal of Faculty of Economics and Administrative Sciences, 18*(3), 89–110.

Yürek, Y. T., Bulut, M., Özyörük, B., & Özcan, E. (2021). Evaluation of the hybrid renewable energy sources using sustainability index under uncertainty. *Sustainable Energy. Grids and Networks, 28*, 100527.

ADDITIONAL READING

Bayraktar, A. (2018). Energy transition in Turkey. *Turkish Policy Quarterly, 17*(3), 19–26.

Karatas, M., Sulukan, E., & Karacan, I. (2018). Assessment of Turkey's energy management performance via a hybrid multi-criteria decision-making methodology. *Energy, 153*, 890–912. doi:10.1016/j.energy.2018.04.051

Kilickaplan, A., Bogdanov, D., Peker, O., Caldera, U., Aghahosseini, A., & Breyer, C. (2017). An energy transition pathway for Turkey to achieve 100% renewable energy powered electricity, desalination and non-energetic industrial gas demand sectors by 2050. *Solar Energy, 158*, 218–235. doi:10.1016/j.solener.2017.09.030

Şahin, U. (2021). Future of renewable energy consumption in France, Germany, Italy, Spain, Turkey and UK by 2030 using optimized fractional nonlinear grey Bernoulli model. *Sustainable Production and Consumption, 25*, 1–14. doi:10.1016/j.spc.2020.07.009 PMID:32835048

Saygın, H., Oral, H. V., & Kardaşlar, S. (2020). Environmental assessment of renewable energy scenarios for a sustainable future in Turkey. *Energy & Environment, 31*(2), 237–255. doi:10.1177/0958305X19855992

Solangi, Y. A., Longsheng, C., Shah, S. A. A., Alsanad, A., Ahmad, M., Akbar, M. A., Gumaei, A., & Ali, S. (2020). Analyzing renewable energy sources of a developing country for sustainable development: An integrated fuzzy based-decision methodology. *Processes (Basel, Switzerland), 8*(7), 825. doi:10.3390/pr8070825

Temiz Dinç, D., & Akdoğan, E. C. (2019). Renewable energy production, energy consumption and sustainable economic growth in Turkey: A VECM approach. *Sustainability*, *11*(5), 1273. doi:10.3390u11051273

Yüksel, S., & Ubay, G. G. (2020). Identifying the influencing factors of renewable energy consumption in Turkey with MARS methodology. *Journal of Economics Business and Finance Research*, 2(1), 1–14.

KEY TERMS AND DEFINITIONS

Biomass: Biomass power is generated from burning material derived from living beings like plants and animals such as plants, wood, and garbage.

Energy Dependency: It refers to humanity's reliance on either primary or secondary energy for energy consumption. It shows the degree of reliance a country has on imports to satisfy its energy demands.

Geothermal Power: Geothermal power is generated from the heat hot water pools found at various temperatures and depths under the ground.

Global Warming: It is the increase in the air and water temperatures which causing change in the climate. It is caused by carbon emissions mainly produced from fossil fuel consumption.

Hydropower: Hydropower is the oldest renewable energy resource which generates electricity by harnessing the natural flow of flowing water.

Multi-Criteria Decision-Making (MCDM) Methods: These methods are used when numerous criteria (or objectives) must be examined simultaneously to rank or choose between the alternatives being assessed.

Renewable Energy: Renewable energy resources are used to produce electricity as an alternative to fossil fuels. They cause much less environmental problems compared to fossil fuels.

Solar Power: Solar power is the conversion of sunlight into thermal or electrical energy. Solar power is the most environmentally friendly and widely used renewable energy source currently accessible.

Wind Power: The kinetic energy generated by moving air is utilized to generate electricity in the wind. Wind turbines convert this into electrical energy.

Chapter 10
The Selection of a Most Feasible Wind Turbine Alternative Under Multi-Criteria Framework

Ezgi Demir
Sumer Robotics Engineering and Consultancy Ltd., UK

Figen Balo
Fırat University, Turkey

ABSTRACT

Because of the effects of nuclear and fossil-based energy on the environment, economics, and security in the world, the need for alternative energy sources has grown steadily and dramatically during the last years. An increasing attraction in renewable power sources, due to rising energy expenses and country-level tax inducement, is driving the research to advance a sequence of improving unified resolutions and novel energy generation equipment. The novel wind turbine installation and the novel wind farm building are critical procedures for long time energy generation. In this chapter, a comprehensive analysis, which combines ARAS and ENTROPY methods, is structured to choose appropriate turbines when improving a wind power plant. The various wind turbine brands were evaluated on different classes (financial, customer satisfaction, environmental, and technical). Data on wind turbines is acquired from 2 MW wind turbine manufacturers.

DOI: 10.4018/978-1-6684-2472-8.ch010

INTRODUCTION

By diverse kinds of energy resources (for example, coal, wood), power has been obtained over the centuries. Meanwhile, there is growing worry over environmental/global generation of waste. This has prompted on a long-term energy supply, which entails more efficient energy usage, reduced pollution, and implicitly, lower energy consumption. These factors have led to a greater focus on short-term stored energy supplies, the most developed of which being wind energy. The pressure differential generated through the sun's uneven heating of the land and sea creates wind energy. With fast growth during the last years, wind energy technology has emerged as the most potential option to existing energy systems (Lee, Chen & Kang, 2009). One of the solutions for completing the Kyoto Protocol and combating global warming has been wind energy (Gamboa, Munda, 2007). Many nations have invested in wind energy which today satisfies 4 percent of global electricity demand and is increasing in value every day. To produce electric, a wind energy station incurs 3 chief kinds of expenses: financing, capital, and running expenses. The finance expenses are the expenses of obtaining the essential money for running and constructing a wind energy station; the capital expenses are the expenses of installing and connecting the wind energy station to the grid; and the running expenses are the expenses of maintaining and operating the wind energy station. The onshore designs have a high capital expense, ranging from 75 percent to 90 percent of total expense, with wind turbines accounting for 64 percent of entire expense for a generally 5MW onshore design. Because of better manufacturing processes, automation, and mass production, the capital expense of making wind turbines has continuously decreased over the last twenty years.

The most essential components of these systems are wind turbines, which transform wind motion energy into electrical energy. As a result, for long-term operation, greater accuracy and attention must be paid to wind turbine selection. Nonetheless, the proper wind turbine selection is extremely critical because the expenses of the wind turbines make up the plurality of the entire expense for a wind energy station design. Furthermore, the appropriateness of wind turbines for a certain site may have an important impact on their capacity factor.

The selection and evaluation of sustainable energy options is a multi criteria decision-making issue as poly criteria, several may still be in oppose, must be simultaneously considered. Multi criteria decision-making methodologies, like "VlšeKriterijumska Optimizacija I Kompromisno Resenje (VIKOR)", "preference ranking organization methodology" for enrichment of evaluations, "analytic network process", "analytical hierarchy process", "technique for order preference by similarity to the ideal solution", "multi-attribute utility theory", "the elimination and choice translating reality English", and "multi-objective decision making" have been utilized in the assessment of sustainable energy designs (San Cristóbal, 2011). The previous implementations of multi criteria decision-making on sustainable energy include sustainable energy planning projects, hydro-site selection projects, wind energy station plans, geothermal designs and solar energy farm designs and the like (San Cristóbal, 2011). Wang et al. conducted a thorough evaluation of multi criteria decision-making as a tool for making sustainable energy decisions (Wang, Jing, Zhang and Zhao, 2009). Methodologies in diverse steps of multi criteria decision-making for renewable energy were researched, containing criteria weighting, criteria selection, final aggregation, and evaluation. For wind energy station location selection, Janke used Geographical Information Systems and Multi Criteria Decision Analysis (JR. Janke, 2010). Utilizing local meteorological circumstances and a fuzzy multi criteria approach, Al-Yahyai and Charabi assessed the appropriateness of a location for a wind turbine (S. Al-Yahyai, Y. Charabi., 2015). Minguez et al. used "Technique for Order of Preference by Similarity to Ideal Solution" to investigate the most optimal support structure choices for a 5.5MW wind turbine. They looked at engineering, financial, and ecological factors to assess the system's efficacy (Lozano-Minguez, Kolios, Brennan, 2011), (Sanchez-Lozano, García-Cascales, Lamata, 2014). The complete evaluation modelling, which combines "interpretive structural modeling" and "fuzzy analytical network process", was created by Lee et al. They were able to choose appropriate turbines for a wind farm using this approach (Lee, Hung, Kang, Pearn, 2012). Haaren et al. developed fourteen criteria to locate ideal places, including land use, historical and memorial monuments, slope, and so on. They utilized a weighted approach utilizing Geographical Information Systems to identify suitable wind farm locations in the United Kingdom (Van Haaren, Fthenakis, 2011). For wind farm appropriateness researches, Haralambopoulos et al. employed a Geographical Information Systems multi criteria decision-making methodology (Tegou, Polatidis & Haralambopoulos, 2010).

During the last years, the significance of studies on wind turbine choice has grown. There are several criteria in the wind turbine choice problem, all of which are in distinct magnitudes and units. All the same, there is a scarcity of literature on applying "multi criteria decision making" methodologies to solve the problem of wind turbine choice (Shirgholami, Zangeneh and Bortolini, 2016). One of the generally utilized methodologies in these researches is the "analytical hierarchy process" methodology (Atanassov, 1999), (Smarandache, 2002), (Bagocius, Zavadskas and Turskis, 2014). By utilizing binary comparisons, it's simple to utilize and solves the hierarchy problem. Except this, the "analytical hierarchy process's" scales can be implemented to conceptual and the judgments' consistency can be checked Shirgholami, Zangeneh, Bortolini, 2016). Nonetheless, there are several drawbacks for analytical hierarchy process which include biased judgements and complexity. In spite of its drawbacks, it has been opted through lots of decision-makers and utilized lone or interconnected with the other "multi criteria decision making" methodologies. For example, Bagocius and co-workers utilized a multi-aim function to build a technique for selecting the optimal wind turbine depend on personal preference. The parameters utilized were the investment cost, power ratio, net energy production, maximum capacity. The "analytical hierarchy process" methodology was used to determine these criteria. Utilizing the weighted aggregated sum product assessment approach, the most optimal option was chosen. Shirgholami and co-workers (Shirgholami, Zangeneh and Bortolini, 2016) chosen a wind turbine to be used in an Iranian wind farm plan. They demonstrated how the criteria were developed and reported the relevant literature in great detail. The primary criteria utilized were the supplier performance, environmental impact, cost, and energy, but sub-criteria were also determined. The weighting was made utilizing the "analytical hierarchy process" methodology, and the sub- primary criteria were computed severally as global and local constituents. After it has combined them all, the primary criterion weightiness was obtained and ranked. Another paper that utilized the "analytical hierarchy process" methodology was the study through Balo et al. for the choice a 1.50 MW wind turbine (Sagbansua, Balo, 2017). As criteria, the maximum capacity, rotor diameter, hub height, total cost, energy output, state support, electromagnetic impact, noise, integration capability, and satisfaction level of the system were utilized. In several researches, "analytical hierarchy process" has been compared to the other acknowledged "multi criteria decision making" methodologies. For example, Kolios and co-workers employed several decision making techniques to choose the wind turbine back structure (Kolios, Mytilinou, Minguez and Salonitis, 2016). The "elimination and choice translating reality English" approaches, preferred "ranking technique" for enrichment evaluation, the "weighted sum" method, the "analytical hierarchy process", the "weighted product" method, and the technique for order of preference by similarity to ideal solution were compared in their research. The optimal conclusions were obtained by the use of stochastic algorithms created through "Monte Carlo simulation". Khan and Rehman utilized the weighted sum methodology to identify the optimal wind turbine in contrast to the "options weighted entire" methodology (Rehman and Khan, 2017). The power ratio, wind speed ratio, rotor diameter, center height, and wind cutting speed were utilized as criteria. C++ software was used to find a solution. The wind turbines' selection as a decision-making problem necessitates several judgments on socio-economic, technical, and ecological problems (Cavallaro and Ciraolo, 2005). Because of the subjectivity, ambiguity and vagueness of human judgement, multi criteria decision making methodologies have been used to energy systems and coupled with fuzzy set theory (Cavallaro and Ciraolo,2005), (Wang, Jing, Zhang and Zhao, 2009), (Suganthi, Inıyan and Samuel, 2005). The novel approach to imprecise assessment and decision environments technique, which is based on fuzzy connections and has low dependency on preference weight input, is one of the approaches based on fuzzy relationships (Cavallaro and Ciraolo, 2005). Ciraolo and Cavallaro evaluated wind power stations using the fuzzy-new F- novel approach to imprecise assessment and decision environments method to imprecise evaluation and decision settings (Cavallaro and Ciraolo, 2005). An option was obtained with 4 diverse designs. As criteria, the maintenance and operation costs, investment costs, fuel saving, energy generation capacities, CO2 absorption, technological development, realization time, noise, aesthetics, social acceptability, and ecosystem effect were utilized. In order to confirm the modifications in the sequence, a sensitivity analysis was undertaken to assess the robustness of the data obtained. Lee et al. employed the "analytical network" technique in another research that incorporated fuzzy logic (Lee, Hung, Kang and Pearn, 2012). The assessment of wind turbines was based on fuzzy logic and structural modelling interpretation. Khan used autonomous decision-making techniques and fuzzy logic for a range of wind turbines in a case study encompassing 500 - 750 kW turbines (Khan, 2015). As criteria, the power ratio, wind speed ratio, wind cutting speed, and rotor diameter were utilized. In their paper, the unified and-or aggregation operator was introduced and used in consequence the respecification. The 2 elements with important effects on this choice were the output percentage and central height. All the same, the modifications in the operator- correlated characteristic value had an impact on the conclusion. Thus, Khan and Rehman researched fuzzy logic sourched wind turbine choice through utilizing a more efficient fuzzy operator (Rehman and Khan, 2016). The research was based through fuzzy arithmetic mean operator and made through fuzzy numbers. As criteria, the center height, rotor diameter, power ratio, wind speed ratio, and wind cutting speed were utilized. Among twenty diverse wind turbines, the most performance wind turbine was chosen. To surpass the these researches' limitation induced through proper fuzzy operator choice, Khan and Rehman (Rehman and Khan, 2019) researched a 2-stage multi criteria decision making methodology fuzzy goal programing based for wind turbine choice. As criteria for the power production, the rated power, rotor diameter, hub height, wind speed rated, and cut-in wind speed were utilized.

In Central and Southeast Europe, Afgan et al. implemented "adoption, substitution, progress, innovation, deterioration model" to compute the precedence classification among a variety of gas transportation system alternatives (Afgan, Carvalho, Pilavachi and Martins, 2008). For a few integrated power and heat systems, Pilavachi and co-workers suggested a multi criteria decisions methodology with an agglomeration function depend on the weight factors' statistical evaluation to compute maintainability surds (Pilavachi, Roumpeas, Minett and Afgan, 2006). In Spain, Cristóbal implemented the "Analytic Hierarchy Process"to determine the attribute notional significance weights and utilized the "VlšeKriterijumska Optimizacija I Kompromisno Resenje (VIKOR)" methodology to choose the best sustainable energy design (San Cristóbal,

2011). Lee ranked and assessed the building energy efficiency from the multiple-objective throughput perspective through implementing fuzzy integral and fuzzy measure, the multiple-attribute decision making method (Lee, 2010). In Turkey, Onut and co-authors accepted the "Analytical Network Processs" to assess the most proper energy sources for the fabricating technology (Önüt, Tuzkaya and Saadet, 2008). Karaaslan and Aydın have been listed renewable energy sources by COPRAS and MULTIMOORA. Ilgın and Alkan studied the factors that prevent the widespread use of renewable energy sources in Turkey were examined by DEMATEL method. Lee and co-workers introduced a "Multi Criteria Decision Analysis" methodology, with the combination of "Analytic Hierarchy Process" and the "benefits, opportunities, costs and risks" model, to aid choose a proper wind energy plant design (Lee, Chen and Kang, 2009). Patlitzianas and co-workers existed an interconnected "Multi-Criteria Decision Making" method, depend on sequenced significance mean, to incorporate qualifying decisions on numerous threats and opportunities elements for evaluating the renewable energy manufacturer environment in European Union membership states (Patlitzianas, Ntotas, Doukas and Psarras, 2007). Nobre and co-workers utilized a multi-criteria geo-spatial analysis method, depend on "geographic information systems" technic, to determine the most proper site for positionning a wave power plant (Nobre, Pacheco, Jorge, Lopes, Gato and 2009). Kolios and co-workers ensured a systematical method depend on the "Technique For Order Of Preference By Similarity To Ideal Solution" for evaluation and classification of diverse existing off-shore wind turbine back formations (Kolios, Collu, Chahardehi and Brennan, 2010). Lee and co-workers improved a notional modelling for produce policy in the photovoltaic cell energy technology, and "the benefits, opportunities, costs and risks", "interpretive structural model", and "fuzzy analytical network process" approach are combined to analyze proper strategical elements (Chen, Kang and Lee,2010). Chen and co-workers devised a "fuzzy analytic hierarchy process" methodology related to the "the benefits, opportunities, costs and risks" approach to assess wind - solar farm designs (Lee, Chen and Kang, 2011).

The sustainable energy assessment issue is happening more caution nowadays; all the same, the utilizes of "Multi-Criteria Decision Making" methods with the imprecise consideration and fuzzy data to work on the complicated problem are quite restricted. Because the wind farm's construction is a complex work and the most proper wind turbine selection is necessary for the wind farm's next operation, a systematical "Multi-Criteria Decision Making" modelling for assessing diverse wind turbine frameworks is essential for fulfillment the most appropriate decision. In the writers' comprehension, this article is the first one that investigates the criteria relationships in the decision-making process through integrating ENTROPY and ARAS to assess diverse wind turbine frameworks. Depend on the assessment conclusions, the company can choose the most proper wind turbine framework to be structured in its novel wind energy station.Ateş and Topal studied the place of establishment of a solar power plant with Entropy Based Topsis, Aras and Moosra Methods: example of Kop region (Ateş and Topal, 2021). Arsu evaluated of financial performance by the Entropy-Based Aras Method: An application on businesses in Bist Electricity, Gas and Steam Industry (Arsu, 2021). In this research, multi criteria decision- making is utilized to select the most feasible 2 MW wind turbine for a wind farm planning. A comparison of six different wind turbine trademarks is carried out. Each wind turbine trademark is evaluated depend on a common set of criteria. Within these base prime criteria, many important sub criteria are presented; likewise, sub-alternatives are signified for each wind turbine trademark. The wind turbine with the best performance is chosen from amongst a number of prominent 2 MW wind turbine trademarks. The necessary information was received from the manufacturer of the wind turbine. In this study 16 criteria have been handled by a consultant company in a real time. This data is handled from real data in the company. These are qualitative criteria and they have been gathered from in the real time study. The Entropy method has been provided weighting of results obtained from real-time data. On the other hand, Aras method has been provided sorting by using real-time data. Entropy based Aras method has been used, since real-time data has been used and this has been providing analysis without the need for decision maker.

ENTROPY AND ARAS IN WIND TURBINE SELECTION

Entropy Method

Step 1: Applying a positive transform to data containing negative values

In this method, Z-score standardization is applied to the criteria data Xij values. It is expressed by the following mathematical expression.

$$Z_{ij} = \frac{X_{ij} - \overline{X}_j}{\sigma_j} \tag{1}$$

Here $\overline{X}_j$ and σj respectively, j. are the mean and standard deviations of the criterion. Then the data is made positive by making coordinate transformation:

$$Z'_{ij} = Z_{ij} + A, \; A > |\min Z_{ij}| \tag{2}$$

In the decision matrix, Z'ij values are now written instead of Xij criterion values.

For Z'ij values, it shows the correspondence of i = 1,2,3,.....,m alternative values to j = 1,2,3,...., n criterion values.

Step 2. Conversion of criteria into benefit or cost analysis

$$R_{ij} = \frac{X_{ij}}{\max X_{ij}} \text{(benefit criteria), (i =1,2,3,..,m number of alternatives)} \quad (3)$$

$$R_{ij} = \frac{\min X_{ij}}{X_{ij}} \text{(cost criteria). (j=1,2,3,....,n number of alternatives)} \quad (4)$$

Step 3. Normalizing the decision matrix

$$P_{ij} = \frac{R_{ij}}{\sum_{i=1}^{m} R_{ij}}, \forall j \quad (5)$$

i = alternatives

j = criteria,

P_{ij} = normalized values,

R_{ij} = Converted values by benefit or cost status.

Step 4. Calculation of Entropy Values

$$E_j = -k. \sum_{i=1}^{m} P_{ij} .In(P_{ij}), \forall j \quad (6)$$

k (entropy) 0≤ Ej ≤ 1 is the entropy value that provides the expression

$$k = \frac{1}{In(m)}; \text{ m, the number of alternatives} \quad (7)$$

Pij = stands for normalized values

Step 5. Calculation of degrees of difference

The D_j value, which represents the degree of difference of the information for each criterion, is calculated as follows.

$$Dj = 1 - Ej \text{ , } j=1,2,\ldots,n \text{ index of criteria} \tag{8}$$

Step 6. Calculation of weights

The significance weights (Wj) of the kits are calculated as follows by normalizing the degree of difference (Dj)

$$W_j = \frac{D_j}{\sum_{j=1}^{n} D_j} \text{, j=1,2,\ldots,n index of criteria} \tag{9}$$

ARAS Method

Step 1: Creating the Decision Matrix

$$X = \begin{bmatrix} x_{01} & \cdots & x_{0j} & \cdots & x_{0n} \\ \vdots & & \ddots & & \vdots \\ x_{m1} & \cdots & x_{mj} & \cdots & x_{mn} \end{bmatrix} i = 0,1,\ldots,m \; j= 0,1, \ldots, n \tag{10}$$

Step 2: Generating the Normalized Decision Matrix

$$\overline{x_{ij}} = \frac{x_{ij}}{\sum_{i=0}^{m} x_{ij}} \tag{11}$$

$$x_{ij}{}^{*} = \frac{1}{x_{ij}} \tag{12}$$

$$\overline{x_{ij}} = \frac{x_{ij}{}^{*}}{\sum_{i=0}^{m} x_{ij}{}^{*}} \tag{13}$$

$$\bar{X} = \begin{bmatrix} \bar{x}_{01} & \cdots & \bar{x}_{0j} & \cdots & \bar{x}_{0n} \\ \vdots & & \ddots & & \vdots \\ \bar{x}_{m1} & \cdots & \bar{x}_{mj} & \cdots & \bar{x}_{mn} \end{bmatrix} \text{i} = 0,1,\ldots,\text{m j} = 0,1, \ldots, \text{n} \quad (14)$$

Step 3: Generating a Weighted Normalized Decision Matrix

$$\sum_{j=1}^{n} w_j = 1 \quad (15)$$

$$\widehat{X_{ij}} = \overline{X_{ij}} \,.\, W_j \quad (16)$$

$$\widehat{X_{ij}} = \begin{bmatrix} \hat{x}_{01} & \cdots & \hat{x}_{0j} & \cdots & \hat{x}_{0n} \\ \vdots & & \ddots & & \vdots \\ \hat{x}_{m1} & \cdots & \hat{x}_{mj} & \cdots & \hat{x}_{mn} \end{bmatrix} \text{i} = 0,1,\ldots,\text{m j} = 0,1, \ldots, \text{n} \quad (17)$$

Step 4: Calculation of Optimality Function Value

$$S_i = \sum_{j=1}^{n} \widehat{X_{ij}} \text{, i} = 0,1,2,\ldots,\text{m} \quad (18)$$

$$K_i = \frac{S_i}{S_0} \text{, i}=0,1,\ldots,\text{m} \quad (19)$$

In this study, it has been aimed to apply a multi-criteria decision-making methodology for wind turbine selection. For this purpose, criteria affecting wind turbine selection have been determined. There are 16 criteria that affect the wind turbine selection. The Entropy method, which allows to determine the weights of the criteria, has been chosen from the multi-criteria decision-making methods. Then, the data of 6 wind turbine corresponding to the criteria have been taken and sorted. Sorting has been done with the ARAS method, which is one of the ranking methods.

In the first stage, the initial decision matrix was created with the entropy method. The initial decision matrix has been shown in Table 1.

Table 1. The important parameters of wind turbine (The Initial Decision Matrix)

Turbine Name	Support of Government(MAX)	Total Cost (million$) (MIN)	Frequency (Hz) (MAX)	Nominal Rotor Diameter (m) (MAX)	Nominal Wind speed (8 m/s)(MAX)	Turbine Output (V) (MAX)	Capacity Factor(MAX)	Cut-Out Wind Speed (m/s)(MAX)	Cut-In Wind Speed(MAX)	Annual Output (8 m/s) in KWh(MAX)	Hub Height (m) (MAX)	Service support(MAX)	Reliability(MAX)	Spare part(MAX)	Max. sound power (dB) (MIN)	Electromagnetic effects(MIN)
T1	0.316	3.89	55	116	10.0	690	32.9	25.0	3.0	9327000	87	4	3	4	105	13.5
T2	0.330	4.00	55	114	12.0	690	32.4	25.0	3.0	9269000	99.3333	3	4	2	101.6	13.5
T3	0.440	3.82	55	110	11.75	690	32	20	3.0	9074000	87.5	2	2	3	107.6	13.5
T4	0.315	3.45	50	103	10	690	30.2	22	3.0	8498000	85	6	6	5	98.6	11.5
T5	0.300	3.67	55	92.5	12.5	690	30.7	24	3.0	8676000	81.3333	5	5	6	102.9	10
T6	0,45	4,25	50	98	15	690	31.1	25	3	8825000	79.3333	1	1	1	103.5	12.5

Table 2. Conversion of the criteria cost and benefit analysis

Turbine Name	Support of Government(MAX)	Total Cost (million$) (MIN)	Frequency (Hz) (MAX)	Nominal Rotor Diameter (m) (MAX)	Nominal Wind speed (8 m/s)(MAX)	Turbine Output (V) (MAX)	Capacity Factor(MAX)	Cut-Out Wind Speed (m/s)(MAX)	Cut-In Wind Speed(MAX)	Annual Output (8 m/s) in KWh(MAX)	Hub Height (m) (MAX)	Service support(MAX)	Reliability(MAX)	Spare part(MAX)	Max. sound power (dB) (MIN)	Electromagnetic effects(MIN)
T1	0,702222222	0,88688946	1	1	0,666667	1	1	1	1	1	0,875839	0,666667	0,5	0,666667	0,939048	0,740741
T2	0,733333333	0,8625	1	0,982759	0,8	1	0,984802	1	1	0,9937815	1	0,5	0,666667	0,333333	0,970472	0,740741
T3	0,977777778	0,903141361	1	0,948276	0,783333	1	0,972644	0,8	1	0,9728745	0,880872	0,333333	0,333333	0,5	0,916357	0,740741
T4	0,7	1	0,909091	0,887931	0,666667	1	0,917933	0,88	1	0,9111183	0,855705	1	0,833333	0,833333	1	0,869565
T5	0,666666667	0,940054496	1	0,797414	0,833333	1	0,933131	0,96	1	0,9302026	0,818792	0,833333	0,833333	1	0,958212	1
T6	1	0,811764706	0,909091	0,844828	1	1	0,945289	1	1	0,9461778	0,798658	0,166667	0,166667	0,166667	0,952657	0,8

Table 3. Normalized decision matrix

Turbine Name	Support of Government(MAX)	Total Cost (million$) (MIN)	Frequency (Hz) (MAX)	Nominal Rotor Diameter (m) (MAX)	Nominal Wind speed (8 m/s) (MAX)	Turbine Output (V)(MAX)	Capacity Factor(MAX)	Cut-Out Wind Speed (m/s)(MAX)	Cut-In Wind Speed(MAX)	Annual Output (8 m/s) in KWh(MAX)	Hub Height (m)(MAX)	Service support(MAX)	Reliability(MAX)	Spare part(MAX)	Max. sound power (dB) (MIN)	Electromagnetic effects(MIN)
T1	0,146908415	0,164106591	0,171875	0,18311	0,140351	0,166667	0,173798	0,177305	0,166667	0,1737875	0,167469	0,190476	0,142857	0,190476	0,16369	0,151425
T2	0,153417015	0,15959366	0,171875	0,179953	0,168421	0,166667	0,171157	0,177305	0,166667	0,1727068	0,191209	0,142857	0,190476	0,095238	0,169168	0,151425
T3	0,20455602	0,16711378	0,171875	0,173639	0,164912	0,166667	0,169044	0,141844	0,166667	0,1690734	0,168431	0,095238	0,095238	0,142857	0,159735	0,151425
T4	0,146443515	0,185036128	0,15625	0,162589	0,140351	0,166667	0,159535	0,156028	0,166667	0,1583409	0,163619	0,285714	0,285714	0,238095	0,174315	0,17776
T5	0,139470014	0,173944044	0,171875	0,146014	0,175439	0,166667	0,162176	0,170213	0,166667	0,1616576	0,156561	0,238095	0,238095	0,285714	0,167031	0,204424
T6	0,209205021	0,150205798	0,15625	0,154696	0,210526	0,166667	0,164239	0,177305	0,166667	0,1644338	0,152711	0,047619	0,047619	0,047619	0,166062	0,163539

Since the unit values are different in the decision matrices in Table 1. Some of the criteria are maximum and some are minimum. So we need to do cost and benefit analysis. The converted cost and benefit tables have been shown in Table 2.

The matrix should be normalized. Therefore, the variables have been normalized as the second step of the Entropy method. The normalized decision matrix has been shown in Table 3.

In the third step, entropy values have been found according to each criterion. Then, the weighted entropy values and the weights of the criteria have been found in Table 4.

After the weights of the criteria have been found, the problem of choosing the best wind turbine has started. Values found in criterion weighting have been used in the selection problem of wind turbine. The initial decision matrix of the ARAS method has been discussed as in Table 5.

Table 4. Criteria of Entropy Weights

	VALUES OF ENTROPY	1-ENTROPY VALUE		Wj
K1	0.317	0.683	K1	0.063
K2	0.312	0.688	K2	0.063
K3	0.312	0.688	K3	0.063
K4	0.313	0.687	K4	0.063
K5	0.315	0.685	K5	0.063
K6	0.311	0.689	K6	0.063
K7	0.312	0.688	K7	0.063
K8	0.313	0.687	K8	0.063
K9	0.311	0.689	K9	0.063
K10	0.312	0.688	K10	0.063
K11	0.313	0.687	K11	0.063
K12	0.356	0.644	K12	0.059
K13	0.356	0.644	K13	0.059
K14	0.356	0.644	K14	0.059
K15	0.312	0.688	K15	0.063
K16	0.313	0.687	K16	0.063
TOPLAM		10.865		

Table 5. Initial decision matrix

Turbine Name	Support of Government(MAX)	Total Cost (million$) (MIN)	Frequency (Hz) (MAX)	Nominal Rotor Diameter (m) (MAX)	Nominal Wind speed (8 m/s) (MAX)	Turbine Output (V)(MAX)	Capacity Factor(MAX)	Cut-Out Wind Speed (m/s) (MAX)	Cut-In Wind Speed(MAX)	Annual Output (8 m/s) in KWh(MAX)	Hub Height (m)(MAX)	Service support(MAX)	Reliability(MAX)	Spare part(MAX)	Max. sound power (dB) (MIN)	Electromagnetic effects(MIN)
T1	0.316	3.89	55	116	10	690	32.9	25	3	9327000	87	4	3	4	105	13.5
T2	0.33	4	55	114	12	690	32.4	25	3	9269000	99.33333	3	4	2	101.6	13.5
T3	0.44	3.82	55	110	11.75	690	32	20	3	9074000	87.5	2	2	3	107.6	13.5
T4	0.315	3.45	50	103	10	690	30.2	22	3	8498000	85	6	6	5	98.6	11.5
T5	0.3	3.67	55	92.5	12.5	690	30.7	24	3	8676000	81.33333	5	5	6	102.9	10
T6	0.45	4.25	50	98	15	690	31.1	25	3	8825000	79.33333	1	1	1	103.5	12.5
Values of Wi	0.063	0.063	0.063	0.063	0.063	0.063	0.063	0.063	0.063	0.063	0.063	0.059	0.059	0.059	0.063	0.063
Optimum	0.45	3.45	55	116	15	690	32.9	25	3	9327000	99.33333	6	6	6	98.6	10

The criteria have been divided into two as benefit-oriented and cost-oriented. It is expected that the benefit-oriented criteria will be maximum. And also cost-oriented criteria have been expected to be minimal. Therefore, the transformed decision matrix has been given in Table 6.

According to the next step of the ARAS method, the transformed decision matrix should also be normalized due to the unit difference. For this reason, the normalization process has been performed in Table 7.

In the next step, the normalized matrix has been weighted. The weighted matrix has been given in Table 8.

In the last step of the ARAS method, the optimum values have been obtained and ranked. These values have been shown in Table 9.

CONCLUSION

Climate change and global warming have raised human awareness of the need of environmental preservation and moved the focus of technological improvement to low-carbon sustainable power. Commercial wind power plants are presently operational in over eighty nations, and there are several advantages to building wind power plants in both developing and developed countries. These advantages contain stable energy prices, increased energy security, financial improvement to create jobs and to attract investment, less dependence on imported fuels, $CO2$ emissions reductions, and air quality improved.

This article purposes to obtain the most fesible 2MW wind turbine brand depend on diverse criteria present in the literature. In this study Entropy-based ARAS method has been used for the selection of wind turbine. For this purpose, the criteria in the selection of wind turbines have been weighted by the Entropy method. In this context, it has been seen that the least important criterion weights have been found to be "Service support", "Reliability" and "Spare part". The other criteria have the same result and the most important ones. Then, in order to find the best wind turbine, sorting has been done with the ARAS method. The most suitable alternative order can be listed as T5, T4, T1, T2, T3 and T6. The study is quite comprehensive within the scope of criterion weighting of the wind turbine. Therefore, it is important. Objective data have been used in the study. The Entropy-based Aras method has been applied to the wind turbine selection system for the first time. For further studies, it has been planning to benefit from fuzzy decision making methods by inviting leading names in the sector as decision makers.

Table 6. Transformed decision matrix

Turbine Name	Support of Government(MAX)	Total Cost (million$) (MIN)	Frequency (Hz) (MAX)	Nominal Rotor Diameter (m) (MAX)	Nominal Wind speed (8 m/s)(MAX)	Turbine Output (V)(MAX)	Capacity Factor(MAX)	Cut-Out Wind Speed (m/s)(MAX)	Cut-In Wind Speed(MAX)	Annual Output (8 m/s) in KWh(MAX)	Hub Height (m)(MAX)	Service support(MAX)	Reliability(MAX)	Spare part(MAX)	Max. sound power (dB) (MIN)	Electromagnetic effects(MIN)
T1	0.316	0.257	55	116	10	690	32.9	25	3	9327000	87	4	3	4	0.010	0.074
T2	0.33	0.25	55	114	12	690	32.4	25	3	9269000	99.33333	3	4	2	0.010	0.074
T3	0.44	0.262	55	110	11.75	690	32	20	3	9074000	87.5	2	2	3	0.010	0.074
T4	0.315	0.290	50	105	10	690	30.2	22	3	8498000	85	6	6	5	0.010	0.087
T5	0.3	0.272	55	92,5	12.5	690	30.7	24	3	8676000	81.33333	5	5	6	0.010	0.100
T6	0.45	0.235	50	98	15	690	31.1	25	3	8825000	79.33333	1	1	1	0.010	0.080
SUM	2.601	1.856	375	749.5	86.25	4830	222.2	166	21	62996000	618.8333	27	27	27	0.068	0.589
Optimum	0.45	0.290	55	115	15	690	32.9	25	3	9327000	99.33333	6	6	6	0.010	0.1

Table 7. Normalized matrix

Turbine Name	Support of Government (MAX)	Total Cost (million$) (MIN)	Frequency (Hz) (MAX)	Nominal Rotor Diameter (m) (MAX)	Nominal Wind speed (8 m/s) (MAX)	Turbine Output (V) (MAX)	Capacity Factor (MAX)	Cut-Out Wind Speed (m/s) (MAX)	Cut-In Wind Speed (MAX)	Annual Output (8 m/s) in KWh (MAX)	Hub Height (m) (MAX)	Service support (MAX)	Reliability (MAX)	Spare part (MAX)	Max. sound power (dB) (MIN)	Electromagnetic effects (MIN)
T1	0.121	0.138	0.147	0.155	0.116	0.143	0.148	0.151	0.143	0.148	0.141	0.148	0.111	0.148	0.139	0.126
T2	0.127	0.135	0.147	0.152	0.139	0.143	0.146	0.151	0.143	0.147	0.161	0.111	0.148	0.074	0.144	0.126
T3	0.169	0.141	0.147	0.148	0.136	0.143	0.144	0.120	0.143	0.144	0.141	0.074	0.074	0.111	0.136	0.126
T4	0.121	0.156	0.133	0.137	0.116	0.143	0.136	0.133	0.143	0.135	0.137	0.222	0.222	0.185	0.148	0.148
T5	0.115	0.147	0.147	0.123	0.145	0.143	0.138	0.145	0.143	0.138	0.131	0.185	0.185	0.222	0.142	0.170
T6	0.173	0.128	0.133	0.131	0.174	0.143	0.140	0.151	0.143	0.140	0.128	0.037	0.037	0.037	0.141	0.136
Optimum	0.173	0.156	0.147	0.155	0.174	0.143	0.148	0.151	0.143	0.148	0.161	0.222	0.222	0.222	0.148	0.170

Table 8. Weighted matrix

Turbine Name	Support of Government (MAX)	Total Cost (million$) (MIN)	Frequency (Hz) (MAX)	Nominal Rotor Diameter (m) (MAX)	Nominal Wind speed (8 m/s) (MAX)	Turbine Output (V)(MAX)	Capacity Factor (MAX)	Cut-Out Wind Speed (m/s) (MAX)	Cut-In Wind Speed (MAX)	Annual Output (8 m/s) in KWh (MAX)	Hub Height (m) (MAX)	Service support (MAX)	Reliability (MAX)	Spare part (MAX)	Max. sound power (dB) (MIN)	Electromagnetic effects (MIN)
T1	0.008	0.009	0.009	0.010	0.007	0.009	0.009	0.010	0.009	0.009	0.009	0.009	0.007	0.009	0.009	0.008
T2	0.008	0.009	0.009	0.010	0.009	0.009	0.009	0.010	0.009	0.009	0.010	0.007	0.009	0.004	0.009	0.008
T3	0.011	0.009	0.009	0.009	0.009	0.009	0.009	0.008	0.009	0.009	0.009	0.004	0.004	0.007	0.009	0.008
T4	0.008	0.010	0.008	0.009	0.007	0.009	0.009	0.008	0.009	0.009	0.009	0.013	0.013	0.011	0.009	0.009
T5	0.007	0.009	0.009	0.008	0.009	0.009	0.009	0.009	0.009	0.009	0.008	0.011	0.011	0.013	0.142	0.011
T6	0.011	0.008	0.008	0.008	0.011	0.009	0.009	0.010	0.009	0.009	0.008	0.002	0.002	0.002	0.009	0.009
Optimum	0.011	0.010	0.009	0.010	0.011	0.009	0.009	0.010	0.009	0.009	0.010	0.013	0.013	0.013	0.009	0.011

Table 9. Optimization function and sorting

	Si	Ki	Ranking
Optimum	0.167	1	
T1	0.139	0.832	3
T2	0.137	0.823	4
T3	0.132	0.788	5
T4	0.150	0.900	2
T5	0.151	0.902	1
T6	0.124	0.744	6

REFERENCES

Afgan, N. H., Carvalho, M. G., Pilavachi, P. A., & Martins, N. (2008). Evaluation of natural gas supply options for Southeast and Central Europe: Part 2. Multi-criteria assessment. *Energy Conversion and Management*, *2008*(49), 2345–2353. doi:10.1016/j.enconman.2008.01.024

Al-Yahyai, S., & Charabi, Y. (2015). Assessment of large-scale wind energy potential in the emerging city of Duqm (Oman). *Renewable & Sustainable Energy Reviews*, *47*, 438–447. doi:10.1016/j.rser.2015.03.024

Arsu, T. (2021, March). Finansal Performansin Entropi Tabanli Aras Yöntemi İle Değerlendirilmesi: Bist Elektrik, Gaz Ve Buhar Sektöründeki İşletmeler Üzerine Bir Uygulama. *Hacettepe Üniversitesi İktisadi ve İdari Bilimler Fakültesi Dergisi*, *39*(1), 15–32. doi:10.17065/huniibf.740393

Ateş, S., & Topal, A. (2021). Entropi temelli topsis, aras ve moosra yöntemleri ile güneş enerji santrali kuruluş yeri seçimi: kop bölgesi örneği. *Uluslararası Yönetim İktisat ve İşletme Dergisi*, *17*(4), 1099–1119. doi:10.17130/ijmeb.869594

Bagocius, V., Zavadskas, E. K., & Turskis, Z. (2014). Multi-person selection of the best wind turbine based on the multi-criteria integrated additive-multiplicative utility function. *Journal of Civil Engineering and Management*, *20*(4), 590–599. doi:10.3846/13923730.2014.932836

Cavallaro, F., & Ciraolo, L. (2005). A multi-criteria approach to evaluate wind energy plants on an Italian island. *Energy Policy*, *33*(2), 235–244. doi:10.1016/S0301-4215(03)00228-3

Chen, H. H., Kang, H. Y., & Lee, A. H. I. (2010). Strategic selection of suitable projects for hybrid solar-wind power generation systems. *Renewable & Sustainable Energy Reviews*, *14*(1), 413–441. doi:10.1016/j.rser.2009.08.004

Gamboa, G., & Munda, G. (2007). The problem of wind farm location a social multi-criteria evaluation framework. *Energy Policy*, *35*(3), 1564–1583. doi:10.1016/j.enpol.2006.04.021

Ilgın, M. A., & Alkan, E. (2020). Yenilenebilir Enerji Kaynaklarının Türkiye'de Yaygın Kullanımını Engelleyen Faktörlerin Çok Kriterli Karar Verme Teknikleri ve Kalite Evi ile Analiz Edilmesi. *International Journal of Engineering Research and Development*, *12*(1), 1–12. doi:10.29137/umagd.469519

Intuitionistic Fuzzy Sets. (1999). *Intuitionistic Fuzzy Sets Atanassov KT35*. Physica.

Janke, J. R. (2010). Multi-criteria GIS modeling of wind and solar farms in Colorado. *Renewable Energy*, *35*(10), 2228–2234. doi:10.1016/j.renene.2010.03.014

Karaaslan, A., & Aydın, S. (2020). Yenilenebilir Enerji Kaynaklarının Çok Kriterli Karar Verme Teknikleri ile Değerlendirilmesi: Türkiye Örneği. *Atatürk Üniversitesi İktisadi ve İdari Bilimler Dergisi*, *34*(4), 1351–1375. doi:10.16951/atauniiibd.749466

Khan, S. A. (2015). An automated decision-making approach for assortment of wind turbines-A case study of turbines in the range of 500 kW to 750 kW. *International Journal of Computing and Network Technology*, *3*, 75–82.

Kolios, A., Collu, M., Chahardehi, A., Brennan, F. P., & Patel, M. H. (2010).A multi-criteria decision making method to compare support structures for offshore wind turbines. *In: European Wind Energy Conference*, Warsaw.

Kolios, A., Mytilinou, V., Minguez, M. L., & Salonitis, K. (2016). A comparative study of multiple-criteria decision making methods under stochastics inputs. *Energies*, *9*(7), 566. doi:10.3390/en9070566

Lee, Hung, Kang, & Pearn. (2012). A wind turbine evaluation model under a multi-criteria decision making environment. *Energy Conversion and Management*, *64*, 289–300.

Lee, A. H., Chen, H. H., & Kang, H. Y. (2009). Multi-criteria decision making on strategic selection of wind farms. *Renewable Energy*, *34*(1), 120–126. doi:10.1016/j.renene.2008.04.013

Lee, A. H. I., Chen, H. H., & Kang, H. Y. (2011). A model to analyze strategic products for photovoltaic silicon thin-film solar cell power industry. *Renewable & Sustainable Energy Reviews*, *15*(2), 1271–1283. doi:10.1016/j.rser.2010.10.008

Lee, A. H. I., Hung, M. C., Kang, H. Y., & Pearn, W. L. (2012). A wind turbine evaluation model under a multi-criteria decision-making environment. *Energy Conversion and Management*, *64*, 289–300. doi:10.1016/j.enconman.2012.03.029

Lee, W. S. (2010). Evaluating and ranking energy performance of office buildings using fuzzy measure and fuzzy integral. *Energy Conversion and Management*, *51*(1), 197–203. doi:10.1016/j.enconman.2009.09.012

Lozano-Minguez, E., Kolios, A. J., & Brennan, F. P. (2011). Multi-criteria assessment of offshore wind turbine support structures. *Renewable Energy*, *36*(11), 2831–2837. doi:10.1016/j.renene.2011.04.020

Nobre, A., Pacheco, M., Jorge, R., Lopes, M. F. P., & Gato, L. M. C. (2009). Geo-spatial multi-criteria analysis for wave energy conversion system deployment. *Renewable Energy*, *34*(1), 97–111. doi:10.1016/j.renene.2008.03.002

Önüt, S., Tuzkaya, U. R., & Saadet, N. (2008). Multiple criteria evaluation of current energy resources for Turkish manufacturing industry. *Energy Conversion and Management*, *49*(6), 1480–1492. doi:10.1016/j.enconman.2007.12.026

Patlitzianas, K. D., Ntotas, K., Doukas, H., & Psarras, J. (2007). Assessing the renewable energy producers' environment in EU accession member states. *Energy Conversion and Management*, *48*(3), 890–897. doi:10.1016/j.enconman.2006.08.014

Pilavachi, P. A., Roumpeas, C. P., Minett, S., & Afgan, N. H. (2006). Multi-criteria evaluation for CHP system options. *Energy Conversion and Management*, *47*(20), 3519–3529. doi:10.1016/j.enconman.2006.03.004

Rehman, S., & Khan, S. A. (2016). Fuzzy logic based multi-criteria wind turbine selection strategy e a case study of qassim, Saudi arabia. *Energies*, *9*(11), 872. doi:10.3390/en9110872

Rehman, S., & Khan, S. A. (2017). Multi-criteria wind turbine selection using weighted sum approach. *International Journal of Advanced Computer Science and Applications*, *8*(6), 128–132. doi:10.14569/IJACSA.2017.080616

Rehman, S., & Khan, S. A. (2019). Goal programming-based two-tier multi-criteria decision-making approach for wind turbine selection. *Applied Artificial Intelligence*, *33*(1), 27–53. doi:10.1080/08839514.2018.1525525

Sagbansua, L., & Balo, F. (2017). Multicriteria decision making for 1.5 MW wind turbine selection. *Procedia Computer Science*, *111*, 413–419. doi:10.1016/j.procs.2017.06.042

San Cristóbal, J. R. (2011). Multi-criteria decision-making in the selection of a renewable energy project in Spain: The Vikor method. *Renewable Energy*, *36*(2), 498–502. doi:10.1016/j.renene.2010.07.031

Sanchez-Lozano, J. M., García-Cascales, M. S., & Lamata, M. T. (2014). Identification and selection of potential sites for onshore wind farms development in Region of Murcia, Spain. *Energy*, *73*, 311–324. doi:10.1016/j.energy.2014.06.024

Shirgholami, Z., Zangeneh, S. N., & Bortolini, M. (2016). Decision system to support the practitiones in the wind farm design: A case study for Iran mainland. *Sustainable Energy Technologies and Assessments*, *16*, 1–10. doi:10.1016/j.seta.2016.04.004

Smarandache, F. (2002). Neutrosophy, A new branch of philosophy. *International Journal (Toronto, Ont.)*, *8*(3), 297–384.

Suganthi, L., Inıyan, S., & Samuel, A. A. (2015). Applications of fuzzy logic in renewable energy systcms-A review. *Renewable & Sustainable Energy Reviews*, *48*, 585–607. doi:10.1016/j.rser.2015.04.037

Tegou, L. I., Polatidis, H., & Haralambopoulos, D. A. (2010). Environmental management framework for wind farm siting: Methodology and case study. *Journal of Environmental Management*, *91*(11), 2134–2147. doi:10.1016/j.jenvman.2010.05.010 PMID:20541310

Van Haaren, R. V., & Fthenakis, V. (2011). GIS-based wind farm site selection using spatial multi-criteria analysis (SMCA): Evaluating the case for New York State. *Renewable & Sustainable Energy Reviews*, *15*(7), 3332–3340. doi:10.1016/j.rser.2011.04.010

Wang, J., Jing, Y., Zhang, C., & Zhao, J. (2009). Review on multi-criteria decision analysis aid in sustainable energy decision-making. *Renewable & Sustainable Energy Reviews*, *13*(9), 2263–2278. doi:10.1016/j.rser.2009.06.021

ADDITIONAL READING

Bayraktar, A. (2018). Energy transition in Turkey. *Turkish Policy Quarterly*, *17*(3), 19–26.

Karatas, M., Sulukan, E., & Karacan, I. (2018). Assessment of Turkey's energy management performance via a hybrid multi-criteria decision-making methodology. *Energy*, *153*, 890–912. doi:10.1016/j.energy.2018.04.051

Şahin, U. (2021). Future of renewable energy consumption in France, Germany, Italy, Spain, Turkey and UK by 2030 using optimized fractional nonlinear grey Bernoulli model. *Sustainable Production and Consumption*, *25*, 1–14. doi:10.1016/j.spc.2020.07.009 PMID:32835048

Solangi, Y. A., Longsheng, C., Shah, S. A. A., Alsanad, A., Ahmad, M., Akbar, M. A., Gumaei, A., & Ali, S. (2020). Analyzing renewable energy sources of a developing country for sustainable development: An integrated fuzzy based-decision methodology. *Processes (Basel, Switzerland)*, *8*(7), 825. doi:10.3390/pr8070825

Temiz Dinç, D., & Akdoğan, E. C. (2019). Renewable energy production, energy consumption and sustainable economic growth in Turkey: A VECM approach. *Sustainability*, *11*(5), 1273. doi:10.3390u11051273

Yüksel, S., & Ubay, G. G. (2020). Identifying the influencing factors of renewable energy consumption in Turkey with MARS methodology. *Journal of Economics Business and Finance Research*, *2*(1), 1–14.

KEY TERMS AND DEFINITIONS

Climate Change: It includes both global warming driven by human-induced emissions of greenhouse gases and the resulting large-scale shifts in weather patterns.

Energy Supply: It is the delivery of fuels or transformed fuels to point of consumption. It potentially encompasses the extraction, transmission, generation, distribution, and storage of fuels. It is also sometimes called energy flow.

Global Warming: It is the current rise in temperature of the air and oceans. It happens because humans burn coal, oil, and natural gas, and cut down forests.

Low-Carbon Power: It is electricity produced with substantially lower greenhouse gas emissions than conventional fossil fuel power generation.

Multi-Criteria Decision-Making (MCDM) Methods: These methods are used when numerous criteria (or objectives) must be examined simultaneously to rank or choose between the alternatives being assessed.

Wind Power: Or wind energy is the use of wind turbines to generate electricity.

Wind Turbine: It is a device that converts the wind's kinetic energy into electrical energy.

Compilation of References

Abdel-Basset, M., Gamal, A., Chakrabortty, R. K., & Ryan, M. (2021). Development of a hybrid multi-criteria decision-making approach for sustainability evaluation of bioenergy production technologies: A case study. *Journal of Cleaner Production, 290*, 125805. doi:10.1016/j.jclepro.2021.125805

Abdel-Basset, M., Gamal, A., Chakrabortty, R. K., & Ryan, M. J. (2021). Evaluation approach for sustainable renewable energy systems under uncertain environment: A case study. *Renewable Energy, 168*, 1073–1095. doi:10.1016/j.renene.2020.12.124

Abdel-Basset, M., Mohamed, R., Zaied, A. E. N. H., Gamal, A., & Smarandache, F. (2020). Solving the supply chain problem using the best-worst method based on a novel Plithogenic model. In *Optimization Theory Based on Neutrosophic and Plithogenic Sets* (pp. 1–19). Academic Press. doi:10.1016/B978-0-12-819670-0.00001-9

Abu-Taha, R. (2011, July). Multi-criteria applications in renewable energy analysis: A literature review. In *2011 Proceedings of PICMET'11. Technology Management in the Energy Smart World (PICMET)* (pp. 1-8). IEEE.

Adua, L., Zhang, K. X., & Clark, B. (2021). Seeking a handle on climate change: Examining the comparative effectiveness of energy efficiency improvement and renewable energy production in the United States. *Global Environmental Change, 70*, 102351. doi:10.1016/j.gloenvcha.2021.102351

Afgan, N. H., Carvalho, M. G., Pilavachi, P. A., & Martins, N. (2008). Evaluation of natural gas supply options for Southeast and Central Europe: Part 2. Multi-criteria assessment. *Energy Conversion and Management, 2008*(49), 2345–2353. doi:10.1016/j.enconman.2008.01.024

Agarwal, S., Kant, R., & Shankar, R. (2020). Evaluating solutions to overcome humanitarian supply chain management barriers: A hybrid fuzzy Swara–fuzzy Waspas approach. *International Journal of Disaster Risk Reduction, 101838*. Advance online publication. doi:10.1016/j.ijdrr.2020.101838

Ahmad, F. B., Zhang, Z., Doherty, W. O. S., & O'Hara, I. M. (2015). A multi-criteria analysis approach for ranking and selection of microorganisms for the production of oils for biodiesel production. *Bioresource Technology, 190*, 264–273. doi:10.1016/j.biortech.2015.04.083 PMID:25958151

Aigner, D. J., & Chu, S. F. (n.d.). On estimating the industry production function. *The American Economic Review*, *58*(4), 826–839.

Aigner, D., Lovell, C. K., & Schmidt, P. (1977). Formulation and estimation of stochastic frontier production function models. *Journal of Econometrics*, *6*(1), 21–37. doi:10.1016/0304-4076(77)90052-5

Ajansı, S. K. (2016). *Project of Research*. Planning and Feasibility Studies on Small-scale Biogas Plants.

Aksoy, A. (2019). Integrated model for renewable energy planning in Turkey. *International Journal of Green Energy*, *16*(1), 34–48. doi:10.1080/15435075.2018.1531872

Al Garni, H., Kassem, A., Awasthi, A., Komljenovic, D., & Al-Haddad, K. (2016). A multicriteria decision making approach for evaluating renewable power generation sources in Saudi Arabia. *Sustainable Energy Technologies and Assessments*, *16*, 137–150. doi:10.1016/j.seta.2016.05.006

Albayrak, Ö. K. (2020). Multi criteria decision making techniques used in evaluation of renewable energy resources and analysis of evaluation criteria: 2017-2020. *Ataturk University Journal of Economics and Administrative Sciences*, *34*(4), 1287–1310. doi:10.16951/atauniiibd.717808

Algarin, R. A., Llanos, A. P., & Castro, A. O. (2017). An analytic hierarchy process based approach for evaluating renewable energy sources. *International Journal of Energy Economics and Policy*, *7*(4), 38–47.

Ali, S., & Waewsak, J. (2019). *GIS-MCDM approach to scrutinize the suitable sites for a biomass power plant in southernmost provinces of Thailand*. Academic Press.

Ali, E. B., Agyekum, E. B., & Adadi, P. (2021). Agriculture for Sustainable Development: A SWOT-AHP Assessment of Ghana's Planting for Food and Jobs Initiative. *Sustainability*, *13*(2), 628. doi:10.3390u13020628

Alkan, Ö., & Albayrak, Ö. K. (2020). Ranking of renewable energy sources for regions in Turkey by fuzzy entropy based fuzzy Copras and fuzzy Multimoora. *Renewable Energy*, *162*, 712–726. doi:10.1016/j.renene.2020.08.062

Alniak, M. O. (2011). *Advanced Technologies Workshop (IATW 2020) Proceedings Book*. Academic Press.

Al-Yahyai, S., & Charabi, Y. (2015). Assessment of large-scale wind energy potential in the emerging city of Duqm (Oman). *Renewable & Sustainable Energy Reviews*, *47*, 438–447. doi:10.1016/j.rser.2015.03.024

Amin, N., Lung, C., & Sopian, K. A. (2009). Practical field study of various solar cells on their performance in Malaysia. *Renewable Energy Journal*, *34*(8), 1939–1946.

Andiloro, S., Romeo, G., Marciano, C., Zimbone, S. M., & Zema, D. A. (2018). A local multi-criteria assessment of alternative feeds for biogas plants in Calabria (Southern Italy). *Chemical Engineering Transactions*, *65*, 859–864. doi:10.3303/CET1865144

Anish Kumar, K., Senthil Kumar, P., Madhusudanan, S., Pasupathy, V., Vignesh, P. R., & Sankaranarayanan, A. R. (2017). A simplified model for evaluating best biodiesel production method: Fuzzy analytic hierarchy process approach. *Sustainable Materials and Technologies*, *12*, 18–22. doi:10.1016/j.susmat.2017.03.002

Arabi, M., Yaghoubi, S., & Tajik, J. (2019). Algal biofuel supply chain network design with variable demand under alternative fuel price uncertainty: A case study. *Computers & Chemical Engineering*, *130*, 106528. doi:10.1016/j.compchemeng.2019.106528

Aragonés-Beltrán, P., Chaparro-González, F., Pastor-Ferrando, J. P., & Rodríguez-Pozo, F. (2010). An ANP-based approach for the selection of photovoltaic solar power plant investment projects. *Renewable & Sustainable Energy Reviews*, *14*(1), 249–264. doi:10.1016/j.rser.2009.07.012

Arce, M. E. (2015). The influence of parameters variability on biomass selection for energy use. *Energy, Sustainability and Society*, *5*(1), 1–10. doi:10.118613705-015-0042-z

Archetti, C., Bertazzi, L., Laporte, G., & Speranza, M. G. (2007). A Branch-and-Cut Algorithm for a Vendor-Managed Inventory-Routing Problem. *Transportation Science*, *41*(3), 382–391. doi:10.1287/trsc.1060.0188

Arsu, T. (2021, March). Finansal Performansin Entropi Tabanli Aras Yöntemi İle Değerlendirilmesi: Bist Elektrik, Gaz Ve Buhar Sektöründeki İşletmeler Üzerine Bir Uygulama. *Hacettepe Üniversitesi İktisadi ve İdari Bilimler Fakültesi Dergisi*, *39*(1), 15–32. doi:10.17065/huniibf.740393

Arsu, T., & Ayçin, E. (2021). Evaluation of third-party reverse logistics service provider selection criteria with fuzzy Swara method. *Journal of Yasar University*, *16*(63), 1282–1300.

Asakereh, A., Omid, M., Alimardani, R., & Sarmadian, F. (2014). Developing a GIS-based fuzzy AHP model for selecting solar energy sites in shodirwan region in Iran. *Int J Adv Sci Technol.*, *68*, 37–48. doi:10.14257/ijast.2014.68.04

Ateş, S., & Topal, A. (2021). Entropi temelli topsis, aras ve moosra yöntemleri ile güneş enerji santrali kuruluş yeri seçimi: kop bölgesi örneği. *Uluslararası Yönetim İktisat ve İşletme Dergisi*, *17*(4), 1099–1119. doi:10.17130/ijmeb.869594

Aworanti, O. A., Agarry, S. E., & Ogunleye, O. O. (2017). Biomethanization of Cattle Manure, Pig Manure and Poultry Manure Mixture in Co-digestion with Waste of Pineapple Fruit and Content of Chicken-Gizzard- Part I: Kinetic and Thermodynamic Modelling Studies. *The Open Biotechnology Journal*, *11*(1), 36–53. doi:10.2174/1874070701711010036

Azevedo, S. G., Sequeira, T., Santos, M., & Mendes, L. (2019). Biomass-related sustainability: A review of the literature and interpretive structural modeling. *Energy*, *171*, 1107–1125. doi:10.1016/j.energy.2019.01.068

Babazadeh, R., Razmi, J., Rabbani, M., & Pishvaee, M. S. (2017). An integrated data envelopment analysis–mathematical programming approach to strategic biodiesel supply chain network design problem. *Journal of Cleaner Production*, *147*, 694–707. doi:10.1016/j.jclepro.2015.09.038

Badea, G., Naghiu, G.S., Safirescu, C., Mureşan, D., Badea, F. & Megyesi, E. (2014). Choosing The Optimal Multi-Junctions Photovoltaic Cells for Application in The Field of Concentrated Photovoltaic. *Computer Applications in Environmental Sciences and Renewable Energy*, 144-150.

Badri, M. A. (1999). Combining the analytic hierarchy process and goal programming for global facility location-allocation problem. *International Journal of Production Economics*, *62*(3), 237–248. doi:10.1016/S0925-5273(98)00249-7

Bagocius, V., Zavadskas, E. K., & Turskis, Z. (2014). Multi-person selection of the best wind turbine based on the multi-criteria integrated additive-multiplicative utility function. *Journal of Civil Engineering and Management*, *20*(4), 590–599. doi:10.3846/13923730.2014.932836

Balezentiene, L., Streimikiene, D., & Balezentis, T. (2013). Fuzzy decision support methodology for sustainable energy crop selection. *Renewable & Sustainable Energy Reviews*, *17*, 83–93. doi:10.1016/j.rser.2012.09.016

Bal, H., & Örkçü, H. H. (2005). Combining the discriminant analysis and the data envelopment analysis in view of multiple criteria decision making: A new model. *Gazi University Journal of Science*, *18*(3), 355–364.

Balin, A., & Baracli, H. (2017). A fuzzy multi-citeria decision making methadology based upon the interval Type-2 fuzzy sets for evaluating renewable energy alternatives in Turkey. *Technological and Economic Development of Economy*, *23*(5), 742–763. doi:10.3846/20294913.2015.1056276

Balo, F., & Sagbansua, L. (2016). The selection of the best solar panel for the photovoltaic system design by using AHP. *Energy Procedia*, *100*, 50–53. doi:10.1016/j.egypro.2016.10.151

Baltagi, B. H., & Griffin, J. M. (1988). A general index of technical change. *Journal of Political Economy*, *96*(1), 20–41. doi:10.1086/261522

Baltazar, B. M., Remolador, M., Sevilla, K. H., Saladaga, I., Ang, M. R. C. O., & Inocencio, L. C. V. (2016). *Locating potential site for biomass power plant development in Central Luzon, Philippines using LANDSAT-based suitability map*. Academic Press.

Barin, A., Canha, L. N., Magnago, K. M., Matos, M. A., & Wottrich, B. (2011). A novel fuzzy-based methodology for biogas fuelled hybrid energy systems decision making. In K. Gopalakrishnan, S. K. Khaitan, & S. Kalogirou (Eds.), Studies in Fuzziness and Soft Computing (Vol. 269, pp. 183–198). Academic Press.

Barros, C. (2008). Efficiency analysis of hydroelectric generating plants: A case study for Portugal. *Energy Economics*, *1*(1), 59–75. doi:10.1016/j.eneco.2006.10.008

Barry, F., Sawadogo, M., Bologo, M., Ouédraogo, I. W. K., & Dogot, T. (2021). Key barriers to the adoption of biomass gasification in Burkina Faso. *Sustainability (Switzerland)*, *13*(13). Advance online publication. doi:10.3390u13137324

Bastan, M., Nahand, P. K., Korlou, S., & Hamid, M. (2019). *Selection of a biomass product using a hybrid approach of BW-PROMETHEE*. Academic Press.

Baykasoğlu, A., Kaplanoğlu, V., Durmuşoğlu, Z. D. U., & Şahin, C. (2013). Integrating fuzzy DEMATEL and fuzzy hierarchical TOPSIS methods for truck selection. *Expert Systems with Applications*, *40*(3), 899–907. doi:10.1016/j.eswa.2012.05.046

Baysal, M. E., & Cetin, N. C. (2018). Priority ranking for energy resources in Turkey and investment planning for renewable energy resources. *Complex & Intelligent Systems*, *4*(4), 261–269. doi:10.100740747-018-0075-y

Ben Yosef, G., Navon, A., Poliak, O., Etzion, N., Gal, N., Belikov, J., & Levron, Y. (2021). Frequency stability of the Israeli power grid with high penetration of renewable sources and energy storage systems. *Energy Reports*, *7*, 6148–6161. doi:10.1016/j.egyr.2021.09.057

Bhushan, S., Rana, M. S., Bhandari, M., Sharma, A. K., Simsek, H., & Prajapati, S. K. (2021). Enzymatic pretreatment of algal biomass has different optimal conditions for biogas and bioethanol routes. *Chemosphere*, *284*(131264), 1–10. doi:10.1016/j.chemosphere.2021.131264 PMID:34216928

Bilgiç, S., Torğul, B., & Paksoy, T. (2021). Evaluation of renewable energy resources with Bwm for sustainable energy management. *Journal of Productivity*, (2), 95–110.

Billig, E., & Thrän, D. (2016). Evaluation of biomethane technologies in Europe – Technical concepts under the scope of a Delphi-Survey embedded in a multi-criteria analysis. *Energy*, *114*, 1176–1186. doi:10.1016/j.energy.2016.08.084

Binswanger, M. (2001). Technological progress and sustainable development: What about the rebound effect? *Ecological Economics*, *36*(1), 119–132. doi:10.1016/S0921-8009(00)00214-7

Boafo-Mensah, G., Neba, F. A., Tornyeviadzi, H. M., Seidu, R., Darkwa, K. M., & Kemausuor, F. (2021). Modelling the performance potential of forced and natural-draft biomass cookstoves using a hybrid Entropy-TOPSIS approach. *Biomass and Bioenergy*, *150*, 106106. doi:10.1016/j.biombioe.2021.106106

Böhler, L., Krail, J., Görtler, G., & Kozek, M. (2020). Fuzzy model predictive control for small-scale biomass combustion furnaces. *Applied Energy*, *276*, 115339. doi:10.1016/j.apenergy.2020.115339

Bojić, S., Đatkov, Đ., Brcanov, D., Georgijević, M., & Martinov, M. (2013). Location allocation of solid biomass power plants: Case study of Vojvodina. *Renewable & Sustainable Energy Reviews*, *26*, 769–775. doi:10.1016/j.rser.2013.06.039

Boran, F. E. (2018). A new approach for evaluation of renewable energy resources: A case of Turkey. *Energy Sources. Part B, Economics, Planning, and Policy*, *13*(3), 196–204. doi:10.1080/15567249.2017.1423414

Boran, F. E., Boran, K., & Menlik, T. (2012). The evaluation of renewable energy technologies for electricity generation in Turkey using intuitionistic fuzzy TOPSIS. *Energy Sources. Part B, Economics, Planning, and Policy*, *7*(1), 81–90. doi:10.1080/15567240903047483

Bottani, E., & Rizzi, A. (2006). A fuzzy TOPSIS methodology to support outsourcing of logistics services. *Supply Chain Management*, *11*(4), 294–308. doi:10.1108/13598540610671743

BP. (2021). *BP Statistical Review of World Energy*. https://www.bp.com/content/dam/bp/business-sites/en/global/corporate/pdfs/energy-economics/statistical-review/bp-stats-review-2020-renewable-energy.pdf

Brand, C., Tran, M., & Anable, J. (2012). The UK transport carbon model: An integrated life cycle approach to explore low carbon futures. *Energy Policy*, *41*, 107–124. doi:10.1016/j.enpol.2010.08.019

Broeck, V., & Meeusen, W. (1977). Efficiency estimation from Cobb-Douglas production functions with composed error. *International Economic Review*, *18*(2), 435–444. doi:10.2307/2525757

Brooks, S. P., & Morgan, B. J. T. (1995). Optimization Using Simulated Annealing. *The Statistician*, *44*(2), 241–257. doi:10.2307/2348448

Bruce, T. (2011). *Investigation of Cost and Performance Characteristic of Photovoltaic Panels*. University of Southern Queensland Faculty of Engineering and Surveying.

Buyukozkan, G., & Guleryuz, S. (2014). A new GDM based AHP framework with linguistic interval fuzzy preference relations for renewable energy planning. *Journal of Intelligent & Fuzzy Systems*, *27*(6), 3181–3195. doi:10.3233/IFS-141275

Büyüközkan, G., & Güleryüz, S. (2016). An integrated Dematel-Anp approach for renewable energy resources selection in Turkey. *International Journal of Production Economics*, *182*, 435–448. doi:10.1016/j.ijpe.2016.09.015

Büyüközkan, G., & Güleryüz, S. (2017). Evaluation of Renewable Energy Resources in Turkey using an integrated MCDM approach with linguistic interval fuzzy preference relations. *Energy*, *123*, 149–163. doi:10.1016/j.energy.2017.01.137

Büyüközkan, G., Havle, C. A., & Feyzioğlu, O. (2021). An integrated SWOT based fuzzy AHP and fuzzy MARCOS methodology for digital transformation strategy analysis in airline industry. *Journal of Air Transport Management*, *97*, 102142. doi:10.1016/j.jairtraman.2021.102142

Buyukozkan, G., Karabulut, Y., & Guler, M. (2018a). Strategic renewable energy source selection for turkey with hesitant fuzzy MCDM method. In *Energy Management—Collective and Computational Intelligence with Theory and Applications* (pp. 229–250). Springer.

Buyukozkan, G., Karabulut, Y., & Mukul, E. (2018b). A novel renewable energy selection model for United Nations' sustainable development goals. *Energy*, *165*, 290–302. doi:10.1016/j.energy.2018.08.215

Büyüközkan, G., Mukul, E., & Kongar, E. (2021). Health tourism strategy selection via SWOT analysis and integrated hesitant fuzzy linguistic AHP-MABAC approach. *Socio-Economic Planning Sciences*, *74*, 100929. doi:10.1016/j.seps.2020.100929

Çabukoglu, E., Georges, G., Küng, L., Pareschi, G., & Boulouchos, K. (2018). Battery electric propulsion: An option for heavy-duty vehicles? Results from a Swiss case-study. *Transportation Research Part C, Emerging Technologies*, *88*, 107–123. doi:10.1016/j.trc.2018.01.013

Cao, J. X., Wang, X., & Gao, J. (2021). A two-echelon location-routing problem for biomass logistics systems. *Biosystems Engineering*, *202*, 106–118. doi:10.1016/j.biosystemseng.2020.12.007

Cao, J. X., Zhang, Z., & Zhou, Y. (2021). A location-routing problem for biomass supply chains. *Computers & Industrial Engineering*, *152*(107017), 1–11. doi:10.1016/j.cie.2020.107017

Çapik, M., Yılmaz, A. O., & Çavuşoğlu, İ. (2012). Present situation and potential role of renewable energy in Turkey. *Renewable Energy*, *46*, 1–13. doi:10.1016/j.renene.2012.02.031

Cavallaro, F. A. (2010). Comparative assessment of thin-film photovoltaic production processes using the ELECTRE III method. *Energy Policy*, *38*, 463–474.

Cavallaro, F., & Ciraolo, L. (2005). A multi-criteria approach to evaluate wind energy plants on an Italian island. *Energy Policy*, *33*(2), 235–244. doi:10.1016/S0301-4215(03)00228-3

Cebi, S., Ilbahar, E., & Atasoy, A. (2016). A fuzzy information axiom based method to determine the optimal location for a biomass power plant: A case study in Aegean Region of Turkey. *Energy*, *116*, 894–907. doi:10.1016/j.energy.2016.10.024

Çelikbilek, Y., & Tüysüz, F. (2016). An integrated grey based multi-criteria decision making approach for the evaluation of renewable energy sources. *Energy*, *115*, 1246–1258. doi:10.1016/j.energy.2016.09.091

Çepelioğullar, Ö., & Pütün, A. E. (2014). A pyrolysis study for the thermal and kinetic characteristics of an agricultural waste with two different plastic wastes. *Waste Management & Research*, *32*(10), 971–979. doi:10.1177/0734242X14542684 PMID:25062939

Chachuli, F. S. M., Ludin, N. A., Mat, S., & Sopian, K. (2020). Renewable energy performance evaluation studies using the data envelopment analysis (DEA): A systematic review. *Journal of Renewable and Sustainable Energy*, *12*(6), 062701. doi:10.1063/5.0024750

Chan, C. C., Bouscayrol, A., & Chen, K. (2009). Electric, hybrid, and fuel-cell vehicles: Architectures and modeling. *IEEE Transactions on Vehicular Technology*, *59*(2), 589–598. doi:10.1109/TVT.2009.2033605

Chandna, R., Saini, S., & Kumar, S. (2021). Fuzzy AHP based performance evaluation of massive online courses provider for online learners. *Materials Today: Proceedings*, *46*, 11103–11112. doi:10.1016/j.matpr.2021.02.255

Chang, D.-Y. (1996). Applications of the extent analysis method on fuzzy AHP. *European Journal of Operational Research*, *95*(3), 649–655. doi:10.1016/0377-2217(95)00300-2

Charnes, A., Cooper, W. W., & Rhodes, E. (1978). Measuring the efficiency of decision making units. *European Journal of Operational Research*, *2*(6), 429–444. doi:10.1016/0377-2217(78)90138-8

Chen, H. C., & Yang, C. H. (2014). A Multi-Criterion Analysis of Cross-Strait Co-Opetitive Strategy in the Crystalline Silicon Solar Cell Industry. Mathematical Problems in Engineering.

Cheng, W., Zhang, Y., & Wang, P. (2020). Effect of spatial distribution and number of raw material collection locations on the transportation costs of biomass thermal power plants. *Sustainable Cities and Society, 55*, 102040. doi:10.1016/j.scs.2020.102040

Chen, H. H., Kang, H. Y., & Lee, A. H. I. (2010). Strategic selection of suitable projects for hybrid solar-wind power generation systems. *Renewable & Sustainable Energy Reviews, 14*(1), 413–441. doi:10.1016/j.rser.2009.08.004

Chen, X. (2016). Economic potential of biomass supply from crop residues in China. *Applied Energy, 166*, 141–149. doi:10.1016/j.apenergy.2016.01.034

Chien, T., & Hu, J. (2007). Renewable energy and macro economic efficiency of OECD and non-OECD economies. *Energy Policy, 35*(7), 3606–3615. doi:10.1016/j.enpol.2006.12.033

Chu, T. C. (2002). Facility location selection using fuzzy Topsis under group decisions. *International Journal of Uncertainty, Fuzziness and Knowledge-based Systems, 10*(06), 687–701. doi:10.1142/S0218488502001739

Cobuloglu, H. I., & Büyüktahtakin, I. E. (2014). *A multi-criteria approach for biomass crop selection under fuzzy environment*. Academic Press.

Cobuloglu, H. I., & Büyüktahtakin, I. E. (2015). A stochastic multi-criteria decision analysis for sustainable biomass crop selection. *Expert Systems with Applications, 42*(15-16), 6065–6074. doi:10.1016/j.eswa.2015.04.006

Coelli, T. J. (1995). Estimators and hypothesis tests for a stochastic frontier function: A Monte Carlo analysis. *Journal of Productivity Analysis, 6*(3), 247–268. doi:10.1007/BF01076978

Coelli, T. J., Rao, D. S. P., O'Donnell, C. J., & Battese, G. E. (2005). *An Introduction to Efficiency and Productivity Analysis*. Springer.

Coffey, L., & Claudio, D. (2021). In defense of group fuzzy AHP: A comparison of group fuzzy AHP and group AHP with confidence intervals. *Expert Systems with Applications, 178*, 114970. doi:10.1016/j.eswa.2021.114970

Colak, M., & Kaya, I. (2017). Prioritization of renewable energy alternatives by using an integrated fuzzy MCDM model: A real case application for Turkey. *Renewable & Sustainable Energy Reviews, 80*, 840–853. doi:10.1016/j.rser.2017.05.194

Costa, F. R., Ribeiro, C. A. A. S., Marcatti, G. E., Lorenzon, A. S., Teixeira, T. R., Domingues, G. F., Castro, N. L. M., Santos, A. R., Soares, V. P., Menezes, S. J. M. C., Mota, P. H. S., Telles, L. A. A., & Carvalho, J. R. D. (2020). GIS applied to location of bioenergy plants in tropical agricultural areas. *Renewable Energy, 153*, 911–918. doi:10.1016/j.renene.2020.01.050

Curiel-Esparza, J., Reyes-Medina, M., Martin-Utrillas, M., Martinez-Garcia, M. P., & Canto-Perello, J. (2019). Collaborative elicitation to select a sustainable biogas desulfurization technique for landfills. *Journal of Cleaner Production, 212*, 1334–1344. doi:10.1016/j.jclepro.2018.12.095

Cutz, L., Haro, P., Santana, D., & Johnsson, F. (2016). Assessment of biomass energy sources and technologies: The case of Central America. *Renewable & Sustainable Energy Reviews, 58*, 1411–1431. doi:10.1016/j.rser.2015.12.322

Dablanc, L. (2009). Freight transport for development toolkit: Urban freight (No. 57971). The World Bank.

Dağdeviren, M., Akay, D., & Kurt, M. (2004). Analytical hierarchy process for job evaluation and application *J. Fac. Eng. Arch. Gazi Univ., 19*(2), 131–138.

Dağdeviren, M., Yavuz, S., & Kılınç, N. (2009). Weapon selection using the AHP and TOPSIS methods under fuzzy environment. *Expert Systems with Applications, 36*(4), 8143–8151. doi:10.1016/j.eswa.2008.10.016

Dağıstan, H. (2008). Our renewable energy and geothermal resources. In *Proceedings 5th World Water Forum*. Sözkesen Publishing.

Damgacı, E., Boran, K., & Boran, F. E. (2017). Evaluation of Turkey's renewable energy using intuitionistic fuzzy Topsis method. *Journal of Polytechnic, 20*(3), 628–637.

Damgaci, E., Boran, K., & Boran, F. E. (2017). Evaluation of Turkey's renewable energy using intuitionistic fuzzy TOPSIS method. *Journal of Polytechnic, 20*(3), 629–637.

Darshini, D., Dwivedi, P., & Glenk, K. (2013). Capturing stakeholders´ views on oil palm-based biofuel and biomass utilisation in Malaysia. *Energy Policy, 62*, 1128–1137. doi:10.1016/j.enpol.2013.07.017

Davtalab, M., & Alesheikh, A. A. (2018). Spatial optimization of biomass power plant site using fuzzy analytic network process. *Clean Technologies and Environmental Policy, 20*(5), 1033 1046. doi:10.100710098-018-1531-5

de Carlo, F., & Schiraldi, M. M. (2013). *Sustainable choice of the location of a biomass plant: An application in Tuscany*. Academic Press.

de Clercq, D., Wen, Z., & Fan, F. (2017). Performance evaluation of restaurant food waste and biowaste to biogas pilot projects in China and implications for national policy. *Journal of Environmental Management, 189*, 115–124. doi:10.1016/j.jenvman.2016.12.030 PMID:28012386

Delivand, M. K., Cammerino, A. R. B., Garofalo, P., & Monteleone, M. (2015). Optimal locations of bioenergy facilities, biomass spatial availability, logistics costs and GHG (greenhouse gas) emissions: A case study on electricity productions in South Italy. *Journal of Cleaner Production, 99*, 129–139. doi:10.1016/j.jclepro.2015.03.018

Demirtas, O. (2013). Evaluating the best renewable energy technology for sustainable energy planning. *International Journal of Energy Economics and Policy, 3*, 23–33.

den Herder, M., Kurttila, M., Leskinen, P., Lindner, M., Haatanen, A., Sironen, S., Salminen, O., Juusti, V., & Holma, A. (2017). Is enhanced biodiversity protection conflicting with ambitious bioenergy targets in eastern Finland? *Journal of Environmental Management*, *187*, 54–62. doi:10.1016/j.jenvman.2016.10.065 PMID:27883939

Derse, O., & Yontar, E. (2020). Determination of the most appropriate renewable energy source by Swara-Topsis method. *Journal of Industrial Engineering*, *31*(3), 389–419.

Derse, O., & Yontar, E. (2020). Determination of the most appropriate renewable energy source by SWARA-TOPSIS Method. *Journal of Industrial Engineering*, *31*(3), 389–419.

Derse, O., & Yontar, E. (2020). Determination of the Most Appropriate Renewable Energy Source by SWARA-TOPSIS Method. *Journal of Industrial Engineering*, *31*(3), 389–419.

Deveci, K., Cin, R., & Kagızman, A. (2020). A modified interval valued intuitionistic fuzzy CODAS method and its application to multi-criteria selection among renewable energy alternatives in Turkey. *Applied Soft Computing*, *96*, 106660. doi:10.1016/j.asoc.2020.106660

Dhanalakshmi, C. S., Madhu, P., Karthick, A., Mathew, M., & Vignesh Kumar, R. (2020). A comprehensive MCDM-based approach using TOPSIS and EDAS as an auxiliary tool for pyrolysis material selection and its application. *Biomass Conversion and Biorefinery*. Advance online publication. doi:10.100713399-020-01009-0

Dimitrakopoulos, G. J., Uden, L., & Varlamis, I. (2020). *The future of intelligent transport systems*. Elsevier.

Dincer, I. (2000). Renewable energy and sustainable development: A crucial review. *Renewable & Sustainable Energy Reviews*, *4*(2), 157–175. doi:10.1016/S1364-0321(99)00011-8

Djaković, D. D., Gvozdenac-Urošević, B. D., & Vasić, G. M. (2016). Multi-criteria analysis as a support for national energy policy regarding the use of biomass: Case study of Serbia. *Thermal Science*, *20*(2), 371–380. doi:10.2298/TSCI150602190D

Dogan, H., & Uludag, A. S. (2018). Evaluation of renewable energy alternatives and selection of suitable facility location: A study in Turkey. *The International Journal of Economic and Social Research*, *14*(2), 157–180.

Durairaj, S., Sathiya Sekar, K., & Ilangkumaran, M., RamManohar, M., Thyalan, B., Yuvaraj, E., & Ramesh, S. (2014). Multi-criteria decision model for biodiesel selection in an electrical power generator based on FAHP-GRA-TOPSIS. *International Journal of Research in Engineering and Technology*, *3*(23), 226–233. doi:10.15623/ijret.2014.0323050

Ecer, F. (2021). An analysis of the factors affecting wind farm site selection through Fucom subjective weighting method. *Pamukkale University Journal of Engineering Sciences*, *27*(1), 24–34. doi:10.5505/pajes.2020.93271

Ecer, F., Pamucar, D., Mardani, A., & Alrasheedi, M. (2021). Assessment of renewable energy resources using new interval rough number extension of the level based weight assessment and combinative distance-based assessment. *Renewable Energy*, *170*, 1156–1177. doi:10.1016/j.renene.2021.02.004

Economic, Social and Environmental Council Rapport. (2018). *L'économie bleue: pilier d'un nouveau modèle de développement du Maroc* [The blue economy: pillar of a new development model for Morocco]. Available online: https://www.cese.ma/media/2020/11/Rapport-AS38-VF-2.pdf

Efeoglu, R., & Pehlivan, C. (2018). The effects of energy consumption and current deficit on economic growth in Turkey. *Journal of Political Economic Theory*, *2*(1), 104–105.

Electric and hybrid car in Morocco. (2019). *Study on sustainable mobility in Morocco*. Group Sunergia. Available online: https://groupe-sunergia.com/wp-content/uploads/2019/11/Potentiel-3.png

El-Halwagi, A. M., Rosas, C., Ponce-Ortega, J. M., Jiménez-Gutiérrez, A., Mannan, M. S., & El-Halwagi, M. M. (2013). Multiobjective optimization of biorefineries with economic and safety objectives. *AIChE Journal. American Institute of Chemical Engineers*, *59*(7), 2427–2434. doi:10.1002/aic.14030

Elhasbi, A., Jami, J., & Kammas, S. (2015). Sustainable transport in morocco: What contingency factors for which maturity level? Case of road haulage" RH" service providers in the region of Casablanca metropolis. *European Scientific Journal*, *11*(29).

Energy Atlas. (2021). *Hidroelecticity power plants*. https://www.enerjiatlasi.com/hidroelektrik/

Erdal, L. (2012). Renewable energy investments and potential for green jobs in Turkey. *Journal of Social and Human Sciences*, *4*(1), 171–181.

Erdogan, M., & Kaya, I. (2015). An integrated multi-criteria decision-making methodology based on Type-2 fuzzy sets for selection among energy alternatives in Turkey. *Iranian Journal of Fuzzy Systems*, *12*(1), 1–25.

Eroglu, H., & Sahin, R. (2020). A neutrosophic VIKOR method-based decision-making with an improved distance measure and score function: Case study of selection for renewable energy alternatives. *Cognitive Computation*, *12*(6), 1338–1355. doi:10.100712559-020-09765-x

Ertay, T., Kahraman, C., & Kaya, I. (2013). Evaluation of renewable energy alternatives using MACBETH and fuzzy AHP multicriteria methods: The case of Turkey. *Technological and Economic Development of Economy*, *19*(1), 38–62. doi:10.3846/20294913.2012.762950

Ertuğrul, İ. (2011). Fuzzy group decision making for the selection of facility location. *Group Decision and Negotiation*, *20*(6), 725–740. doi:10.100710726-010-9219-1

ETKB. (2021a). *Solar*. https://enerji.gov.tr/eigm-yenilenebilir-enerji-kaynaklar-gunes

ETKB. (2021b). *Wind energy potential atlas*. https://repa.enerji.gov.tr/REPA/bolgeler/TURKIYE-GENELI.pdf

ETKB. (2021c). *Geothermal.* https://enerji.gov.tr/bilgi-merkezi-enerji-

ETKB. (2021d). *Biomass.* https://enerji.gov.tr/bilgi-merkezi-enerji-biyokutle

European Parliament and of the Council. (2007). *Establishing a Framework for the Approval of Motor Vehicles and their Trailers, and of Systems, Components and Separate Technical Units Intended for such Vehicles.* Available online: https://www.legislation.gov.uk/eudr/2007/46/annex/I/division/2/2020-01-31

European strategy for low emission mobility, (2016). *Annexe 1 historical development activity inn transport activity and energy use and emissions.* Available online: https://ec.europa.eu/transparency/documents-register/detail?ref=COM(2016)501&lang=en

Ezbakhe, F., & Pérez-Foguet, A. (2021). Decision analysis for sustainable development: The case of renewable energy planning under uncertainty. *European Journal of Operational Research, 291*(2), 601–613. doi:10.1016/j.ejor.2020.02.037

Fan, Y. T., Zhang, G. S., Guo, X. Y., Xing, Y., & Fan, M. H. (2006). Biohydrogen-production from beer lees biomass by cow dung compost. *Biomass and Bioenergy, 30*(5), 493–496. doi:10.1016/j.biombioe.2005.10.009

Farrell, M. J. (1957). The measurement of productive efficiency. *Journal of the Royal Statistical Society. Series A (General), 120*(3), 253–290. doi:10.2307/2343100

Fehrenbach, H., Giegrich, J., Reinhardt, G., Schmitz, J., Sayer, U., Gretz, M., . . . Lanje, K. (2008). *Criteria for a sustainable use of bioenergy on a global scale.* Retrieved from Germany: https://www.osti.gov/etdeweb/servlets/purl/21240931

Feng, W., & Figliozzi, M. (2013). An economic and technological analysis of the key factors affecting the competitiveness of electric commercial vehicles: A case study from the USA market. *Transportation Research Part C, Emerging Technologies, 26*, 135–145. doi:10.1016/j.trc.2012.06.007

Firouzi, S., Allahyari, M. S., Isazadeh, M., Nikkhah, A., & Van Haute, S. (2021). Hybrid multi-criteria decision-making approach to select appropriate biomass resources for biofuel production. *The Science of the Total Environment, 770*, 144449. doi:10.1016/j.scitotenv.2020.144449 PMID:33513499

Franco, C., Bojesen, M., Hougaard, J. L., & Nielsen, K. (2015). A fuzzy approach to a multiple criteria and Geographical Information System for decision support on suitable locations for biogas plants. *Applied Energy, 140*, 304–315. doi:10.1016/j.apenergy.2014.11.060

Fu, H.-H., Chen, Y.-Y., & Wang, G.-J. (2020). Using a Fuzzy Analytic Hierarchy Process to Formulate an Effectual Tea Assessment System. *Sustainability, 12*(15), 6131. doi:10.3390u12156131

Galang, W. N., Tabañag, I. D., & Loretero, M. (2021). GIS-based biomass energy sustainability analysis using analytical hierarchy process: A case study in Medellin, Cebu. *International Journal of Renewable Energy Development, 10*(3), 551–561. doi:10.14710/ijred.0.33260

Galvez, D., Rakotondranaivo, A., Morel, L., Camargo, M., & Fick, M. (2015). Reverse logistics network design for a biogas plant: An approach based on MILP optimization and Analytical Hierarchical Process (AHP). *Journal of Manufacturing Systems*, *37*, 616–623. doi:10.1016/j.jmsy.2014.12.005

Gamboa, G., & Munda, G. (2007). The problem of wind farm location a social multi-criteria evaluation framework. *Energy Policy*, *35*(3), 1564–1583. doi:10.1016/j.enpol.2006.04.021

Gandhi, P., Kunwar, P., Pareek, N., Mathur, S., Lizasoain, J., Gronauer, A., ... Vivekanand, V. (2018). Multicriteria Decision Model and Thermal Pretreatment of Hotel Food Waste for Robust Output to Biogas: Case Study from City of Jaipur, India. *BioMed Research International*, *13*. Advance online publication. doi:10.1155/2018/9416249 PMID:30306090

Gautam, S., & LeBel, L. (2016). *A multi-criteria decision making approach to locate a terminal in bioenergy supply chains*. Canadian Institute of Forestry.

Gençoğlu, M. T. (2002). Importance of renewable energy resources for Turkey. *Fırat University Journal of Science and Engineering Sciences*, *14*(2), 57–64.

Gendreau, M., Guertin, F., Potvin, J.-Y., & Taillard, É. (1999). Parallel Tabu Search for Real-Time Vehicle Routing and Dispatching. *Transportation Science*, *33*(4), 381–390. doi:10.1287/trsc.33.4.381

George, J., Arun, P., & Muraleedharan, C. (2021). Region-specific biomass feedstock selection for gasification using multi-attribute decision-making techniques. *International Journal of Sustainable Engineering*, *14*(5), 1101–1109. doi:10.1080/19397038.2020.1790058

Geri, F., Sacchelli, S., Bernetti, I., & Ciolli, M. (2018). Urban-rural bioenergy planning as a strategy for the sustainable development of inner areas: A GIS-based method to chance the forest chain. *Green Energy Technology*, *0*, 539–550.

Ghenai, C., Albawab, M., & Bettayeb, M. (2020). Sustainability indicators for renewable energy systems using multi-criteria decision-making model and extended Swara/Aras hybrid method. *Renewable Energy*, *146*, 580–597. doi:10.1016/j.renene.2019.06.157

Ghorabaee, M. K., Amiri, M., Zavadskas, E. K., & Antucheviciene, J. (2018). A new hybrid fuzzy MCDM approach for evaluation of construction equipment with sustainability considerations. *Archives of Civil and Mechanical Engineering*, *18*(1), 32–49. doi:10.1016/j.acme.2017.04.011

Ghose, D., Naskar, S., & Uddin, S. (2019). *Q-GIS-MCDA based approach to identify suitable biomass facility location in Sikkim*. doi:10.1109/ICACCP.2019.8882978

Gil, D. R. G., Costa, M. A., Lopes, A. L. M., & Mayrink, V. D. (2017). Spatial statistical methods applied to the 2015 Brazilian energy distribution benchmarking model: Accounting for unobserved determinants of inefficiencies. *Energy Economics*, *64*, 373–383. doi:10.1016/j.eneco.2017.04.009

Giurca, I., Aşchilean, I., Safirescu, C. O., & Mureşan, D. (2014). Choosing Photovoltaic Panels Using The Promethee Method. *Proceedings of the 8th International Management Conference "Management Challenges For Sustainable Development"*.

Goepel, K. D. (2018). Implementation of an online software tool for the analytic hierarchy process (AHP-OS). *International Journal of the Analytic Hierarchy Process*, *10*(3). Advance online publication. doi:10.13033/ijahp.v10i3.590

Göhlich, D., Nagel, K., Syré, A. M., Grahle, A., Martins-Turner, K., Ewert, R., Miranda Jahn, R., & Jefferies, D. (2021). Integrated approach for the assessment of strategies for the decarbonization of urban traffic. *Sustainability*, *13*(2), 839. doi:10.3390u13020839

Gökgöz, F., & Güvercin, M. T. (2018). Energy security and renewable energy efficiency in EU. *Renewable & Sustainable Energy Reviews*, *96*, 226–239. doi:10.1016/j.rser.2018.07.046

González-Cruz, L. A., Morales-Mendoza, L. F., Aguilar-Lasserre, A. A., Azzaro-Pantel, C., Martínez-Isidro, P., & Meza-Palacios, R. (2021). Optimal ecodesign selection for biodiesel production in biorefineries through multicriteria decision making. *Clean Technologies and Environmental Policy*, *23*(8), 2337–2356. Advance online publication. doi:10.100710098-021-02141-9

Guenounou, A., Malek, A., & Aillerie, M. (2016). Comparative performance of PV panels of different technologies over one year of exposure: Application to a coastal Mediterranean region of Algeria. *Energy Conversion and Management*, *114*, 356–363.

Güldeş, M., Atici, U., & Şahin, C. (2022). Fuzzy Resource-Constrained Project Scheduling Under Learning Considerations. Intelligent and Fuzzy Techniques for Emerging Conditions and Digital Transformation, Cham.

Guler, D., Charisoulis, G., Buttenfield, B. P., & Yomralioglu, T. (2021). Suitability modeling and sensitivity analysis for biomass energy facilities in Turkey. *Clean Technologies and Environmental Policy*, *23*(7), 2183–2199. doi:10.100710098-021-02126-8

Guo, J. X., & Zhu, K. W. (2021). *Operation management of hybrid biomass power plant considering environmental constraints*. Sustainable Production and Consumption., doi:10.1016/j.spc.2021.09.017

Gupta, S., Dangayach, G. S., Singh, A. K., & Rao, P. N. (2015). Analytic Hierarchy Process (AHP) Model for Evaluating Sustainable Manufacturing Practices in Indian Electrical Panel Industries. *Procedia: Social and Behavioral Sciences*, *189*, 208–216.

Gurtu, A. (2019). A pioneering approach to reducing fuel cost and carbon emissions from transportation. *Transportation Journal*, *58*(4), 309–322. doi:10.5325/transportationj.58.4.0309

Hajinajaf, N., Mehrabadi, A., & Tavakoli, O. (2021). Practical strategies to improve harvestable biomass energy yield in microalgal culture: A review. *Biomass and Bioenergy*, *145*(105941), 1–11. doi:10.1016/j.biombioe.2020.105941

Halkos, G. E., & Tzeremes, N. G. (2012). Analyzing the Greek renewable energy sector: A data envelopment analysis approach. *Renewable & Sustainable Energy Reviews*, *16*(5), 2884–2893. doi:10.1016/j.rser.2012.02.003

Halog, A. (2011). Sustainable development of bioenergy sector: An integrated methodological framework. *International Journal of Multicriteria Decision Making*, *1*(3), 338–361. doi:10.1504/IJMCDM.2011.041193

Hamid, N. N. A., Martinez-Hernandez, E., & Lim, J. S. (2017). Technological screening of algae-based biorefinery for sustainable biofuels production using analytic hierarchy process (AHP) with feature scaling normalisation. *Chemical Engineering Transactions*, *61*, 1369–1374. doi:10.3303/CET1761226

Haugen, H. H., Halvorsen, B. M., & Eikeland, M. S. (2015).. . *Simulation of Gasification of Livestock Manure with Aspen Plus.*, *119*, 271–277. doi:10.3384/ecp15119271

Hepbaşlı, A., & Utlu, Z. (2004). Evaluating the energy utilization efficiency of Turkey's renewable energy sources during 2001. *Renewable & Sustainable Energy Reviews*, *8*(3), 237–255. doi:10.1016/j.rser.2003.11.001

Herrera-Seara, M. A., Aznar Dols, F., Zamorano, M., & Alameda-Hernández, E. (2010). Optimal location of a biomass power plant in the province of granada analyzed by multi-criteria evaluation using appropriate geographic information system according to the analytic hierarchy process. *Renewable Energy and Power Quality Journal*, *1*(8), 813–818. doi:10.24084/repqj08.484

Hoicka, C. E., Conroy, J., & Berka, A. L. (2021). Reconfiguring actors and infrastructure in city renewable energy transitions: A regional perspective. *Energy Policy*, *158*, 112544. doi:10.1016/j.enpol.2021.112544

Hollander, A. L., & Wright, R. E. (1980). Impact of tahanids on cattle: Blood meal size and preferred feeding sites12. *Journal of Economic Entomology*, *73*(3), 431–433. doi:10.1093/jee/73.3.431

Hong, J. D., & Mwakalonge, J. L. (2020). Biofuel logistics network scheme design with combined data envelopment analysis approach. *Energy*, *209*, 118342. Advance online publication. doi:10.1016/j.energy.2020.118342

Hosseinabadi, A. A. R., Rostami, N. S. H., Kardgar, M., Mirkamali, S., & Abraham, A. (2017). A new efficient approach for solving the capacitated Vehicle Routing Problem using the Gravitational Emulation Local Search Algorithm. *Applied Mathematical Modelling*, *49*, 663–679. doi:10.1016/j.apm.2017.02.042

Hsiao, W., Hu, J., Hsiao, C., & Chang, M. (2019). Energy efficiency of the Baltic Sea countries: An application of stochastic frontier analysis. *Energies*, *12*(1), 104. doi:10.3390/en12010104

Iasimone, F., Seira, J., Panico, A., De Felice, V., Pirozzi, F., & Steyer, J.-P. (2021). Insights into bioflocculation of filamentous cyanobacteria, microalgae and their mixture for a low-cost biomass harvesting system. *Environmental Research*, *199*(111359), 1–10. doi:10.1016/j.envres.2021.111359 PMID:34022232

IEA. (2016). *Energy and air pollution: World energy outlook special report 2016*. Available online: http://pure.iiasa.ac.at/id/eprint/13467/1/WorldEnergyOutlookSpecialReport2016EnergyandAirPollution.pdf

IEA. (2021). *World Energy Outlook 2021.* https://iea.blob.core.windows.net/assets/888004cf-1a38-4716-9e0c-3b0e3fdbf609/WorldEnergyOutlook2021.pdf

Ilbahar, E., Kahraman, C., & Cebi, S. (2021). Location selection for waste-to-energy plants by using fuzzy linear programming. *Energy*, *234*, 121189. Advance online publication. doi:10.1016/j.energy.2021.121189

Ilbahar, E., Kahraman, C., & Cebi, S. (2022). Risk assessment of renewable energy investments: A modified failure mode and effect analysis based on prospect theory and intuitionistic fuzzy AHP. *Energy*, *239*, 121907. doi:10.1016/j.energy.2021.121907

Ilgın, M. A., & Alkan, E. (2020). Yenilenebilir Enerji Kaynaklarının Türkiye'de Yaygın Kullanımını Engelleyen Faktörlerin Çok Kriterli Karar Verme Teknikleri ve Kalite Evi ile Analiz Edilmesi. *International Journal of Engineering Research and Development*, *12*(1), 1–12. doi:10.29137/umagd.469519

İmre, Ş., Çelebi, D., & Koca, F. (2021). Understanding barriers and enablers of electric vehicles in urban freight transport: Addressing stakeholder needs in Turkey. *Sustainable Cities and Society*, *68*, 102794. doi:10.1016/j.scs.2021.102794

International Energy Agency. (2009). *Transport and CO2 moving toward sustainability. Paris IEA declaration.* General United Nations Publications. Available online: https://www.iea.org/news/transport-energy-and-co2-moving-toward-sustainability

Intuitionistic Fuzzy Sets. (1999). *Intuitionistic Fuzzy Sets Atanassov KT35.* Physica.

Ioannou, K., Tsantopoulos, G., Arabatzis, G., Andreopoulou, Z., & Zafeiriou, E. (2018). A spatial decision support system framework for the evaluation of biomass energy production locations: Case study in the regional unit of drama, greece. *Sustainability*, *10*(2), 531. doi:10.3390u10020531

IRENA. (2020). *Global renewables outlook 2050.* https://irena.org/-/media/Files/IRENA/Agency/Publication/2020/Apr/IRENA_GRO_Summary_2020.pdf?la=en&hash=1F18E445B56228AF8C4893CAEF147ED0163A0E47

Jacob, D. J., & Winner, D. A. (2009). Effect of climate change on air quality. *Atmospheric Environment*, *43*(1), 51–63. doi:10.1016/j.atmosenv.2008.09.051

Janke, J. R. (2010). Multi-criteria GIS modeling of wind and solar farms in Colorado. *Renewable Energy*, *35*(10), 2228–2234. doi:10.1016/j.renene.2010.03.014

Jayathilakan, K., Sultana, K., Radhakrishna, K., & Bawa, A. S. (2012). Utilization of byproducts and waste materials from meat, poultry and fish processing industries: A review. *Journal of Food Science and Technology*, *49*(3), 278–293. doi:10.100713197-011-0290-7 PMID:23729848

Jeong, J. S. (2018). Biomass feedstock and climate change in agroforestry systems: Participatory location and integration scenario analysis of biomass power facilities. *Energies*, *11*(6), 1404. Advance online publication. doi:10.3390/en11061404

Jeong, J. S., & Ramírez-Gómez, A. (2017). A multi-criteria GIS-based assessment to optimize biomass facility sites with parallel environment - A case study in Spain. *Energies, 10*(12), 2095. Advance online publication. doi:10.3390/en10122095

Jeong, J. S., & Ramírez-Gómez, Á. (2018). Optimizing the location of a biomass plant with a fuzzy-DEcision-MAking Trial and Evaluation Laboratory (F-DEMATEL) and multi-criteria spatial decision assessment for renewable energy management and long-term sustainability. *Journal of Cleaner Production, 182*, 509–520. doi:10.1016/j.jclepro.2017.12.072

jeotermal

Jha, D. K., & Shrestha, R. (2006). Measuring efficiency of hydropower plants in Nepal using data envelopment analysis. *IEEE Transactions on Power Systems, 21*(4), 1502–1511. doi:10.1109/TPWRS.2006.881152

Jin, W., Pastor-Pérez, L., Yu, J., Odriozola, J. A., Gu, S., & Reina, T. R. (2020). Cost-effective routes for catalytic biomass upgrading. *Current Opinion in Green and Sustainable Chemistry, 23*, 1–9. doi:10.1016/j.cogsc.2019.12.008

Johnson, L., Lippke, B., & Oneil, E. (2012). Modeling Biomass Collection and Woods Processing Life-Cycle Analysis*. *Forest Products Journal, 62*(4), 258–272. doi:10.13073/FPJ-D-12-00019.1

Júnior, E. S., Colmenero, J. C., & Junior, A. B. (2021). Biomass selection method to produce biogas with a multi-criteria approach. *Waste and Biomass Valorization, 12*(6), 3169–3177. doi:10.100712649-020-01231-x

Kaa, G., Rezaei, J., Kamp, L., & Winter, A. (2014). Photovoltaic technology selection: A fuzzy MCDM approach. *Renewable & Sustainable Energy Reviews, 32*, 662–670. doi:10.1016/j.rser.2014.01.044

Kabak, M., & Dagdeviren, M. (2014). Prioritization of renewable energy sources for Turkey by using a hybrid MCDM methodology. *Energy Conversion and Management, 79*, 25–33. doi:10.1016/j.enconman.2013.11.036

Kahraman, C., Cebi, S., & Kaya, I. (2010). Selection among renewable energy alternatives using fuzzy axiomatic design: The case of Turkey. *Journal of Universal Computer Science, 16*(1), 82–102.

Kahraman, C., & Kaya, I. (2010). A fuzzy multicriteria methodology for selection among energy alternatives. *Expert Systems with Applications, 37*(9), 6270–6281. doi:10.1016/j.eswa.2010.02.095

Kale, R. D., Mallesh, B. C., Kubra, B., & Bagyaraj, D. J. (1992). Influence of vermicompost application on the available macronutrients and selected microbial populations in a paddy field. *Soil Biology & Biochemistry, 24*(12), 1317–1320. doi:10.1016/0038-0717(92)90111-A

Kammas, S., & Zendal, S. (2017). La logistique urbaine durable «LUD»: Concepts, état des lieux à Tanger (nord du Maroc), vers un modèle conceptuel de mise en œuvre dans les pays en développement [Sustainable urban logistics "LUD": Concepts, inventory in Tangier (northern Morocco), towards a conceptual model for implementation in developing countries]. *Revue des Etudes et Recherche en Logistique et Développement, 2*, 1–25.

Karaaslan, A., & Aydın, S. (2020). Evaluation of renewable energy resources with multi criteria decision making techniques: Evidence from Turkey. *Ataturk University Journal of Economics and Administrative Sciences*, *34*(4), 1351–1375. doi:10.16951/atauniiibd.749466

Karaca, C., & Ulutas, A. (2018). The selection of appropriate renewable energy source for Turkey by using Entropy and WASPAS methods. *Ege Academic Review*, *18*(3), 483–494.

Karaca, C., & Ulutaş, A. (2018). The selection of appropriate renewable energy source for Turkey by using Entropy and Waspas methods. *Ege Academic Review*, *18*(3), 483–494. doi:10.21121/eab.2018341150

Karaca, C., Ulutaş, A., & Eşgünoğlu, M. (2017). Determination of renewable energy source in turkey by Copras and analysis of the employment-enhancing effect of renewable energy investments. *Maliye Dergisi*, *172*, 111–132.

Kara, I., & Bektas, T. (2006). Integer linear programming formulations of multiple salesman problems and its variations. *European Journal of Operational Research*, *174*(3), 1449–1458. doi:10.1016/j.ejor.2005.03.008

Karakaş, E., & Yildiran, O. V. (2019). Evaluation of renewable energy alternatives for Turkey via modified fuzzy AHP. *International Journal of Energy Economics and Policy*, *9*(2), 31–39. doi:10.32479/ijeep.7349

Karakul, A. K. (2020). Selection of renewable energy source using fuzzy Ahp method. *Bingöl University Journal of Social Sciences Institute*, *10*(19), 127–150. doi:10.29029/busbed.640162

Karakuş, C. B., Demiroğlu, D., Çoban, A., & Ulutaş, A. (2020). Evaluation of GIS-based multi-criteria decision-making methods for sanitary landfill site selection: The case of Sivas city, Turkey. *Journal of Material Cycles and Waste Management*, *22*(1), 254–272. doi:10.100710163-019-00935-0

Karam, A., Hussein, M., & Reinau, K. H. (2021). Analysis of the barriers to implementing horizontal collaborative transport using a hybrid fuzzy Delphi-AHP approach. *Journal of Cleaner Production*, *321*, 128943. doi:10.1016/j.jclepro.2021.128943

Karasan, A., & Kahraman, C. (2020). Selection of the most appropriate renewable energy alternatives by using a novel interval-valued neutrosophic ELECTRE I method. *Informatica (Vilnius)*, *31*(2), 225–248. doi:10.15388/20-INFOR388

Karatop, B., Taşkan, B., Adar, E., & Kubat, C. (2021). Decision analysis related to the renewable energy investments in Turkey based on a Fuzzy AHP-EDAS-Fuzzy FMEA approach. *Computers & Industrial Engineering*, *151*, 106958. doi:10.1016/j.cie.2020.106958

Karayilmazlar, S. S., Cabuk, Y.; & Kurt, R. (2011). Biyokütlenin Türkiye'de Enerji Üretiminde Degerlendirilmesi. *Bartin Orman Fakültesi Dergisi*, *13*(19), 63–75.

Karl, T. R., & Trenberth, K. E. (2003). Modern Global Climate Change. *Science*, *302*(5651), 1719–1723. doi:10.1126cience.1090228 PMID:14657489

Kayahan Karakul, A. (2020). Selection of renewable energy source using fuzzy AHP Method. *Journal of Social Sciences Institute*, *10*(19), 127–150.

Kaya, T., & Kahraman, C. (2011). Multicriteria decision making in energy planning using a modified fuzzy TOPSIS methodology. *Expert Systems with Applications*, *38*(6), 6577–6585. doi:10.1016/j.eswa.2010.11.081

Keleş, M. K., Özdağoğlu, A., & Işıldak, B. (2021). An Application with Multi-Criteria Decision-Making Methods for the Evaluation of Airports from Passengers' View. *Ankara Hacı Bayram Veli University Journal of the Faculty of Economics and Administrative Sciences*, *23*(2), 419–456.

Kengpol, A., Rontlaong, P., & Tuominen, M. (2013). A decision support system for selection of solar power plant locations by applying fuzzy AHP and TOPSIS: An empirical study. *J Software Eng Appl*, *6*(9), 470–481. doi:10.4236/jsea.2013.69057

Khan, D., Nouman, M., Popp, J., Khan, M., Ur Rehman, F., & Olah, J. (2021). Link between technically derived energy efficiency and ecological footprint: Empirical evidence from the Asean region. *Energies*, *14*(13), 13. doi:10.3390/en14133923

Khang, D. S., Promentilla, M. A. B., Tan, R. R., Abe, N., Tuan, P. D., & Razon, L. F. (2016). Multi-criteria approach to assess stakeholders preferences for selection of biodiesel feedstock in Vietnam. *International Journal of Business and Systems Research*, *10*(2-4), 306–331. doi:10.1504/IJBSR.2016.075738

Khan, S. A. (2015). An automated decision-making approach for assortment of wind turbines-A case study of turbines in the range of 500 kW to 750 kW. *International Journal of Computing and Network Technology*, *3*, 75–82.

Kheybari, S., Javdanmehr, M., Rezaie, F. M., & Rezaei, J. (2021). Corn cultivation location selection for bioethanol production: An application of BWM and extended PROMETHEE II. *Energy*, *228*, 120593. doi:10.1016/j.energy.2021.120593

Kheybari, S., Mahdi Rezaie, F., & Rezaei, J. (2021). Measuring the importance of decision-making criteria in biofuel production technology selection. *IEEE Transactions on Engineering Management*, *68*(2), 483–497. doi:10.1109/TEM.2019.2908037

Kheybari, S., Rezaie, F. M., Naji, S. A., & Najafi, F. (2019). Evaluation of energy production technologies from biomass using analytical hierarchy process: The case of Iran. *Journal of Cleaner Production*, *232*, 257–265. doi:10.1016/j.jclepro.2019.05.357

Khishtandar, S., Zandieh, M., & Dorri, B. (2017). A multi criteria decision making framework for sustainability assessment of bioenergy production technologies with hesitant fuzzy linguistic term sets: The case of Iran. *Renewable & Sustainable Energy Reviews*, *77*, 1130–1145. doi:10.1016/j.rser.2016.11.212

Khorasaninejad, E., Fetanat, A., & Hajabdollahi, H. (2016). Prime mover selection in thermal power plant integrated with organic Rankine cycle for waste heat recovery using a novel multi criteria decision making approach. *Applied Thermal Engineering*, *102*, 1262–1279.

Kigozi, R., Aboyade, A. O., & Muzenda, E. (2014). *Technology selection of biogas digesters for OFMSW via multi-criteria decision analysis*. Academic Press.

Kim, K. T., Lee, D. J., Park, S. J., Zhang, Y., & Sultanov, A. (2015). Measuring the efficiency of the investment for renewable energy in Korea using data envelopment analysis. *Renewable & Sustainable Energy Reviews*, *47*, 694–702. doi:10.1016/j.rser.2015.03.034

Kingwell, R., & Abadi, A. (2014). Cereal straw for bioenergy production in an Australian region affected by climate change. *Biomass and Bioenergy*, *61*, 58–65. doi:10.1016/j.biombioe.2013.11.026

Kirkpatrick, S., Gelatt, C. D. Jr, & Vecchi, M. P. (1983). Optimization by Simulated Annealing. *Science*, *220*(4598), 671–680. doi:10.1126cience.220.4598.671 PMID:17813860

Koç, E., & Kaya, K. (2014). Energy resources–state of renewable energy. *Engineer and Machine*, *56*(668), 36–47.

Kodde, D. A., & Palm, F. (1986). Wald criteria for jointly testing equality and inequality restrictions. *Econometrica*, *54*(5), 1243–1248. doi:10.2307/1912331

Kohlheb, N., & Krausmann, F. (2009). Land use change, biomass production and HANPP: The case of Hungary 1961–2005. *Ecological Economics*, *69*(2), 292–300. doi:10.1016/j.ecolecon.2009.07.010

Kokkinos, K., Karayannis, V., & Moustakas, K. (2021). Optimizing microalgal biomass feedstock selection for nanocatalytic conversion into biofuel clean energy, using fuzzy multi-criteria decision making processes. *Frontiers in Energy Research*, *8*, 622210. Advance online publication. doi:10.3389/fenrg.2020.622210

Kolios, A., Collu, M., Chahardehi, A., Brennan, F. P., & Patel, M. H. (2010).A multi-criteria decision making method to compare support structures for offshore wind turbines. *In: European Wind Energy Conference*, Warsaw.

Kolios, A., Mytilinou, V., Minguez, M. L., & Salonitis, K. (2016). A comparative study of multiple-criteria decision making methods under stochastics inputs. *Energies*, *9*(7), 566. doi:10.3390/en9070566

Külekçi, Ö. C. (2009). Place of geothermal energy in the content of renewable energy sources and it's importance for Turkey. *Ankara University Journal of Environmental Sciences*, *2*(2), 83–91.

Kuleli Pak, B., Albayrak, Y. E., & Erensal, Y. C. (2015). Renewable energy perspective for Turkey using sustainability indicators. *International Journal of Computational Intelligence Systems*, *8*(1), 187–197.

Kuleli, P. B., Albayrak, Y. E., & Erensal, Y. C. (2015). Renewable energy perspective for Turkey using sustainability indicators. *International Journal of Computational and Intelligent Systems*, *8*(1), 187–197.

Kumar, A., & Jayanti, S. (2021). A land-use-constrained, generation–transmission model for electricity generation through solar photovoltaic technology: A case study of south India. *Clean Technologies and Environmental Policy*, *23*(9), 2757–2774. doi:10.100710098-021-02202-z

Kumar, A., Sah, B., Singh, A. R., Deng, Y., He, X., Kumar, P., & Bansal, R. C. (2017). A review of multi criteria decision making (MCDM) towards sustainable renewable energy development. *Renewable & Sustainable Energy Reviews*, *69*, 596–609. doi:10.1016/j.rser.2016.11.191

Kurka, T. (2013). Application of the analytic hierarchy process to evaluate the regional sustainability of bioenergy developments. *Energy*, *62*, 393–402. doi:10.1016/j.energy.2013.09.053

Lau, L. J. (1986). Functional forms in econometric model building. Handbook of Econometrics, 3, 1515-1566.

Lee, Hung, Kang, & Pearn. (2012). A wind turbine evaluation model under a multi-criteria decision making environment. *Energy Conversion and Management*, 64, 289–300.

Lee, A. H. I., Chen, H. H., & Kang, H. Y. (2011). A model to analyze strategic products for photovoltaic silicon thin-film solar cell power industry. *Renewable & Sustainable Energy Reviews*, *15*, 1271–1283.

Lee, A. H. I., Hung, M. C., Kang, H. Y., & Pearn, W. L. (2012). A wind turbine evaluation model under a multi-criteria decision-making environment. *Energy Conversion and Management*, *64*, 289–300. doi:10.1016/j.enconman.2012.03.029

Lee, A. H., Chen, H. H., & Kang, H. Y. (2009). Multi-criteria decision making on strategic selection of wind farms. *Renewable Energy*, *34*(1), 120–126. doi:10.1016/j.rcncnc.2008.04.013

Lee, D.-Y., Thomas, V. M., & Brown, M. A. (2013). Electric Urban Delivery Trucks: Energy Use, Greenhouse Gas Emissions, and Cost-Effectiveness. *Environmental Science & Technology*, *47*(14), 8022–8030. doi:10.1021/es400179w PMID:23786706

Lee, H. C., & Chang, C. T. (2018). Comparative analysis of MCDM methods for ranking renewable energy sources in Taiwan. *Renewable & Sustainable Energy Reviews*, *92*, 883–896. doi:10.1016/j.rscr.2018.05.007

Lee, W. S. (2010). Evaluating and ranking energy performance of office buildings using fuzzy measure and fuzzy integral. *Energy Conversion and Management*, *51*(1), 197–203. doi:10.1016/j.enconman.2009.09.012

Lerche, N., Wilkens, I., Schmehl, M., Eigner-Thiel, S., & Geldermann, J. (2019). Using methods of Multi-Criteria Decision Making to provide decision support concerning local bioenergy projects. *Socio-Economic Planning Sciences*, *68*, 100594. doi:10.1016/j.seps.2017.08.002

Lerkkasemsan, N., & Achenie, L. E. K. (2014). Pyrolysis of biomass – fuzzy modeling. *Renewable Energy*, *66*, 747–758. doi:10.1016/j.renene.2014.01.014

Leśniak, A., Kubek, D., Plebankiewicz, E., Zima, K., & Belniak, S. (2018). Fuzzy AHP application for supporting contractors' bidding decision. *Symmetry*, *10*(11), 642. doi:10.3390ym10110642

Lewis, S. M., Gross, S., Visel, A., Kelly, M., & Morrow, W. (2015). Fuzzy GIS-based multi-criteria evaluation for US Agave production as a bioenergy feedstock. *Global Change Biology. Bioenergy*, *7*(1), 84–99. doi:10.1111/gcbb.12116

Liano, T., Dosal, E., Lindorfer, J., & Finger, D. C. (2021). Application of multi-criteria decision-making tools for assessing biogas plants: A case study in reykjavik, Iceland. *Water (Basel)*, *13*(16), 2150. doi:10.3390/w13162150

Li, J.-M., Li, A.-H., Varbanov, P. S., & Liu, Z.-Y. (2017). Distance potential concept and its applications to the design of regional biomass supply chains and solving vehicle routing problems. *Journal of Cleaner Production*, *144*, 426–436. doi:10.1016/j.jclepro.2016.12.166

Li, M., & Tao, W. (2017). Review of methodologies and polices for evaluation of energy efficiency in high energy-consuming industry. *Applied Energy*, *187*, 203–215. doi:10.1016/j.apenergy.2016.11.039

Lim, S. L., Wu, T. Y., Lim, P. N., & Shak, K. P. Y. (2015). The use of vermicompost in organic farming: Overview, effects on soil and economics. *Journal of the Science of Food and Agriculture*, *95*(6), 1143–1156. doi:10.1002/jsfa.6849

Li, S., Wang, Z., Wang, X., Zhang, D., & Liu, Y. (2019). Integrated optimization model of a biomass feedstock delivery problem with carbon emissions constraints and split loads. *Computers & Industrial Engineering*, *137*(106013), 1–18. doi:10.1016/j.cie.2019.106013

Liserre, M., Sauter, T., & Hung, J. Y. (2010). Future Energy Systems: Integrating Renewable Energy Sources into the Smart Power Grid Through Industrial Electronics. *IEEE Industrial Electronics Magazine*, *4*(1), 18–37. https://doi.org/10.1109/MIE.2010.935861

Liu, L., Wang, J., Wang, F., & Yang, X. (2021). The impact of the planting of forest biomass energy plants under the embedded Internet of Things technology on the biodiversity of the local environmental ecology. *Environmental Technology & Innovation, 24*, 2-16. doi:10.1016/j.eti.2021.101894

Li, X., Chen, J., Sun, X., Zhao, Y., Chong, C., Dai, Y., & Wang, C. H. (2021). Multi-criteria decision making of biomass gasification-based cogeneration systems with heat storage and solid dehumidification of desiccant coated heat exchangers. *Energy*, *233*, 121122. Advance online publication. doi:10.1016/j.energy.2021.121122

Lopes, A. L. M., & Mesquita, R. B. (2015). *Tariff regulation of electricity distribution: A comparative analysis of regulatory benchmarking models*. In The 14th European Workshop on Efficiency and Productivity Analysis 2015, Helsinki, Finland.

Lozano-Minguez, E., Kolios, A. J., & Brennan, F. P. (2011). Multi-criteria assessment of offshore wind turbine support structures. *Renewable Energy*, *36*(11), 2831–2837. doi:10.1016/j.renene.2011.04.020

Lu, Z., Sun, X., Wang, Y., & Xu, C. (2019). Green supplier selection in straw biomass industry based on cloud model and possibility degree. *Journal of Cleaner Production*, *209*, 995–1005. doi:10.1016/j.jclepro.2018.10.130

Madhu, P., Dhanalakshmi, C. S., & Mathew, M. (2020). Multi-criteria decision-making in the selection of a suitable biomass material for maximum bio-oil yield during pyrolysis. *Fuel*, *277*, 118109. doi:10.1016/j.fuel.2020.118109

Madhu, P., Nithiyesh Kumar, C., Anojkumar, L., & Matheswaran, M. (2018). Selection of biomass materials for bio-oil yield: A hybrid multi-criteria decision making approach. *Clean Technologies and Environmental Policy*, *20*(6), 1377–1384. doi:10.100710098-018-1545-z

Madugu, F., & Collu, M. (2014). *Techno-economic modelling analysis of microalgae cultivation for biofuels and co-products*. doi:10.2495/EQ141022

Maghsoodi, A. I., Maghsoodi, A. I., Mosavi, A., Rabczuk, T., & Zavadskas, E. K. (2018). Renewable energy technology selection problem using integrated H-Swara-Multimoora approach. *Sustainability*, *10*(12), 4481. doi:10.3390u10124481

Malladi, K. T., & Sowlati, T. (2018). Biomass logistics: A review of important features, optimization modeling and the new trends. *Renewable & Sustainable Energy Reviews*, *94*, 587–599. https://doi.org/10.1016/j.rser.2018.06.052

Maman Ali, M. M., & Yu, Q. (2021). Assessment of the impact of renewable energy policy on sustainable energy for all in West Africa. *Renewable Energy*, *180*, 544–551. doi:10.1016/j.renene.2021.08.084

Martín-Gamboa, M., Iribarren, D., García-Gusano, D., & Dufour, J. (2017). A review of life-cycle approaches coupled with data envelopment analysis within multi-criteria decision analysis for sustainability assessment of energy systems. *Journal of Cleaner Production*, *150*, 164–174. doi:10.1016/j.jclepro.2017.03.017

Martinkus, N., Latta, G., Brandt, K., & Wolcott, M. (2018). A multi-criteria decision analysis approach to facility siting in a wood-based depot-and-biorefinery supply chain model. *Frontiers in Energy Research*, *6*(NOV), 124. Advance online publication. doi:10.3389/fenrg.2018.00124

Martinkus, N., Latta, G., Rijkhoff, S. A. M., Mueller, D., Hoard, S., Sasatani, D., Pierobon, F., & Wolcott, M. (2019). A multi criteria decision support tool for biorefinery siting: Using economic, environmental, and social metrics for a refined siting analysis. *Biomass and Bioenergy*, *128*, 105330. doi:10.1016/j.biombioe.2019.105330

Mazaheri, N., Akbarzadeh, A. H., Madadian, E., & Lefsrud, M. (2019). Systematic review of research guidelines for numerical simulation of biomass gasification for bioenergy production. *Energy Conversion and Management*, *183*, 671–688. https://doi.org/10.1016/j.enconman.2018.12.097

McCollum, D., Gomez Echeverri, L., Riahi, K., & Parkinson, S. (2017). *Sdg7: Ensure access to affordable, reliable, sustainable and modern energy for all.* Available online: http://pure.iiasa.ac.at/id/eprint/14621/1/SDGs-interactions-7-clean-energy.pdf

McKendry, P. (2002). Energy production from biomass (part 1): Overview of biomass. *Bioresource Technology*, *83*(1), 37–46. https://doi.org/10.1016/S0960-8524(01)00118-3

Meidiana, C., Nurfitriya, I. D., & Sari, K. E. (2018). Multi-criteria evaluation for determination of anaerobic di-gester location in rural area. *International Journal of Recent Technology and Engineering*, *7*(4), 153–157.

Memona, L. R., Harijana, K., Mirjata, N. H., & Nixonb, J. D. (2018). *A multi-criteria analysis of options for power generation from biomass in Pakistan*. Academic Press.

Menegaki, A. N. (2013). Growth and renewable energy in Europe: Benchmarking with data envelopment analysis. *Renewable Energy*, *60*, 363–369. doi:10.1016/j.renene.2013.05.042

MFA. (2015). *Turkey's energy profile and strategy*. https://www.mfa.gov.tr/turkeys-energy-strategy.en.mfa

Midtgård, O. M., & Sætre, T. O. (2006). Seasonal variations in yield for different types of PV modules measured under real life conditions in northern Europe. *21st European photovoltaic solar energy conference,* 2383–6.

Mishra, A. R., Rani, P., Pandey, K., Mardani, A., Streimikis, J., Streimikiene, D., & Alrasheedi, M. (2020). Novel multi-criteria intuitionistic fuzzy Swara–Copras approach for sustainability evaluation of the bioenergy production process. *Sustainability*, *12*(10), 4155. doi:10.3390u12104155

Misra, A. K. (2014). Climate change and challenges of water and food security. *International Journal of Sustainable Built Environment*, *3*(1), 153–165. https://doi.org/10.1016/j.ijsbe.2014.04.006

Mojaver, P., Khalilarya, S., & Chitsaz, A. (2020). Multi-objective optimization and decision analysis of a system based on biomass fueled SOFC using couple method of entropy/VIKOR. *Energy Conversion and Management*, *203*, 112260. doi:10.1016/j.enconman.2019.112260

Mostafaeipour, A., Sarikhani, S., Sedaghat, A., & Arabnia, H. R. (2017). *Location planning of bioethanol plants from agricultural crop residues for fuel cells using DEA*. Academic Press.

Mostafaeipour, A., Sedaghat, A., Hedayatpour, M., & Jahangiri, M. (2020). Location planning for production of bioethanol fuel from agricultural residues in the south of Caspian Sea. *Environmental Development*, *33*, 100500. doi:10.1016/j.envdev.2020.100500

Moufad, I., & Jawab, F. (2018). The Determinants of the performance of the urban freight transport-An empirical Analysis. In *International Colloquium on Logistics and Supply Chain Management (LOGISTIQUA)* (pp. 99-104). IEEE.

Moulogianni, C., & Bournaris, T. (2017). Biomass production from crops residues: Ranking of agro-energy regions. *Energies*, *10*(7), 1061. Advance online publication. doi:10.3390/en10071061

Mourmouris, J. C., & Potolias, C. (2013). A multi-criteria methodology for energy planning and developing renewable energy sources at a regional level: A case study Thassos, Greece. *Energy Policy*, *52*, 522–530. doi:10.1016/j.enpol.2012.09.074

Mukeshimana, M. C., Zhao, Z. Y., Ahmad, M., & Irfan, M. (2021). Analysis on barriers to biogas dissemination in Rwanda: AHP approach. *Renewable Energy, 163*, 1127–1137. doi:10.1016/j.renene.2020.09.051

Mutlu, N. T., Karaca, C., & Öztürk, H. H. (2020). Developments in Biomass Gasification Technology. *Journal of Agricultural Machinery. Science, 15*(2), 53–59.

Myllyviita, T., Holma, A., Antikainen, R., Lähtinen, K., & Leskinen, P. (2012). Assessing environmental impacts of biomass production chains-Application of life cycle assessment (LCA) and multi-criteria decision analysis (MCDA). *Journal of Cleaner Production, 29-30*, 238–245. doi:10.1016/j.jclepro.2012.01.019

Myllyviita, T., Leskinen, P., Lähtinen, K., Pasanen, K., Sironen, S., Kähkönen, T., & Sikanen, L. (2013). Sustainability assessment of wood-based bioenergy - A methodological framework and a case-study. *Biomass and Bioenergy, 59*, 293–299. doi:10.1016/j.biombioe.2013.07.010

Naeini, M. A., Zandieh, M., Najafi, S. E., & Sajadi, S. M. (2020). Analyzing the development of the third-generation biodiesel production from microalgae by a novel hybrid decision-making method: The case of Iran. *Energy, 195*, 116895. doi:10.1016/j.energy.2020.116895

Naghiu, G. S., Giurca, I., Achilean, I., & Badea, G. (2016). Multicriterial Analysis on Selecting Solar Radiation Concentration Ration for Photovoltaic Panels Using Electre-Boldur Method. *Procedia Technology, 22*, 773 – 780.

Nahand, P. K. (2019). *Selection of a biomass product using a Hybrid Approach of BW-PROMETHEE*. Academic Press.

Najafi, F., Sedaghat, A., Mostafaeipour, A., & Issakhov, A. (2021). Location assessment for producing biodiesel fuel from Jatropha Curcas in Iran. *Energy, 236*, 121446. doi:10.1016/j.energy.2021.121446

Nalan, Ç. B., Murat, Ö., & Nuri, Ö. (2009). Renewable energy market conditions and barriers in Turkey. *Renewable & Sustainable Energy Reviews, 13*(6), 1428 1436. https://doi.org/10.1016/j.rser.2008.09.001

Nantasaksiri, K., Charoen-amornkitt, P., & Machimura, T. (2021). Integration of multi-criteria decision analysis and geographic information system for site suitability assessment of Napier grass-based biogas power plant in southern Thailand. *Renewable and Sustainable Energy Transition, 100011*, 100011. Advance online publication. doi:10.1016/j.rset.2021.100011

Narwane, V. S., Yadav, V. S., Raut, R. D., Narkhede, B. E., & Gardas, B. B. (2021). Sustainable development challenges of the biofuel industry in India based on integrated MCDM approach. *Renewable Energy, 164*, 298–309. doi:10.1016/j.renene.2020.09.077

Naughton-Treves, L., Kammen, D. M., & Chapman, C. (2007). Burning biodiversity: Woody biomass use by commercial and subsistence groups in western Uganda's forests. *Biological Conservation, 134*(2), 232–241. https://doi.org/10.1016/j.biocon.2006.08.020

Nesterova, N., Quak, H., Balm, S., Roche-Ceraso, I., & Tretvik, T. (2013). State of the art of the electric freight vehicles implementation in city logistics. *FREVUE Project Deliverable D, 1.*

Niu, Y., Zhang, Y., Cao, Z., Gao, K., Xiao, J., Song, W., & Zhang, F. (2021). MIMOA: A membrane-inspired multi-objective algorithm for green vehicle routing problem with stochastic demands. *Swarm and Evolutionary Computation*, *60*(100767), 1–12. https://doi.org/10.1016/j.swevo.2020.100767

Nobre, A., Pacheco, M., Jorge, R., Lopes, M. F. P., & Gato, L. M. C. (2009). Geo-spatial multi-criteria analysis for wave energy conversion system deployment. *Renewable Energy*, *34*(1), 97–111. doi:10.1016/j.renene.2008.03.002

Nwokoagbara, E., Olaleye, A. K., & Wang, M. (2015). Biodiesel from microalgae: The use of multi-criteria decision analysis for strain selection. *Fuel*, *159*, 241–249. doi:10.1016/j.fuel.2015.06.074

O'Shaughnessy, E., Heeter, J., Shah, C., & Koebrich, S. (2021). Corporate acceleration of the renewable energy transition and implications for electric grids. *Renewable & Sustainable Energy Reviews*, *146*, 111160. doi:10.1016/j.rser.2021.111160

Ocal, F. (2013). *Biogas Energy Production and Application for Eskişehir* [Master of Science Thesis]. Department of Mechanical Engineering.

Okello, C., Pindozzi, S., Faugno, S., & Boccia, L. (2014). Appraising bioenergy alternatives in Uganda using strengths, weaknesses, opportunities and threats (SWOT)-Analytical hierarchy process (AHP) and a desirability functions approach. *Energies*, *7*(3), 1171–1192. doi:10.3390/en7031171

Omrani, K., Safaei, A. S., Paydar, M. M., & Nikzad, M. (2020). Pretreatment process selection in a biofuel production line. *International Journal of Industrial Engineering and Production Research*, *31*(1), 51–61. doi:10.22068/ijiepr.31.1.51

Önüt, S., Tuzkaya, U. R., & Saadet, N. (2008). Multiple criteria evaluation of current energy resources for Turkish manufacturing industry. *Energy Conversion and Management*, *49*(6), 1480–1492. doi:10.1016/j.enconman.2007.12.026

Ossei-Bremang, R. N., & Kemausuor, F. (2021). A decision support system for the selection of sustainable biomass resources for bioenergy production. *Environment Systems & Decisions*, *41*(3), 437–454. doi:10.100710669-021-09810-6

Ouyang, X., Chen, J., & Du, K. (2021). Energy efficiency performance of the industrial sector: From the perspective of technological gap in different regions in China. *Energy*, *214*, 118865. doi:10.1016/j.energy.2020.118865

Ozcan, E. C., Unlusoy, S., & Tamer, E. (2017). Evaluation of the renewable energy investments in Turkey using ANP and TOPSIS methods. *Selcuk University Journal of Engineering. Science and Technology*, *5*(2), 204–219.

Özdağoğlu, A., Keleş, M. K., & Işıldak, B. (2021). Cabin crew selection in civil aviation with fuzzy Swara and fuzzy Marcos methods. Gümüşhane University Journal of Social Sciences Institute, 12(2), 284-302.

Ozkale, C., Celik, C., Turkmen, A. C., & Cakmaz, E. S. (2017). Decision analysis application intended for selection of a power plant running on renewable energy sources. *Renewable & Sustainable Energy Reviews*, *70*, 1011–1021. doi:10.1016/j.rser.2016.12.006

Pahla, G., Mamvura, T. A., Ntuli, F., & Muzenda, E. (2017). Energy densification of animal waste lignocellulose biomass and raw biomass. *South African Journal of Chemical Engineering*, *24*, 168–175. https://doi.org/10.1016/j.sajce.2017.10.004

Pan, W., Chen, L., & Zhan, W. (2019). PESTEL analysis of construction productivity enhancement strategies: A case study of three economies. *Journal of Management Engineering*, *35*(1), 05018013. doi:10.1061/(ASCE)ME.1943-5479.0000662

Parajuli, R., Knudsen, M. T., & Dalgaard, T. (2015). Multi-criteria assessment of yellow, green, and woody biomasses: Pre-screening of potential biomasses as feedstocks for biorefineries. *Biofuels, Bioproducts & Biorefining*, *9*(5), 545–566. doi:10.1002/bbb.1567

Pastare, L., Romagnoli, F., Lauka, D., Dzene, I., & Kuznecova, T. (2014). Sustainable use of Macro-Algae for biogas production in Latvian conditions: A preliminary study through an integrated MCA and LCA approach. *Environmental and Climate Technologies*, *13*(1), 44–56. doi:10.2478/rtuect-2014-0006

Pathak, B., Chaudhari, S., & Fulekar, M. H. (2013). Biomass-resource for sustainable development. *Int J Adv Res Technol*, *2*(6), 271–287.

Patlitzianas, K. D., Ntotas, K., Doukas, H., & Psarras, J. (2007). Assessing the renewable energy producers' environment in EU accession member states. *Energy Conversion and Management*, *48*(3), 890–897. doi:10.1016/j.enconman.2006.08.014

Pattanaik, L., Duraivadivel, P., Hariprasad, P., & Naik, S. N. (2020). Utilization and re-use of solid and liquid waste generated from the natural indigo dye production process – A zero waste approach. *Bioresource Technology*, *301*, 122721. doi:10.1016/j.biortech.2019.122721

Pehlken, A., Wulf, K., Grecksch, K., Klenke, T., & Tsydenova, N. (2020). More sustainable bioenergy by making use of regional alternative biomass? *Sustainability (Switzerland)*, *12*(19), 7849. Advance online publication. doi:10.3390u12197849

Pelletier, S., Jabali, O., & Laporte, G. (2016). 50th Anniversary Invited Article—Goods Distribution with Electric Vehicles: Review and Research Perspectives. *Transportation Science*, *50*(1), 3–22. doi:10.1287/trsc.2015.0646

Perçin, S. (2019). An integrated fuzzy Swara and fuzzy AD approach for outsourcing provider selection. *Journal of Manufacturing Technology Management*, *30*(2), 531–552. doi:10.1108/JMTM-08-2018-0247

Perea-Moreno, M. A., Samerón-Manzano, E., & Perea-Moreno, A. J. (2019). Biomass as renewable energy: Worldwide research trends. *Sustainability*, *11*(3), 863. doi:10.3390u11030863

Perpiña, C., Martínez-Llario, J. C., & Pérez-Navarro, Á. (2013). Multicriteria assessment in GIS environments for siting biomass plants. *Land Use Policy*, *31*, 326–335. doi:10.1016/j.landusepol.2012.07.014

Petković, B., Agdas, A. S., Zandi, Y., Nikolić, I., Denić, N., Radenkovic, S. D., Almojil, S. F., Roco-Videla, A., Kojić, N., Zlatković, D., & Stojanović, J. (2021). Neuro fuzzy evaluation of circular economy based on waste generation, recycling, renewable energy, biomass and soil pollution. *Rhizosphere*, *19*, 100418. doi:10.1016/j.rhisph.2021.100418

Pezdevšek Malovrh, Š., Kurttila, M., Hujala, T., Kärkkäinen, L., Leban, V., Lindstad, B. H., Peters, D. M., Rhodius, R., Solberg, B., Wirth, K., Zadnik Stirn, L., & Krč, J. (2016). Decision support framework for evaluating the operational environment of forest bioenergy production and use: Case of four European countries. *Journal of Environmental Management*, *180*, 68–81. doi:10.1016/j.jenvman.2016.05.021 PMID:27208996

Pilavachi, P. A., Roumpeas, C. P., Minett, S., & Afgan, N. H. (2006). Multi-criteria evaluation for CHP system options. *Energy Conversion and Management*, *47*(20), 3519–3529. doi:10.1016/j.enconman.2006.03.004

Piscioneri, I., Sharma, N., Baviello, G., & Orlandini, S. (2000). Promising industrial energy crop, Cynara cardunculus: A potential source for biomass production and alternative energy. *Energy Conversion and Management*, *41*(10), 1091–1105. https://doi.org/10.1016/S0196-8904(99)00135-1

Polat, M. (2021). Türkiye'nin Tarımsal Atık Biyokütle Enerji Potansiyelindeki Degisim. *Toprak Su Dergisi*, 19-24. doi:10.21657/topraksu.692275

Polprasert, C. (2007). *Organic Waste Recycling: Technology and Management* (3rd ed.). IWA Publishing.

Poyraz, P. (2021). *Fault analysis with process phase fuzzy multiccriterial decision making methods in supply chain risk managament* [M.Sc. Thesis]. Karabük University Institute of Graduate Programs Department of Industrial Engineering, Karabük, Turkey.

Prasad, R. (2015). The production function methodology for estimating the value of spectrum. *Telecommunications Policy*, *39*(1), 77–88. doi:10.1016/j.telpol.2014.12.007

Prasad, R. D., & Raturi, A. (2021). Prospects of sustainable biomass-based power generation in a small island country. *Journal of Cleaner Production*, *318*, 128519. doi:10.1016/j.jclepro.2021.128519

Priyanka & Rajneesh. (2017). *A fuzzy VIKOR model for selection of optimal Biomass usage in India*. Academic Press.

Quak, H., & Nesterova, N. (2014). Towards Zero Emission Urban Logistics: Challenges and Issues for Implementation of Electric Freight Vehicles in City Logistics. In C. Macharis, S. Melo, J. Woxenius, & T. V. Lier (Eds.), *Transport and Sustainability* (Vol. 6, pp. 265–294). Emerald Group Publishing Limited.

Quinta-Nova, L., Fernandez, P., & Pedro, N. (2017). *GIS-based suitability model for assessment of forest biomass energy potential in a region of Portugal*. Academic Press.

Rabbani, M., Akbarian-Saravi, N., Ansari, M., & Musavi, M. (2020). A Bi-Objective Vehicle-Routing Problem for Optimization of a Bioenergy Supply Chain by Using NSGA-II Algorithm. *Journal of Quality Engineering and Production Optimization*, *5*(1), 87–102. https://doi.org/10.22070/jqepo.2020.3650.1079

Rahemi, H., Torabi, S. A., Avami, A., & Jolai, F. (2020). Bioethanol supply chain network design considering land characteristics. *Renewable & Sustainable Energy Reviews*, *119*, 109517. Advance online publication. doi:10.1016/j.rser.2019.109517

Raina, G., Mumbai, N., & Hedau, R. (2013). A Novel Technique for PV Panel Performance Prediction. *IJCA Proc. Int. Conf. Work. Emerg. Trends Technol*, *4*, 19–24.

Rani, P., Mishra, A. R., Mardani, A., Cavallaro, F., Štreimikienė, D., & Khan, S. A. R. (2020). Pythagorean fuzzy Swara–Vikor framework for performance evaluation of solar panel selection. *Sustainability*, *12*(10), 4278. doi:10.3390u12104278

Rani, P., Mishra, A. R., Saha, A., & Pamucar, D. (2021). Pythagorean fuzzy weighted discrimination-based approximation approach to the assessment of sustainable bioenergy technologies for agricultural residues. *International Journal of Intelligent Systems*, *36*(6), 2964–2990. Advance online publication. doi:10.1002/int.22408

Rao, B., Mane, A., Rao, A. B., & Sardeshpande, V. (2014). *Multi-criteria analysis of alternative biogas technologies*. Academic Press.

Rao, P. V., & Baral, S. S. (2011). Attribute based specification, comparison and selection of feed stock for anaerobic digestion using MADM approach. *Journal of Hazardous Materials*, *186*(2-3), 2009–2016. doi:10.1016/j.jhazmat.2010.12.108 PMID:21247688

Rasheed, R., Javed, H., Rizwan, A., Yasar, A., Tabinda, A. B., Mahfooz, Y., Wang, Y., & Su, Y. H. (2020). Sustainability and CDM potential analysis of a novel vs conventional bioenergy projects in South Asia by multi-criteria decision-making method. *Environmental Science and Pollution Research International*, *27*(18), 23081–23093. doi:10.100711356-020-08862-6 PMID:32333350

Recanatesi, F., Tolli, M., & Lord, R. (2014). Multi criteria analysis to evaluate the best location of plants for renewable energy by forest biomass: A case study in central Italy. *Applied Mathematical Sciences*, *8*(129), 6447–6458. doi:10.12988/ams.2014.46451

Rehman, S., & Khan, S. A. (2016). Fuzzy logic based multi-criteria wind turbine selection strategy e a case study of qassim, Saudi arabia. *Energies*, *9*(11), 872. doi:10.3390/en9110872

Rehman, S., & Khan, S. A. (2017). Multi-criteria wind turbine selection using weighted sum approach. *International Journal of Advanced Computer Science and Applications*, *8*(6), 128–132. doi:10.14569/IJACSA.2017.080616

Rehman, S., & Khan, S. A. (2019). Goal programming-based two-tier multi-criteria decision-making approach for wind turbine selection. *Applied Artificial Intelligence*, *33*(1), 27–53. doi:10.1080/08839514.2018.1525525

Ren, J., Fedele, A., Mason, M., Manzardo, A., & Scipioni, A. (2013). Fuzzy Multi-actor Multi-criteria Decision Making for sustainability assessment of biomass-based technologies for hydrogen production. *International Journal of Hydrogen Energy*, *38*(22), 9111–9120. doi:10.1016/j.ijhydene.2013.05.074

Ren, J., Manzardo, A., Mazzi, A., Zuliani, F., & Scipioni, A. (2015). Prioritization of bioethanol production pathways in China based on life cycle sustainability assessment and multi-criteria decision-making. *The International Journal of Life Cycle Assessment*, *20*(6), 842–853. doi:10.100711367-015-0877-8

Rentizelas, A., Melo, I. C., Alves, P. N. Junior, Campoli, J. S., & Aparecida do Nascimento Rebelatto, D. (2019). Multi-criteria efficiency assessment of international biomass supply chain pathways using data envelopment analysis. *Journal of Cleaner Production*, *237*, 117690. Advance online publication. doi:10.1016/j.jclepro.2019.117690

Rezk, H., Inayat, A., Abdelkareem, M. A., Olabi, A. G., & Nassef, A. M. (2022). Optimal operating parameter determination based on fuzzy logic modeling and marine predators algorithm approaches to improve the methane production via biomass gasification. *Energy*, *239*, 122072. doi:10.1016/j.energy.2021.122072

Röder, M., Mohr, A., & Liu, Y. (2020). Sustainable bioenergy solutions to enable development in low- and middle-income countries beyond technology and energy access. *Biomass and Bioenergy*, *143*(105876), 1–8. https://doi.org/10.1016/j.biombioe.2020.105876

Rodrigues, C., Rodrigues, A. C., Vilarinho, C., Alves, M., & Alonso, J. M. (2019). Spatial multi-criteria gis-based analysis to anaerobic biogas plant location for dairy waste and wastewater treatment and energy recovery (Barcelos, NW Portugal). Springer Verlag.

Rodríguez, R., Gauthier-Maradei, P., & Escalante, H. (2017). Fuzzy spatial decision tool to rank suitable sites for allocation of bioenergy plants based on crop residue. *Biomass and Bioenergy*, *100*, 17–30. doi:10.1016/j.biombioe.2017.03.007

Romeo, F., & Sangiovanni-Vincentelli, A. (1991). A theoretical framework for simulated annealing. *Algorithmica*, *6*(1), 302–345. https://doi.org/10.1007/BF01759049

Rupf, G. V., Bahri, P. A., de Boer, K., & McHenry, M. P. (2016). Development of a model for identifying the optimal biogas system design in Sub-Saharan Africa. In Z. Kravanja & M. Bogataj (Eds.), *Computer Aided Chemical Engineering* (Vol. 38, pp. 1533–1538). Elsevier.

Sacchelli, S., & Cipollaro, M. (2016). Public perception of bioenergy chain: An integrated evaluation based on semantic differential approach and multi-criteria analysis. *Chemical Engineering Transactions*, *50*, 427–432. doi:10.3303/CET1650072

Saelee, S., Paweewan, B., Tongpool, R., Witoon, T., Takada, J., & Manusboonpurmpool, K. (2014). Biomass type selection for boilers using TOPSIS multi-criteria model. *International Journal of Environmental Sciences and Development*, *5*(2), 181–186. doi:10.7763/IJESD.2014.V5.474

Sagbansua, L., & Balo, F. (2017). Multicriteria decision making for 1.5 MW wind turbine selection. *Procedia Computer Science*, *111*, 413–419. doi:10.1016/j.procs.2017.06.042

Sahebi, I. G., Arab, A., & Toufighi, S. P. (2020). Analyzing the barriers of organizational transformation by using fuzzy Swara. *Journal of Fuzzy Extension & Applications*, *1*(2), 88–103. doi:10.22105/jfea.2020.249191.1010

Şahin, S., Alakoç, N. P., & Keçeci, B. A. (2010). DSS Based Selectıon of Solar Panels for Dıfferent Regıons Of Turkey. *10th International Conference On Clean Energy (Icce-2010).*

Sakthivel, G., Ilangkumaran, M., & Gaikwad, A. (2015). A hybrid multi-criteria decision modeling approach for the best biodiesel blend selection based on ANP-TOPSIS analysis. *Ain Shams Engineering Journal*, *6*(1), 239–256. doi:10.1016/j.asej.2014.08.003

Sakthivel, G., Ilangkumaran, M., Nagarajan, G., Priyadharshini, G. V., Dinesh Kumar, S., Satish Kumar, S., Suresh, K. S., Thirumalai Selvan, G., & Thilakavel, T. (2014). Multi-criteria decision modelling approach for biodiesel blend selection based on GRA-TOPSIS analysis. *International Journal of Ambient Energy*, *35*(3), 139–154. doi:10.1080/01430750.2013.789984

Sakthivel, G., Ilangkumaran, M., Nagarajan, G., & Shanmugam, P. (2013). Selection of best biodiesel blend for IC engines. An integrated approach with FAHP-TOPSIS and FAHP VIKOR. *International Journal of Oil, Gas and Coal Technology*, *6*(5), 581–612. doi:10.1504/IJOGCT.2013.056153

Salah, C. B., Chaabenea, M., & Ammara, M. B. (2008). Multi-criteria fuzzy algorithm for energy management of a domestic photovoltaic panel. *Renewable Energy*, *33*, 993–1001.

Salihoglu, N. K., Teksoy, A., & Altan, K. (2019). Determination of Biogas Production Potential From Cattle And Sheep Wastes: Balikesir Case Study. *Ömer Halisdemir Üniversitesi Mühendislik Bilimleri Dergisi*, *8*(1), 31-47. doi:10.28948/ngumuh.516798

Salminen, E., & Rintala, J. (2002). Anaerobic digestion of organic solid poultry slaughterhouse waste – a review. *Bioresource Technology*, *83*(1), 13–26. doi:10.1016/S0960-8524(01)00199-7

San Cristóbal, J. R. (2011). Multi-criteria decision-making in the selection of a renewable energy project in spain: The Vikor method. *Renewable Energy*, *36*(2), 498–502. doi:10.1016/j.renene.2010.07.031

San Cristóbal, J. R. A. (2011). Multi criteria data envelopment analysis model to evaluate the efficiency of the renewable energy technologies. *Renewable Energy*, *36*(10), 2742–2746. doi:10.1016/j.renene.2011.03.008

Sanaei, S., Chambost, V., & Stuart, P. R. (2018). Systematic assessment of triticale-based biorefinery strategies: Sustainability assessment using multi-criteria decision-making (MCDM). *Biofuels, Bioproducts & Biorefining*, *12*, S73–S86. doi:10.1002/bbb.1482

Sanchez-Lozano, J. M., García-Cascales, M. S., & Lamata, M. T. (2014). Identification and selection of potential sites for onshore wind farms development in Region of Murcia, Spain. *Energy*, *73*, 311–324. doi:10.1016/j.energy.2014.06.024

Sansa, M., Badreddine, A., & Romdhane, T. B. (2021). Sustainable design based on LCA and operations management methods: SWOT, PESTEL, and 7S. In Methods in Sustainability Science (pp. 345-364). Elsevier.

Schievano, A., D'Imporzano, G., & Adani, F. (2009). Substituting energy crops with organic wastes and agro-industrial residues for biogas production. *Journal of Environmental Management*, *90*(8), 2537–2541. https://doi.org/10.1016/j.jenvman.2009.01.013

Schillo, R. S., Isabelle, D. A., & Shakiba, A. (2017). Linking advanced biofuels policies with stakeholder interests: A method building on Quality Function Deployment. *Energy Policy*, *100*, 126–137. doi:10.1016/j.enpol.2016.09.056

Schröder, T., Lauven, L. P., Beyer, B., Lerche, N., & Geldermann, J. (2019). Using PROMETHEE to assess bioenergy pathways. *Central European Journal of Operations Research*, *27*(2), 287–309. doi:10.100710100-018-0590-3

Schulte, J., & Ny, H. (2018). Electric road systems: Strategic stepping stone on the way towards sustainable freight transport? *Sustainability*, *10*(4), 1148. doi:10.3390u10041148

Scott, J. A., Ho, W., & Dey, P. K. (2012). A review of multi-criteria decision-making methods for bioenergy systems. *Energy*, *42*(1), 146–156. doi:10.1016/j.energy.2012.03.074

Scott, J. A., Ho, W., & Dey, P. K. (2013). Strategic sourcing in the UK bioenergy industry. *International Journal of Production Economics*, *146*(2), 478–490. doi:10.1016/j.ijpe.2013.01.027

Şengül, D., Çağıl, G., & Ardalı, Z. (2021). Fuzzy Swara and interval-valued intuitionistic fuzzy analytic hierarchy process application in job evaluation process. *Journal of Management and Economics*, *28*(2), 243–263. doi:10.18657/yonveek.731727

Sengul, U., Eren, M., Shiraz, S. E., Gezder, V., & Sengul, A. B. (2015). Fuzzy TOPSIS method for ranking renewable energy supply systems in Turkey. *Renewable Energy*, *75*, 617–625. doi:10.1016/j.renene.2014.10.045

Şenyiğit, E., & Demirel, B. (2018). The selection of material in dental implant with entropy based sımple additive weighting and analytic hıerarchy process methods. *Sigma: Journal of Engineering & Natural Sciences*, *36*(3), 731–740.

Setiawan, E. A., Kurniawan, K., & Setiawan, A. (2015). Gaussıan approach to compare crystallıne solar panel performance. *International Journal of Technology*, *3*(3), 336–344. doi:10.14716/ijtech.v6i3.1474

Shabani, N., & Sowlati, T. (2013). A mixed integer non-linear programming model for tactical value chain optimization of a wood biomass power plant. *Applied Energy, 104*, 353–361. https://doi.org/10.1016/j.apenergy.2012.11.013

Shahraki Shahdabadi, R., Maleki, A., Haghighat, S., & Ghalandari, M. (2021). Using multi-criteria decision-making methods to select the best location for the construction of a biomass power plant in Iran. *Journal of Thermal Analysis and Calorimetry, 145*(4), 2105–2122. doi:10.100710973-020-10281-1

Shih, H. S., Shyur, H. J., & Lee, E. S. (2007). An extension of TOPSIS for group decision making. *Mathematical and Computer Modelling, 45*(7-8), 801–813. doi:10.1016/j.mcm.2006.03.023

Shin, Y. O., & Zul, I. (2020). *Energy priority estimation model for quantitative analysis of potential bioethanol feedstock*. Academic Press.

Shirgholami, Z., Zangeneh, S. N., & Bortolini, M. (2016). Decision system to support the practitiones in the wind farm design: A case study for Iran mainland. *Sustainable Energy Technologies and Assessments, 16*, 1–10. doi:10.1016/j.seta.2016.04.004

Silva, S., Alçada-Almeida, L., & Dias, L. C. (2014). Biogas plants site selection integrating Multicriteria Decision Aid methods and GIS techniques: A case study in a Portuguese region. *Biomass and Bioenergy, 71*, 58–68. doi:10.1016/j.biombioe.2014.10.025

Sindhu, S., Nehra, V., & Luthra, S. (2017). Investigation of feasibility study of solar plants deployment using hybrid AHP-TOPSIS analysis, Case study of India. *Renew Sustain Energy*, (73), 496–511.

Singh, J., Panesar, B. S., & Sharma, S. K. (2010). A mathematical model for transporting the biomass to biomass based power plant. *Biomass and Bioenergy, 34*(4), 483–488. https://doi.org/10.1016/j.biombioe.2009.12.012

Sinha, A., Shahbaz, M., & Balsalobre, D. (2017). Exploring the relationship between energy usage segregation and environmental degradation in N-11 countries. *Journal of Cleaner Production, 168*, 1217–1229. doi:10.1016/j.jclepro.2017.09.071

Šiškova, J. (2013). Multi-criterion analysis of the risks involved in a biogas plant in relation to the structure and sources of biomass and its application in agricultural companies. *Acta Universitatis Agriculturae et Silviculturae Mendelianae Brunensis, 61*(7), 2843–2850. doi:10.11118/actaun201361072843

Sivaraja, C. M., Sakthivel, G., & Jegadeeshwaran, R. (2020). Selection of optimum bio-diesel fuel blend using fuzzy TOPSIS and fuzzy VIKOR approaches. *International Journal of Oil, Gas and Coal Technology, 23*(2), 261–291. doi:10.1504/IJOGCT.2020.105455

Smarandache, F. (2017). *Plithogeny, plithogenic set, logic, probability, and statistics*. Pons Publishing House.

Smarandache, F. (2002). Neutrosophy, A new branch of philosophy. *International Journal (Toronto, Ont.), 8*(3), 297–384.

Smarandache, F. (2018). Plithogenic Set, an extension of crisp, fuzzy, intuitionistic fuzzy, and neutrosophic sets-revisited. *Neutrosophic Sets Syst.*, *21*, 153–166.

Smyth, B. M., Smyth, H., & Murphy, J. D. (2011). Determining the regional potential for a grass biomethane industry. *Applied Energy*, *88*(6), 2037–2049. doi:10.1016/j.apenergy.2010.12.069

Solangi, Y. A., Tan, Q., Mirjat, N. H., Valasai, G. D., Khan, M. W. A., & Ikram, M. (2019). An integrated Delphi-AHP and fuzzy TOPSIS approach toward ranking and selection of renewable energy resources in Pakistan. *Processes (Basel, Switzerland)*, *7*(2), 118. doi:10.3390/pr7020118

Song, J., Sun, Y., & Jin, L. (2017). PESTEL analysis of the development of the waste-to-energy incineration industry in China. *Renewable & Sustainable Energy Reviews*, *80*, 276–289. doi:10.1016/j.rser.2017.05.066

Sözen, A., Alp, İ., & Kilinc, C. (2012). Efficiency assessment of the hydro-power plants in Turkey by using data envelopment analysis. *Renewable Energy*, *46*, 192–202. doi:10.1016/j.renene.2012.03.021

Sözen, A., Alp, İ., & Özdemir, A. (2010). Assessment of operational and environmental performance of thermal power plants in Turkey by using data envelopment analysis. *Energy Policy*, *38*(10), 6194–6203. doi:10.1016/j.enpol.2010.06.005

Spinelli, R., & Magagnotti, N. (2010). Comparison of two harvesting systems for the production of forest biomass from the thinning of Picea abies plantations. *Scandinavian Journal of Forest Research*, *25*(1), 69–77. https://doi.org/10.1080/02827580903505194

Stamatakis, M., Mandalaki, T. T., & Tsoutsos, T. (2016). Multi-criteria analysis for PV integrated in shading devices for Mediterranean region. *Energy and Building*, *117*, 128–137. doi:10.1016/j.enbuild.2016.02.007

Stephen, J. D., Mabee, W. E., & Saddler, J. N. (2010). Biomass logistics as a determinant of second-generation biofuel facility scale, location and technology selection. *Biofuels, Bioproducts and Biorefining*, *4*(5), 503-518. doi:10.1002/bbb.239

Suganthi, L., Inıyan, S., & Samuel, A. A. (2015). Applications of fuzzy logic in renewable energy systems-A review. *Renewable & Sustainable Energy Reviews*, *48*, 585–607. doi:10.1016/j.rser.2015.04.037

Sugiawan, Y., & Managi, S. (2019). New evidence of energy-growth nexus from inclusive wealth. *Renewable & Sustainable Energy Reviews*, *103*, 40–48. doi:10.1016/j.rser.2018.12.044

Suh, J., & Brownson, J. (2016). Solar farm suitability using geographic information system fuzzy sets and analytic hierarchy processes: Case study of ulleung island, korea. *Energies*, *9*(8), 648. doi:10.3390/en9080648

Sultana, A., & Kumar, A. (2012). Ranking of biomass pellets by integration of economic, environmental and technical factors. *Biomass and Bioenergy*, *39*, 344–355. doi:10.1016/j.biombioe.2012.01.027

Sunil Chaitanya, G., Raj Kumar, M., & Deivanathan, R. (2021). *Multi criteria decision making approach for selection of biodiesel blend using AHP-TOPSIS analysis*. Academic Press.

Su, S., Yu, Z., Zhu, W., & Chang, W. (2020). A comprehensive evaluation and optimal utilization structure of crop straw-based energy production in eastern China. *BioResources*, *15*(2), 2850–2868. doi:10.15376/biores.15.2.2850-2868

Taefi, T. T., Kreutzfeldt, J., Held, T., Konings, R., Kotter, R., Lilley, S., & Nyquist, C. (2016). Comparative analysis of european examples of freight electric vehicles schemes—a systematic case study approach with examples from Denmark, Germany, the Netherlands, Sweden and the UK. In Dynamics in logistics (pp. 495-504). Springer.

Tazzit, S., Ibne Hossain, N. U., Nur, F., Elakramine, F., Jaradat, R., & El Amrani, S. (2021). Selecting a biomass pelleting processing depot using a data driven decision-making approach. *Systems*, *9*(2), 32. Advance online publication. doi:10.3390ystems9020032

TCMB. (2021). *Electronic data delivery system*. https://evds2.tcmb.gov.tr/

Tegou, L. I., Polatidis, H., & Haralambopoulos, D. A. (2010). Environmental management framework for wind farm siting: Methodology and case study. *Journal of Environmental Management*, *91*(11), 2134–2147. doi:10.1016/j.jenvman.2010.05.010 PMID:20541310

TEIAS. (2021). *Turkey electricity generation-transmission statistics for 2020*. https://www.teias.gov.tr/tr-TR/turkiye-elektrik-uretim-iletim-istatistikleri

Thakur, V. (2021). Framework for PESTEL dimensions of sustainable healthcare waste management: Learnings from COVID-19 outbreak. *Journal of Cleaner Production*, *287*, 125562. doi:10.1016/j.jclepro.2020.125562 PMID:33349739

The World Bank. (2021). *World development indicators*. https://databank.worldbank.org/data/reports.aspx?source=world-development-indicators&Type=TABLE&preview=on

Toklu, E. (2017). Biomass energy potential and utilization in Turkey. *Renewable Energy*, *107*, 235–244. https://doi.org/10.1016/j.renene.2017.02.008

Toklu, M. C., & Taskin, H. (2018). A fuzzy hybrid decision model for renewable energy sources selection. *International Journal of Computational and Experimental Science and Engineering*, *4*(1), 6–10. doi:10.22399/ijcesen.399976

Tolga, A. C., & Turgut, Z. K. (2018). Sustainable and renewable energy power plants evaluation by fuzzy TODIM technique. *Alphanumeric Journal*, *6*(1), 49–68. doi:10.17093/alphanumeric.371754

Trivedi, J. K., Sareen, H., & Dhyani, M. (2008). Rapid urbanization - Its impact on mental health: A South Asian perspective. *Indian Journal of Psychiatry*, *50*(3), 161–165. https://doi.org/10.4103/0019-5545.43623

Tsangas, M., Jeguirim, M., Limousy, L., & Zorpas, A. (2019). The application of analytical hierarchy process in combination with PESTEL-SWOT analysis to assess the hydrocarbons sector in Cyprus. *Energies*, *12*(5), 791. doi:10.3390/en12050791

Tsao, Y.-C., Vu, T.-L., & Lu, J.-C. (2021). Pricing, capacity and financing policies for investment of renewable energy generations. *Applied Energy, 303*, 117664. doi:10.1016/j.apenergy.2021.117664

Turcksin, L., Macharis, C., Lebeau, K., Boureima, F., Van Mierlo, J., Bram, S., De Ruyck, J., Mertens, L., Jossart, J.-M., Gorissen, L., & Pelkmans, L. (2011). A multi-actor multi-criteria framework to assess the stakeholder support for different biofuel options: The case of Belgium. *Energy Policy, 39*(1), 200–214. doi:10.1016/j.enpol.2010.09.033

TUREB. (2019). *Turkey wind energy statistics report.* https://tureb.com.tr//lib/uploads/4e77501b714739a9.pdf

Turk, S., & Sahin, G. (2018). Multi-criteria decision-making in the location selection for a solar PV power plant using AHP. *Measurement., 129*, 218–226. doi:10.1016/j.measurement.2018.07.020

Tuysuz, F. (2017). A hybrid multi-criteria analysis approach for the assessment of renewable energy resources under uncertainty. *Alphanumeric Journal, 5*(2), 317-328.

Ubando, A. T., Promentilla, M. A. B., Culaba, A. B., & Tan, R. R. (2016). *Application of spatial analytic hierarchy process in the selection of algal cultivation site for biofuel production: A case study in the Philippines*. Academic Press.

Ubando, A. T., Cuello, J. L., Culaba, A. B., Promentilla, M. A. B., & Tan, R. R. (2014). Multi-criterion evaluation of cultivation systems for sustainable algal biofuel production using Analytic Hierarchy Process and Monte Carlo simulation. *Energy Procedia, 61*, 389–392. doi:10.1016/j.egypro.2014.11.1132

Ubando, A. T., Cuello, J. L., El-Halwagi, M. M., Culaba, A. B., Promentilla, M. A. B., & Tan, R. R. (2016). Application of stochastic analytic hierarchy process for evaluating algal cultivation systems for sustainable biofuel production. *Clean Technologies and Environmental Policy, 18*(5), 1281–1294. doi:10.100710098-015-1073-z

Ulutaş, A., Karakuş, C. B., & Topal, A. (2020). Location selection for logistics center with fuzzy SWARA and CoCoSo methods. *Journal of Intelligent & Fuzzy Systems, 38*, 4693-4709. https://doi:10.3233/JIFS-191400

Ulutaş, A., Topal, A., Karabasevic, D., Stanujkic, D., Popovic, G., & Smarandache, F. (2021, August). Prioritization of logistics risks with plithogenic PIPRECIA method. In *International Conference on Intelligent and Fuzzy Systems* (pp. 663-670). Springer.

Umar, H. S., Girei, A. A., & Yakubu, D. (2017). Comparison of cobb-douglas and Translog frontier models in the analysis of technical efficiency in dry-season tomato production. *Agrosearch, 17*(2), 67. doi:10.4314/agrosh.v17i2.6

Unay, E., Ozkaya, B., & Yoruklu, H. C. (2021). A multi-criteria decision analysis for the evaluation of microalgal growth and harvesting. *Chemosphere, 279*, 130561. doi:10.1016/j.chemosphere.2021.130561 PMID:33892454

Uysal, F. (2011). Graph Theory and Matrix Approach for the Selection of Renewable Energy Alternatives in Turkey. *Istanbul University Econometrics and Statistics e-Journal, 13*, 23-40.

Vafaeipour, M., Zolfani, S. H., Varzandeh, M. H. M., Derakhti, A., & Eshkalag, M. K. (2014). Assessment of regions priority for implementation of solar projects in Iran: New application of a hybrid multi-criteria decision making approach. *Energy Conversion and Management*, *86*, 653–663. doi:10.1016/j.enconman.2014.05.083

van Dael, M., Van Passel, S., Pelkmans, L., Guisson, R., Swinnen, G., & Schreurs, E. (2012). Determining potential locations for biomass valorization using a macro screening approach. *Biomass and Bioenergy*, *45*, 175–186. doi:10.1016/j.biombioe.2012.06.001

van Dyken, S., Bakken, B. H., & Skjelbred, H. I. (2010). Linear mixed-integer models for biomass supply chains with transport, storage and processing. *Energy*, *35*(3), 1338–1350. https://doi.org/10.1016/j.energy.2009.11.017

Van Haaren, R. V., & Fthenakis, V. (2011). GIS-based wind farm site selection using spatial multi-criteria analysis (SMCA): Evaluating the case for New York State. *Renewable & Sustainable Energy Reviews*, *15*(7), 3332–3340. doi:10.1016/j.rser.2011.04.010

Vera, D., Carabias, J., Jurado, F., & Ruiz-Reyes, N. (2010). A Honey Bee Foraging approach for optimal location of a biomass power plant. *Applied Energy*, *87*(7), 2119–2127. doi:10.1016/j.apenergy.2010.01.015

Viana, H., Cohen, W. B., Lopes, D., & Aranha, J. (2010). Assessment of forest biomass for use as energy. GIS-based analysis of geographical availability and locations of wood-fired power plants in Portugal. *Applied Energy*, *87*(8), 2551–2560. doi:10.1016/j.apenergy.2010.02.007

Vida, E., & Tedesco, D. E. A. (2017). The carbon footprint of integrated milk production and renewable energy systems – A case study. *The Science of the Total Environment*, *609*, 1286–1294. https://doi.org/10.1016/j.scitotenv.2017.07.271

Vindiš, P., Muršec, B., Rozman, Č., & Čus, F. (2010). A Multi-Criteria Assessment of Energy Crops for Biogas Production. Strojniski Vestnik. *Jixie Gongcheng Xuebao*, *56*(1).

Vitali, F., Parmigiani, S., Vaccari, M., & Collivignarelli, C. (2013). Agricultural waste as household fuel: Techno-economic assessment of a new rice-husk cookstove for developing countries. *Waste Management, 33*(12), 2762-2770. doi:10.1016/j.wasman.2013.08.026

Vlachokostas, C., Achillas, C., Agnantiaris, I., Michailidou, A. V., Pallas, C., Feleki, E., & Moussiopoulos, N. (2020). Decision support system to implement units of alternative biowaste treatment for producing bioenergy and boosting local bioeconomy. *Energies*, *13*(9), 2306. Advance online publication. doi:10.3390/en13092306

Vlahinić-Dizdarević, N., & Šegota, A. (2021). Total-factor energy efficiency in the EU countries. *Zbornik Radova Ekonomskog Fakulteta U Rijeci: Časopis Za Ekonomsku Teoriju I Praksu, 30*(2), 247-265.

Voivontas, D., Assimacopoulos, D., & Koukios, E. G. (2001). Aessessment of biomass potential for power production: A GIS based method. *Biomass and Bioenergy*, *20*(2), 101–112. https://doi.org/10.1016/S0961-9534(00)00070-2

Voivontas, D., Assimacopoulos, D., & Koukios, E. G. (2001). Assessment of biomass potential for power production: A GIS based method. *Biomass and Bioenergy*, *20*(2), 101–112. doi:10.1016/S0961-9534(00)00070-2

Volkova, A., Latosov, E., & Siirde, A. (2010). Selection of the most appropriate regions for wood fuel based cogeneration plants using multi criteria decision analysis methods. *International Journal of Exergy*, *4*(2).

von Doderer, C. C. C., & Kleynhans, T. E. (2014). Determining the most sustainable lignocellulosic bioenergy system following a case study approach. *Biomass and Bioenergy*, *70*, 273–286. doi:10.1016/j.biombioe.2014.08.014

Wang, C. N., Fu, H. P., Hsu, H. P., Nguyen, V. T., Nguyen, V. T., & Ahmar, A. S. (2021). A model for selecting a biomass furnace supplier based on qualitative and quantitative factors. Computers, Materials, &. *Continua*, *69*(2), 2339–2353. doi:10.32604/cmc.2021.016284

Wang, C. N., Tsai, T. T., & Huang, Y. F. (2019). A model for optimizing location selection for biomass energy power plants. *Processes (Basel, Switzerland)*, *7*(6), 353. doi:10.3390/pr7060353

Wang, J., Jing, Y., Zhang, C., & Zhao, J. (2009). Review on multi-criteria decision analysis aid in sustainable energy decision-making. *Renewable & Sustainable Energy Reviews*, *13*(9), 2263–2278. doi:10.1016/j.rser.2009.06.021

Wang, M., Thoben, K. D., Bernardo, M., & Daudi, M. (2018). Diversity in employment of electric commercial vehicles in urban freight transport: A literature review. *Logistics Research*, *11*(10), 1–13.

Wang, Q., Dong, Z., Li, R., & Wang, L. (2022). Renewable energy and economic growth: New insight from country risks. *Energy*, *238*, 122018. doi:10.1016/j.energy.2021.122018

Wang, S., Jena, U., & Das, K. C. (2018). Biomethane production potential of slaughterhouse waste in the United States. *Energy Conversion and Management*, *173*, 143–157. doi:10.1016/j.enconman.2018.07.059

Wang, Y., Xu, L., & Solangi, Y. A. (2020). Strategic renewable energy resources selection for Pakistan: Based on swot-fuzzy Ahp approach. *Sustainable Cities and Society*, *52*(101861), 1–14. doi:10.1016/j.scs.2019.101861

Wannasiri, W. (2020). The potential of biomass fuel and land suitability for biomass power plant based on gis spatial analysis in the Nakhon Ratchasima Province, Thailand. *Chemical Engineering Transactions*, *78*, 325–330. doi:10.3303/CET2078055

WBCSD. (2004). *Doing Business with the Poor: A field guide*. WBCSD.

Wheeler, J., Páez, M. A., Guillén-Gosálbez, G., & Mele, F. D. (2018). Combining multi-attribute decision-making methods with multi-objective optimization in the design of biomass supply chains. *Computers & Chemical Engineering*, *113*, 11–31. doi:10.1016/j.compchemeng.2018.02.010

Wikipedia. (2021). https://en.wikipedia.org/wiki/Growth_of_photovoltaics#Forecast

Woo, H., Acuna, M., Moroni, M., Taskhiri, M. S., & Turner, P. (2018). Optimizing the location of biomass energy facilities by integrating multi-criteria analysis (MCA) and geographical information systems (GIS). *Forests*, *9*(10), 585. Advance online publication. doi:10.3390/f9100585

Wu, Y., Xu, C., & Zhang, T. (2018). Evaluation of renewable power sources using a fuzzy MCDM based on cumulative prospect theory: A case in China. *Energy*, *147*, 1227–1239. doi:10.1016/j.energy.2018.01.115

Wu, Y., Yan, Y., Wang, S., Liu, F., Xu, C., & Zhang, T. (2019). Study on location decision framework of agroforestry biomass cogeneration project: A case of China. *Biomass and Bioenergy*, *127*, 105289. Advance online publication. doi:10.1016/j.biombioe.2019.105289

Wzorek, M., Junga, R., Yilmaz, E., & Niemiec, P. (2021). Combustion behavior and mechanical properties of pellets derived from blends of animal manure and lignocellulosic biomass. *Journal of Environmental Management*, *290*(112487), 1–8. https://doi.org/10.1016/j.jenvman.2021.112487

Xiang, W., Xue, S., Qin, S., Xiao, L., Liu, F., & Yi, Z. (2018). Development of a multi-criteria decision making model for evaluating the energy potential of Miscanthus germplasms for bioenergy production. *Industrial Crops and Products*, *125*, 602–615. doi:10.1016/j.indcrop.2018.09.050

Xiao, J., Yao, Z., Qu, J., & Sun, J. (2013). Research on an optimal site selection model for desert photovoltaic power plants based on analytic hierarchy process and geographic information system. *Journal of Renewable and Sustainable Energy*, *5*(2), 1–15. doi:10.1063/1.4801451

Xue, J., Ding, J., Zhao, L., Zhu, D., & Li, L. (2022). An option pricing model based on a renewable energy price index. *Energy*, *239*, 122117. doi:10.1016/j.energy.2021.122117

Xu, X., Chen, H. H., Feng, Y., & Tang, J. (2018). The production efficiency of renewable energy generation and its influencing factors: Evidence from 20 countries. *Journal of Renewable and Sustainable Energy*, *10*(2), 025901. doi:10.1063/1.5006844

Yadav, S., Srivatava, A. K., & Singh, R. S. (2015). Selection and ranking of multifaceted criteria for the prioritization of most appropriate conversion technology for biomass to biofuel in Indian perspective using analytic hierarchy process. *International Journal of Advanced Technology in Engineering and Science*, *3*, 869–881.

Yakici Ayan, T., & Pabuccu, H. (2013). Evaluation of the renewable energy investment project with Analytic Hierarchy Process method. *Suleyman Demirel University The Journal of Faculty of Economics and Administrative Sciences*, *18*(3), 89–110.

Yakıcı Ayan, T., & Pabuçcu, H. (2013). Evaluation of the renewable energy investment project with analytic hierarchy process method. *Suleyman Demirel University. The Journal of Faculty of Economics and Administrative Sciences*, *18*(3), 89–110.

Yan, Q., & Tao, J. (2014). Biomass power generation industry efficiency evaluation in China. *Sustainability*, *6*(12), 8720–8735. doi:10.3390u6128720

Yazici, I., Beyca, O. F., Gurcan, O. F., Zaim, H., Delen, D., & Zaim, S. (2020). A comparative analysis of machine learning techniques and fuzzy analytic hierarchy process to determine the tacit knowledge criteria. *Annals of Operations Research*. Advance online publication. doi:10.100710479-020-03697-3

Yenioğlu, Z. A., & Toklu, B. (2021). Performance measurement with stochastic data envelopment analysis: Comparative analysis of Turkish electricity distribution companies. *Journal of Polytechnic of Gazi University*, *24*, 87–101.

Yeo, S. Z., How, B. S., Ngan, S. L., Ng, W. P. Q., Leong, W. D., Lim, C. H., & Lam, H. L. (2020). An integrated approach to prioritise parameters for multi-objective optimisation: A case study of biomass network. *Journal of Cleaner Production*, *274*, 123053. doi:10.1016/j.jclepro.2020.123053

Yergin, D. (2006). Ensuring Energy Security. *Foreign Affairs*, *85*(2), 69–82. https://doi.org/10.2307/20031912

Yilan, G., Kadirgan, M. N., & Çiftçioğlu, G. A. (2020). Analysis of electricity generation options for sustainable energy decision making: The case of Turkey. *Renewable Energy*, *146*, 519–529. doi:10.1016/j.renene.2019.06.164

Yilmaz, S., Ozcalik, H. R., Kesler, S., Dincer, F., & Yelmen, B. (2015). The analysis of different PV power systems for the determination of optimal PV panels and system installation— A case study in Kahramanmaras, Turkey. *Renewable & Sustainable Energy Reviews*, *52*, 1015–1024.

Yücenur, G. N., & Ipekçi, A. (2021). Swara/Waspas methods for a marine current energy plant location selection problem. *Renewable Energy*, *163*, 1287–1298. doi:10.1016/j.renene.2020.08.131

Yürek, Y. T., Bulut, M., Özyörük, B., & Özcan, E. (2021). Evaluation of the hybrid renewable energy sources using sustainability index under uncertainty. *Sustainable Energy. Grids and Networks*, *28*, 100527.

Zadeh, L. A. (1978). Fuzzy sets as a basis for a theory of possibility. *Fuzzy Sets and Systems*, *1*(1), 3–28. doi:10.1016/0165-0114(78)90029-5

Zahid, F., Tahir, A., Khan, H. U., & Naeem, M. A. (2021). Wind farms selection using geospatial technologies and energy generation capacity in Gwadar. *Energy Reports*, *7*, 5857–5870.

Zarbakhshnia, N., Soleimani, H., & Ghaderi, H. (2018). Sustainable third-party reverse logistics provider evaluation and selection using fuzzy Swara and developed fuzzy Copras in the presence of risk criteria. *Applied Soft Computing*, *65*, 307–319. doi:10.1016/j.asoc.2018.01.023

Zeng, Y., Guo, W., & Zhang, F. (2019). Comprehensive evaluation of renewable energy technical plans based on data envelopment analysis. *Energy Procedia*, *158*, 3583–3588. doi:10.1016/j.egypro.2019.01.907

Zeyuan, Y. (2013). Selection of Solar Cell based on TOPSIS Method. *International Conference on Advanced Information Engineering and Education Science (ICAIEES)*, 151-154.

Zhang, L., & Hu, G. (2013). Supply chain design and operational planning models for biomass to drop-in fuel production. *Biomass and Bioenergy*, *58*, 238–250. https://doi.org/10.1016/j.biombioe.2013.08.016

Zhang, L., Xin, H., Yong, H., & Kan, Z. (2019). Renewable energy project performance evaluation using a hybrid multi-criteria decision-making approach: Case study in Fujian, China. *Journal of Cleaner Production*, *206*, 1123–1137. doi:10.1016/j.jclepro.2018.09.059

Zhang, W., Wang, C., Zhang, L., Xu, Y., Cui, Y., Lu, Z., & Streets, D. G. (2018). Evaluation of the performance of distributed and centralized biomass technologies in rural China. *Renewable Energy*, *125*, 445–455. doi:10.1016/j.renene.2018.02.109

Zhang, Z., & Paudel, K. P. (2021). Small-Scale Forest Cooperative Management of the Grain for Green Program in Xinjiang, China: A SWOT-ANP Analysis. *Small-scale Forestry*, *20*(2), 221–233. doi:10.100711842-020-09465-2

Zhao, X. G., & Li, A. (2016). A multi-objective sustainable location model for biomass power plants: Case of China. *Energy*, *112*, 1184–1193. doi:10.1016/j.energy.2016.07.011

Zhou, S., Zhang, Y., & Bao, X. (2012). *Methodology of location selection for biofuel refinery based on fuzzy TOPSIS*. Academic Press.

Zhou, T., Roorda, M. J., MacLean, H. L., & Luk, J. (2017). Life cycle GHG emissions and lifetime costs of medium-duty diesel and battery electric trucks in Toronto, Canada. *Transportation Research Part D, Transport and Environment*, *55*, 91–98. doi:10.1016/j.trd.2017.06.019

Zhou, X. Y., Wang, X. K., Wang, J. Q., Li, J. B., & Li, L. (2020). Decision support framework for the risk ranking of agroforestry biomass power generation projects with picture fuzzy information. *Journal of Intelligent & Fuzzy Systems*, •••, 1–20.

Zhu, L., Li, Z., & Hiltunen, E. (2018). Microalgae Chlorella vulgaris biomass harvesting by natural flocculant: effects on biomass sedimentation, spent medium recycling and lipid extraction. *Biotechnology for Biofuels, 11*(1), 1-10. doi:10.1186/s13068-018-1183-z

Ziolkowska, J. R. (2013). Evaluating sustainability of biofuels feedstocks: A multi objective framework for supporting decision making. *Biomass and Bioenergy*, *59*, 425–440. doi:10.1016/j.biombioe.2013.09.008

Ziolkowska, J. R. (2014). Optimizing biofuels production in an uncertain decision environment: Conventional vs. advanced technologies. *Applied Energy*, *114*, 366–376. doi:10.1016/j.apenergy.2013.09.060

Zubaryeva, A., Zaccarelli, N., Del Giudice, C., & Zurlini, G. (2012). Spatially explicit assessment of local biomass availability for distributed biogas production via anaerobic co-digestion – Mediterranean case study. *Renewable Energy*, *39*(1), 261–270. doi:10.1016/j.renene.2011.08.021

About the Contributors

Figen Balo received the BSc, MSc, and PhD degrees in Mechanical Engineering from Fırat University. She worked at the Ministries of Environment-Urbanization and Education. She is currently professor in Engineering Faculty at Fırat University. She has published 66 papers in scientific journals, 33 book chapters and 100 research papers in conferences. As reviewer, she reviewed more than 1348 article for scholarly journals. As Editor, she completed to 6 scientific books. And also she has been actively serving as Area Editor for 6 different scientific journals. Her main research area is renewable energy, green buildings, building-insulation materials containing natural sources, optimization of energy efficiency at buildings, MCDM for the best selection of equipment at renewable energy systems.

Ayse Topal was born in Turkey in 1983. She received the B.S. degree in business administration from Marmara University, Istanbul, Turkey, in 2006, and the Doctor of Philosophy (Integrated Ph.D.) degree in energy planning and policy from the University of Technology Sydney (UTS), Australia, in 2015. She is currently an Assistant Professor with Nigde Omer Halisdemir University, Turkey. Her research interests are decision making, renewable energy, sustainability, and energy planning. From 2015 to 2019, she supervised and completed one Masters by Research thesis, currently supervising three Masters by Research thesis, published several articles and conference papers, and teaching several courses at bachelor's, master's, and doctorate degrees. The awards she received are UTS International Research Scholarship (IRS), YLSY—International Postgraduate Research Scholarship (Turkish Ministry of Education), Istanbul Chamber of Commerce Scholarship, and Istanbul Metropolitan Municipality Educational Award.

Ezgi Demir is a mathematical engineer from Yıldız Technical University. She completed her master and PhD. degree in the department of operational research from Marmara University. She has been working as Assistant Professor in Gebze Technical University.

* * *

Vildan Ateş is an Assoc. Prof. Dr. in the Management Information Systems Department of Business School at Ankara Yıldırım Beyazıt University in Turkey since March 2017. In 2010, she received her master's degree from the Management Information Systems Department of the Institute of Informatics at Gazi University. She received her Ph.D. degree in 2014 from the same school and the same department. Dr. Ateş worked as an IT Expert at Gazi University from 2009 to 2017. Dr. Ateş is the instructor on undergraduate topics in e-business, production and services management, database management, computer networks and security on graduate topics in management information systems, system analysis and design, qualitative research methods, human computer interaction. Her research interests include information management, IT security and risk management, educational technologies, human computer interaction, and e-commerce. She is married with two children.

Uğur Atici received B.S. degree in industrial engineering from Balıkesir University, Balıkesir, Turkey, in 2002 M.Sc. degree in industrial engineering from Middle Erciyes University, Kayseri, Turkey, in 2010, and Ph.D. degree in machine scheduling, ergonomic risk consideration, and learning effect from Erciyes University, Kayseri, Turkey, in 2020. He worked for 12 years as an engineer in an air maintenance factory. Uğur Atici was awarded the award for outstanding service by the Turkish Air Force. His research areas are human factors, ergonomics in aviation, optimization, multi-criteria decision making, production planning, learning curve, etc. Dr. Uğur made presentations in conferences worldwide, and his many papers are published in international journals and conference proceedings. He is now at the Department of Industrial Engineering, Sivas Republic University.

Zühre Aydın Yenioğlu received her PhD degree in Management Information Systems from the University of Gazi in 2019. She worked in IT and Energy Conversion departments as engineer in the fields of Digital Transformation, IT Project Manager, Data Analyst and Database Administrator between 2007-2022. She continues as a lecturer of Database Management Systems and Information Retrieval courses at the Yıldırım Beyazıt University, Ankara. She writes and presents widely on issues of database management systems, energy data analysis, energy efficiency, operational research, cybersecurity in critical infrastructures, energy forecasting with machine learning. Aydın Yenioğlu, has been working at Turkey Energy Market Regulatory Authority since 2014.

Rim Bakhat is a doctor in management logistics and applied sciences at the faculty of economics and social sciences of the University of Abdelmalek Essaadi, Tangier, Morocco. She is a lecturer of Master Degree (Management Logistics) at the University of Abdelmalek Essaadi Tetouan. Her research study focuses primarily on the logistics and risk assessment in industry and service sectors, where she specializes in data analysis using econometric techniques and decision science. Her dissertation focused on the risk assessment of complex systems in the renewable energy field by using grey data. She is the author of many academic publications (refereed journal articles and conference papers).

Dursun Darendelioğlu is working as a Technology and Design teacher in secondary school. Areas of interest due to its branch, 3d designs, innovative product designs. In addition, he takes part in projects in the field of robotic coding. He is also continuing his postgraduate studies in the field of energy engineering.

Meral Güldeş received her MS in Industrial Engineering from the Cukurova University in 2011. She is studying for her PhD in Industrial Engineering at the Çukurova University. She has been working as research assistant at the Cumhuriyet University since 2006.

Ömer Faruk Gürcan is research assistant at Sivas Cumhuriyet University. He received BSc in Industrial Engineering from Selçuk University and an MSc from İstanbul Technical University. He has been doing Ph.D. in Industrial Engineering at İstanbul Technical University. His research areas include computer vision, machine learning, deep learning.

Murat Kemal Keleş gained his B.S. degree from Wood Affairs Industrial Engineering Department at Hacettepe University in 1994 with honorary degree. Until 2002, he worked as a production and planning engineer in the private sector. In 2002, he became a lecturer at Suleyman Demirel University in Isparta. He received his M.S. and Ph.D. degrees from the Department of Business Administration of Süleyman Demirel University in 2007 and 2014, respectively. He is currently working at Department of Transportation Service Keçiborlu Vocational School of Isparta University of Applied Sciences as a Assist Prof. Dr. His research areas are multi criteria decision making, production management, civil aviation, marketing, planning, entrepreneurship, Technoparks, Technology Transfer Offices, R&D, and Innovation.

Aşkın Özdağoğlu, Associate Professor at Dokuz Eylul University, Faculty of Business, Dept. of Business Administration, Division of Production Management and Marketing. In 1999, he graduated from Dokuz Eylul University Faculty of Economics and Administrative Sciences, Department of Business Administration. He gained MSc. in Production Management in 2003 and PhD degree in 2008. Askin Ozdagoglu is still working as Associate Professor at Dokuz Eylul University Faculty of Business, Division of Production Management and Marketing. He is also the member of Systems Thinking Society in Turkey and Production Researches Society in Turkey. His research interests are multi criteria decision making, fuzzy logic, and system dynamics.

Cenk Şahin received the B.S. degree and M.S. degree in Industrial Engineering from Cukurova University in 2001 and 2004, respectively. He started his Ph.D. studies in Industrial Engineering Department at Cukurova University in September 2004. He worked as a research assistant in the Department of Industrial Engineering at Cukurova University, between December 2002 and January 2010. He is currently working as an Associate Professor in Industrial Engineering Department at Cukurova University since 2018. His main research interests include multi criteria decision making, fuzzy systems,Industry 4.0, the agent and holonic-based approaches to controlling manufacturing systems and the applications of artificial intelligence techniques in manufacturing and service systems.

Yunis Torun has obtained his doctor's degree from the Electrical and Electronics Engineering, Gaziantep University, Turkey in 2010. He has coauthored over 15 journal publications including Machine Learning, Optimization, Biomedical and Control. Dr.Torun currently works in Cumhuriyet University, Sivas, Turkey. His current research focuses on machine learning in robotics, biomedical and physical science.

Melik Ziya Yakut graduated in Computer Systems Education in 2005, and Master degree in Mechanical Education, in 2007, PhD in Mechanical Engineering in 2014, at Suleyman Demirel University, in Isparta, Turkey. He was researcher from 2008 to 2014 at the Research and Application Centre for Renewable Energy Resources at the Suleyman Demirel University. He is currently working as an Assistant Professor in the Faculty of Technology about Mechatronics Engineering and he teaches undergraduate and graduate courses at Mechatronics Engineering at Engineering Faculty, Istanbul Gedik University in part-time. He is currently teaching and researching various subjects like Renewable Energy Sources & Policy, System Modelling (CAE), Computer Programming, Building Information Modelling (BIM), Green and Smart Buildings, Industrial Revolutions and effects of 4th Industrial Revolution, New Technologies, System Control, Polymer Processing. He is a member of CITAD, Research Centre for Territory, Architecture and Design, the Technologies of Architecture research group of Minerva Foundation, University Lusíada, Lisbon, Portugal.

Index

F

G

H

L

M

N

O

P

R

S

T

U

V

W

www.ingramcontent.com/pod-product-compliance
Lightning Source LLC
LaVergne TN
LVHW081334300926
849161LV00008B/1973

* 9 7 8 1 6 6 8 4 2 4 7 2 8 *